Applied Mechanics

(Engineering Mechanics)

This Page has been intentionally left blank

Applied Mechanics

(Engineering Mechanics)

For the Students of
Polytechnic Diploma Courses in Engineering & Technology

Er. R.K. Dhawan
M.I.E., M.E., M.I.S.T.E.
Principal
Ramgarhia Institute of Engineering & Technology
Satnampura, Phagwara
(Punjab Technical University, Jalandhar)

S. CHAND
PUBLISHING
S Chand And Company Limited
(ISO 9001 Certified Company)

S Chand And Company Limited
(ISO 9001 Certified Company)
Head Office: Block B-1, House No. D-1, Ground Floor, Mohan Co-operative Industrial Estate, New Delhi – 110 044 | Phone: 011-66672000
Registered Office: A-27, 2nd Floor, Mohan Co-operative Industrial Estate, New Delhi – 110 044 Phone: 011-49731800
www.**schandpublishing.com**; e-mail: **info@schandpublishing.com**

Branches

Chennai	:	Ph: 23632120; chennai@schandpublishing.com
Guwahati	:	Ph: 2738811, 2735640; guwahati@schandpublishing.com
Hyderabad	:	Ph: 40186018; hyderabad@schandpublishing.com
Jalandhar	:	Ph: 4645630; jalandhar@schandpublishing.com
Kolkata	:	Ph: 23357458, 23353914; kolkata@schandpublishing.com
Lucknow	:	Ph: 4003633; lucknow@schandpublishing.com
Mumbai	:	Ph: 25000297; mumbai@schandpublishing.com
Patna	:	Ph: 2260011; patna@schandpublishing.com

First Edition 2011

ISBN: 978-81-219-3592-0 **Product Code:** H3APM64PHYS10ENAA11O

PRINTED IN INDIA

By Vikas Publishing House Private Limited, Plot 20/4, Site-IV, Industrial Area Sahibabad, Ghaziabad – 201 010 and Published by S Chand And Company Limited, A-27, 2nd Floor, Mohan Co-operative Industrial Estate, New Delhi – 110 044.

PREFACE

This book of **Applied Mechanics** is a sincere attempt of the author to present the subject matter in SI units. It covers the syllabii of the various state boards of technical education in India. The needs and requirements of all kinds of students have been taken into consideration while presenting the subject matter in simple and lucid manner. The theoretical contents have been illustrated by suitable diagrams and problems in an elaborate way. At the end of each chapter, Model Test Questions, Questions for Self Examination, i.e., Objective Type Questions have been given which will serve as quick test for students. Problems for Practice are also given for the students to solve themselves.

The multicolour edition of this treatise has been thoroughly updated, strengthened and streamlined. The multicolour illustrations have been incorporated to enhance the content value and to give the students an idea of what they will be dealing in reality and to abridge the gap between theory and actual practice.

Author expresses his deep gratitude to Principal Yudhvir Saini, Principal Jagroop Singh, Principal S.L. Trehn, Principal Prabhkar, Prof. Jasbir Singh, Kamlesh Dhawan and Seema Dhawan for their valueable suggestions during the various stages in the preparation of this book. Author's warmest thanks are due to Mr. Navin Joshi, Vice-President (Publishing), Mr. Shishir Bhatnagar, Manager (Editorial & Pre-Press) S.Chand & Company Ltd.

Errors and suggestions for the improvement of the book will be gratefully acknowledged and incorporated in the next edition of the book.

Er. R.K. DHAWAN

This Page has been intentionally left blank

CONTENTS

1c

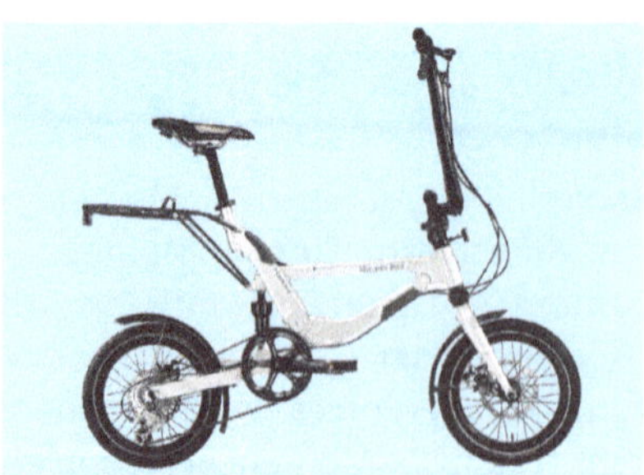

This Page has been intentionally left blank

Chapter 1

Basic Concepts

INTRODUCTION

The science of Mechanics mainly concerned with the knowledge of state of rest or motion of bodies under the action of different forces. It is a common experience that various bodies in this universe are either at rest or in motion. Naturally, to study the behaviour of different bodies, the knowledge of Engineering (Applied) Mechanics is of paramount importance. Although there are only a few fundamental laws and principles of Mechanics yet their applications are so varied and extensive that an engineer has to often depend upon this subject for solving many problems. It is with this view that this chapter deals entirely to explain the basic concepts of Engineering Mechanics. Once the reader has mastery of these basic concepts, he can readily apply them to solve even complicated problems.

1.1. MECHANICS

The branch of science which deals with the study of action of forces on objects is called Mechanics.

In other words, Mechanics deals with the behaviour of forces and their effects on different bodies.

1.2. ENGINEERING MECHANICS (APPLIED MECHANICS)

The branch of engineering which deals with the study of different laws of Mechanics as applied to Engineering problems is called Engineering Mechanics or Applied Mechanics.

1.3. IMPORTANCE AND NECESSITY OF ENGINEERING MECHANICS

The study of Engineering Mechanics has utmost importance for an engineer because the knowledge of this subject enables him to design all types of structures and machines successfully.

Therefore, it is very necessary to study the various laws of Engineering Mechanics to determine the forces acting on the members of framed structure, motion of vehicles, flight of an aeroplanes, projectiles etc. so as to execute the design and construction in the engineering field.

Sir Issac Newton (1643–1727)

Note: *As a matter of fact, scientific history of Engineering Mechanics starts with* ***Sir Issac Newton*** *(1643–1727). He introduced the concept of force and mass and gave Laws of Motion in 1686. Today, knowledge of Engineering mechanics, coupled with the knowledge of other subjects e.g. strength of materials, Theory of machines, Calculus, Vector Algebra etc. to touch its present height.*

1.4. BRANCHES OF ENGINEERING (APPLIED) MECHANICS

Depending upon the state of position of bodies under the action of forces, Engineering Mechanics may be divide into the following two branches :

1. Statics
2. Dynamics
   (a) Kinematics
   (b) Kinetics

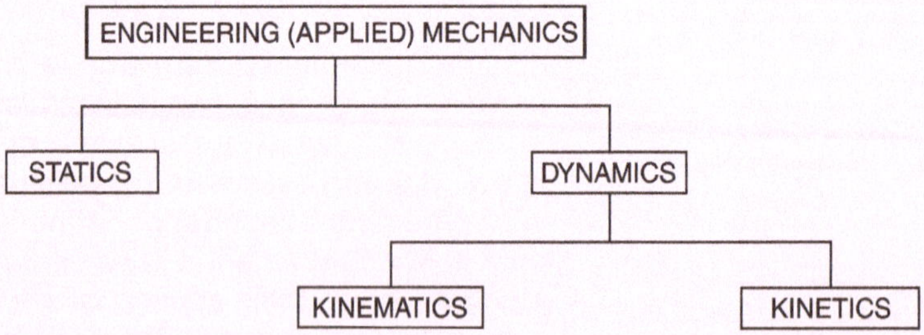

1. Statics : *The branch of Engineering Mechanics which deals with the action of forces on bodies at rest is called statics.*

Infact, statics is the study of the equilibrium of bodies under the action of forces. Here, we are mainly concerned with the conditions of equilibrium of stationary bodies.

For example, the *equilibrium of a book lying on the table* and the study of different forces acting on it are topics concerning to statics.

2. Dynamics : *The branch of Engineering Mechanics which deals with the action of forces on bodies in motion is called Dynamics.*

Infact, Dynamics is the study of the motion of bodies kinematically or kinetically. Here, we are mainly concerned with the study of motion of bodies and the effect of forces acting on them.

For example, working of a watch and the study of different forces acting on its various parts are topics concerning to Dynamics.

Dynamics may be further divided into two parts :

(a) Kinematics
(b) Kinetics

(a) **Kinematics**: *The branch of dynamics which deals with the motion of bodies without any reference to the forces producing motion is called kinematics.*

Kinematics is used to relate displacement, velocity, acceleration and time of the bodies without any reference to the cause of motion.

(b) **Kinetics**: *The branch of dynamics which deals with the motion of bodies with reference to the forces producing motion is called kinetics.*

Kinetics is used to predict the motion of body caused by given forces or to determine the forces required to produce a given motion.

1.5. RIGID BODY

The body, which does not change its shape or size under the action of forces is called rigid body.

Infact, in this universe no body is perfectly rigid. It is because some change takes place in the body under the action of external forces depending upon the shape and size. But this change is very small as compared to the size of the body and therefore, the body is considered as rigid body.

1.6. UNIT

The quantity used as the standard of measurement is called unit.

In the study of Engineering Mechanics, we have to deal with a large number of quantities like length, mass, time, velocity, etc. Therefore, we must be able to measure these quantities. In order to measure a quantity, we select another quantity of same kind as standard. This standard quantity is known as unit.

For example, when we say that the length of a rod is 2 metres, we compare length of the rod with a standard quantity of length called metre. Hence, metre is the unit of length.

1.7. TYPES OF UNITS

The following two types of units are used in practice:

1. Fundamental units
2. Derived units

1. Fundamental units: *The units adopted for measuring the fundamental quantities i.e., mass, length and time are called fundamental units.*

2. Derived units: *The units which are derived from the fundamental units are called derived units.*

For example, units of *area*, *velocity*, *acceleration*, *pressure*, etc.

We know, Area = length × length

Since, the unit of area is derived from the unit of length (fundamental unit), therefore, it is a derived unit.

$$\text{Similarly, Velocity} = \frac{\text{Distance}}{\text{Time}} = \frac{\text{Length}}{\text{Time}}$$

Since, the unit of velocity is derived from the unit of length and time, therefore, it is a derived unit.

1.8. SYSTEM OF UNITS

Following four systems of units are used in Engineering Mechanics:

1. Foot, Pound, Second system i.e., F. P. S. system.
2. Centimetre, Gram, Second system i.e., C. G. S. system.
3. Metre, Kilogram, Second system i.e., M. K.S. system.
4. System International Units system i.e., S. I. system.

1. **F. P. S. system**: *The system in which the fundamental units of length, mass and time are foot, pound and second is called F. P. S. system.*

2. **C. G. S. system**: *The system in which the fundamental units of length, mass and time are centimetre, gram and second is called C. G. S. system.*

3. **M. K. S. system**: *The system in which the fundamental units of length, mass and time are metre, kilogram and second is called M. K. S. system.*

Metre: The international metre may be defined as the shortest distance (at 0°C) between two parallel lines engraved upon the polished surface of the Platinum-iridium bar, kept at the International Bureau of Weights and Measures at Serves near Paris.

Kilogram: The international kilogram may be defined as the mass of the Platinum-iridium cylinder, which is also kept at the International Bureau of Weights and Measures at Sevres near Paris.

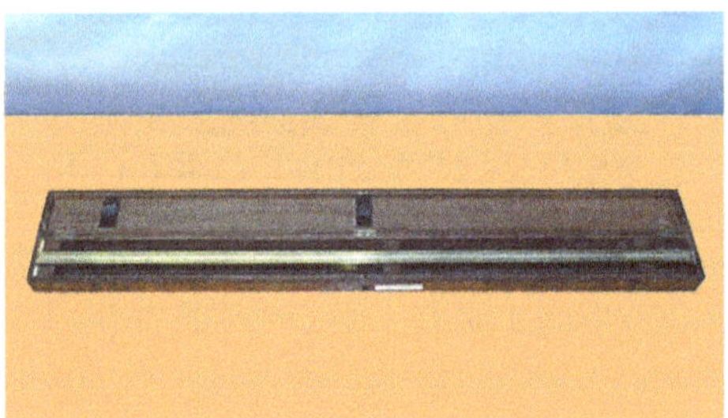

A bar of platinum-iridium metre kept at a temperature of 0°C at the international Bureau of Weights and Measures at Serves in France.

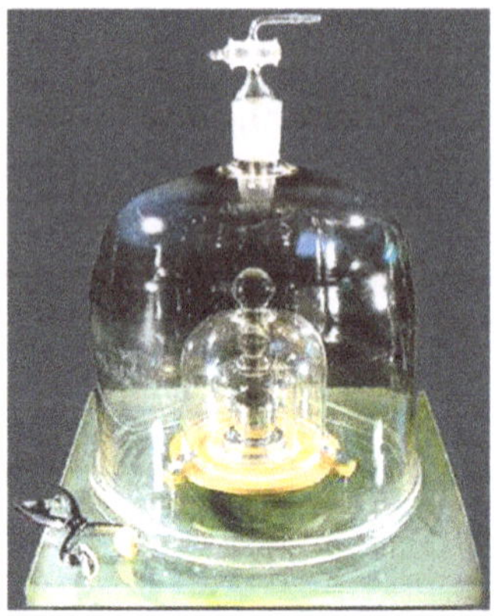

The standard platinum-kilogram is kept at the International Bureau of Weights and Measures at Serves in France.

Second: The fundamental unit of time for all the four systems is second, which is 1(24 × 60 × 60) = 1/86400th of the mean solar day.

A solar day may be defined as the interval of time between the instants at which the sun crosses the meridian on two consecutive days. This value varies throuhout the year. The average of all the solar days, of one year, is called the mean solar day.

4. **S. I. Units system**: *The system which considers three more fundamental units of electric current, temperature and luminous intensity in addition to the fundamental units of length, mass and time is called S. I. units system.*

This system of units was agreed upon at the 11th International General Conference of weights and measures held in Paris in October 1960. It rationalizes the main metric units of measurement and also standardizes their names and symbolic representation.

The recommendations of this conference were endorsed by the International Standards Organisation (I. S. O.) in 1962.

Infact, the SI unit system in the MKS system is absolute units.

The six basic units (SI units) are given below:

S. No.	Physical Quantity	Name of SI units	Symbol
1.	Length	metre	m
2.	Mass	kilogram	kg
3.	Time	second	s
4.	Electric current	ampere	A
5.	Temperature	kelvin	K
6.	Luminous intensity	candela	cd

In addition to these six basic units, two more supplementary units have been introduced in this system.

S. No.	Physical Quantity	Name of SI unit	Symbol
1.	Plane angle	radian	rad
2.	Solid angle	steradian	sr

In SI system, the unit of force is derived from that of mass by considering the force required to impart unit mass (i.e. 1 kg), a unit acceleration (i.e.1 m/sec^2) and given the name newton (N).

Therefore, the unit of force is newton and is the main departing feature of this system from M. K. S. system.

Table 1.1 shows some physical quantities and their derived units used in MKS and SI systems.

TABLE 1.1.
IMPORTANT MKS SYSTEM AND SI SYSTEM UNITS

S. No.	Quantities	MKS System		SI System	
		Unit	Symbol	Unit	Symbol
1.	Area	square metre	m^2	Square metre	m^2
2.	Volume	cubic metre	m^3	cubic metre	m^3
3.	Density	kilogram per cubic metre	kg/m^3	kilogram per cubic metre	kg/m^3
4.	Velocity	metre per second	m/s	metre per second	m/s
5.	Acceleration	metre per second per second	m/s^2	metre per second per second	m/s^2

6.	Force	kologram weight or newton	kgf or N	newton	N
7.	Moment of force	kilogram	kgf-m	newton metre	Nm
8.	Work	kilogram metre	kgf-m	newton metre	Nm
9.	Torque	kilogram metre	kgf-m	newton metre	Nm
10.	Power	kilogram metre per second	kgf-m/s	watt or joule/ second	W or J/s
11.	Energy	kilogram metre	kgf-m	newton metre	Nm
12.	Pressure Intensity or stress	kilogram per square metre	kgf/m^2	newton per square metre	N/m^2
13.	Momentum	kilogram metre per second	kg-m/s	kilogram metre per second	kg m/s
14.	Plane angle	radian	rad	radian	rad
15.	Heat	kilo calorie	kcal	joule	J
16.	Mass	kilogram	kg	kilogram	kg

1.9. SCALAR AND VECTOR QUANTITIES

The various quantities used in Engineering Mechanics may be classified into two types, namely:

1. Scalar quantities and
2. Vector quantities

1. Scalar quantities: *The quantities which possess magnitude only are called scalar quantities.*

These quantities are completely known by their *magnitude*. Length, area, volume, mass, time, speed, etc. are examples of scalar quantities.

2. Vector quantities: *The quantities which possess magnitude and direction are called vector quantities.*

Displacement, acceleration, velocity, force, weight, momentum, impulse, etc. are the examples of vector quantities.

1.10. MASS

The quantity of matter contained in a body is called mass.

The mass of a body is a constant quantity and is independent of place and position of the body. Thus, the mass of a body is the same on the surface of earth, at the centre of earth or on the moon.

Mass is a scalar quantity i.e. it possesses magnitude only. The unit of mass in M.K.S. and S.I. system is kilogram (kg).

1.11. WEIGHT

The force with which a body is attracted towards the centre of earth is called weight.

Weight = Force with which the body is attracted towards earth

= Mass of body × Acceleration due to gravity.

$W = m \times g$ [∵ F = mass × acceleration]

The value of 'g' is usually taken as 9.81 m/sec^2 in M.K.S. system as well as in SI unit unless and until specified.

Weight of a body is not constant at all places. It is becuase the value of 'g' is different at different places. Thus, the weight of a body will be different at the equator and at the poles of earth.

Weight is a vector quantity i.e., it possesses magnitude and direction. The unit of weight are kg wt. in MKS and N in SI system.

DIFFERENCE BETWEEN MASS AND WEIGHT

S. No.	Mass	Weight
1.	It is the qantity of matter contained in a body.	It is the force with which the body is attracted towards the centre of earth.
2.	It is constant at all places.	It is not constant at all places.
3.	It resists motion in the body.	It produces motion in the body.
4.	It is a scalar quantity.	It is vector quantity.
5.	It is measured in kg in M.K.S. and S.I. units.	It is measured in kg wt in M.K.S. and N in S.I. units
6.	It can be measured by ordinary balance.	It is to be measured by spring balance.
7.	It is never zero.	It is zero at the centre of earth.

Problem 1: Calculate the weight of a body on the moon having a mass of 10 kg on the earth. The gravitational acceleration on the moon is 1.7 m/s^2.

Solution: Mass of body, m = 10 kg

Weight on the moon, $W = m \times g$ [$g = 1.7$ m/s^2]

= 10 kg × 1.7 m/s^2

= 17 kg m/s^2

= **17 newton [Ans]**

1.12. VECTOR

A quantity having magnitude, direction and sense is called a vector.

A vector is represented by scaled straight line in a co-ordinate space with a specific inclination and an arrow head.

1.13. IMPORTANT MATHEMATICAL FORMULAE

The following are the important mathematical formulae mostly used in this book:

(a) **Trigonometry :** In a right angled triangle (see Fig. 1.1).

1. $\sin\theta = \frac{AB}{AC}$; $\cos\theta = \frac{CB}{AC}$ and $\tan\theta = \frac{AB}{CB}$

2. $\tan\theta = \frac{\sin\theta}{\cos\theta}$; $\text{cosec}\,\theta = \frac{1}{\sin\theta}$; $\sec\theta = \frac{1}{\cos\theta}$; $\cot\theta = \frac{1}{\tan\theta}$

3. $\sin(A + B) = \sin A \cos B + \cos A \sin B$

4. $\sin(A - B) = \sin A \cos B - \cos A \sin B$

5. $\cos(A + B) = \cos A \cos B + \sin A \sin B$

6. $\cos(A - B) = \cos A \cos B - \sin A \sin B$

7. $\tan(A + B) = \frac{\tan A + \tan B}{1 - \tan A \tan B}$

8. $\tan(A - B) = \frac{\tan A - \tan B}{1 + \tan A \tan B}$

9. $\sin 2A = 2 \sin A \cos A$

10. $\sin^2\theta + \cos^2\theta = 1$

11. $1 + \tan^2\theta = \sec^2\theta$

12. $1 + \cos^2\theta = \cos ec^2\theta$

13. $1 + \cos\theta = 2\cos^2\frac{\theta}{2}$

14. $1 - \cos\theta = 2\sin^2\frac{\theta}{2}$

15. Rules for change of trigonometerical ratios:

$$\sin(-\theta) = -\sin\theta$$

$$\cos(-\theta) = \cos\theta$$

$$\tan(-\theta) = -\tan\theta$$

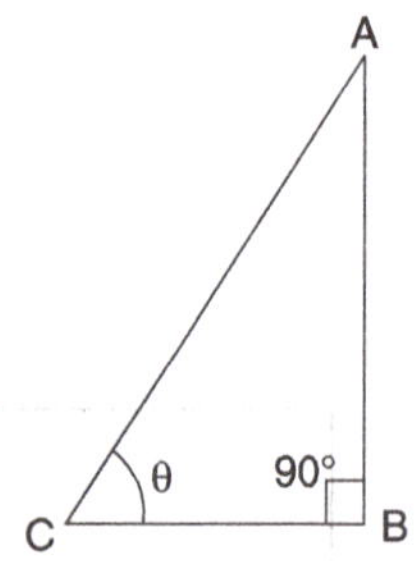

Fig. 1.1.

Change of important T-Ratios

Angle	$-\theta$	$90° - \theta$	$90° + \theta$	$180° - \theta$	$180° + \theta$	$360° - \theta$	$360° + \theta$
sin	$-\sin\theta$	$\cos\theta$	$\cos\theta$	$\sin\theta$	$-\sin\theta$	$-\sin\theta$	$\sin\theta$
cos	$\cos\theta$	$\sin\theta$	$-\sin\theta$	$-\cos\theta$	$-\cos\theta$	$\cos\theta$	$\cos\theta$
tan	$-\tan\theta$	$\cot\theta$	$-\cot\theta$	$-\tan\theta$	$\tan\theta$	$-\tan\theta$	$\tan\theta$

16. In any triangle ABC, (sine formula) as shown in Fig. 1.2.

$$\frac{a}{\sin A} = \frac{b}{\sin B} = \frac{c}{\sin C}$$

where a, b and c are the lengths of three sides and A, B and C are the opposite angles to the sides a, b and c respectively.

17. Solution of quadractic equation.

if $ax^2 + bx + c = 0$

$$\therefore \quad x = \frac{-b \pm \sqrt{b^2 - 4ac}}{2a}$$

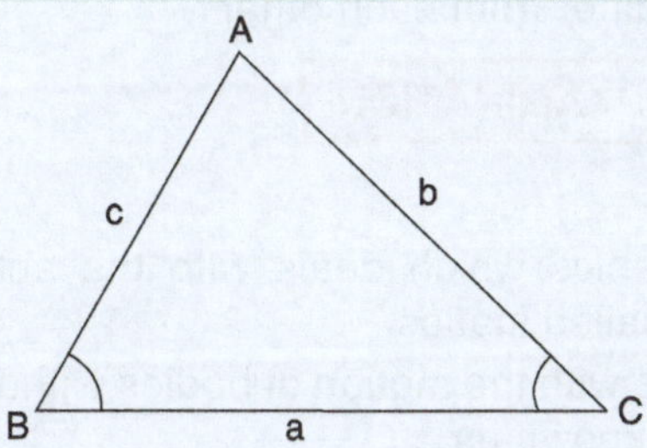

Fig. 1.2.

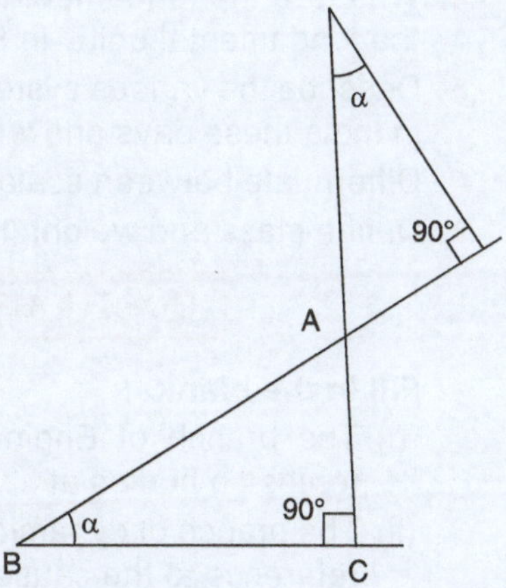

Fig. 1.3.

18. In any triangle ABC (cosine formula) as shown in Fig. 1.2.

$$a^2 = b^2 + c^2 - 2bc\cos A$$

19. The angle between any two lines is equal to the angle between their perpendiculars (see Fig. 1.3)

20. $x^\circ = 1$ (*Any thing raised to the power zero is one*)

$x^m \times x^n = x^{m+n}$ (*In multiplication the powers are added*)

$\frac{x^m}{x^n} = x^{m-n}$ (*In division the powers are subtracted*)

Abbreviations in powers of ten

Sub-multiples			Multiples		
Perfix	**Symbol**	**Power of 10**	**Perfix**	**Symbol**	**Power of 10**
deci	d	10^{-1}	deca	da	10^{1}
centi	c	10^{-2}	hecto	h	10^{2}
milli	m	10^{-3}	kilo	k	10^{3}
micro	μ	10^{-6}	mega	M	10^{6}
nano	n	10^{-9}	giga	G	10^{9}
plco	p	10^{-12}	tera	T	10^{12}
femto	f	10^{-15}	peta	P	10^{15}
atto	a	10^{-18}	exa	E	10^{18}

MODEL TEST QUESTIONS

1. Define Mechanics and Engineering Mechanics. What are the different classes of Engineering Mechanics?
2. Why is the study of Engineering Mechanics so essential for engineers?
3. Differentiate between Kinematics and Kinetics.
4. What are the fundamental and derived units? Why are they so called? State the fundamental units in SI systems.
5. Describe the various systems of units. Which sytems of units is being followed in India these days and why?
6. Differntiate between scalar and vector quantities with five examples of each.
7. Define mass and weight. How do they differ from each other?

QUESTIONS FOR SELF EXAMINATION

1. **Fill in the blanks:**
   (i) The branch of Engineering Mechanics which deals with the action of forces on bodies at .... is called statics.
   (ii) The branch of dynamics which deals with the motion of bodies without any reference to the cause of motion is known as.............. .
   (iii) The body in which the distance between any two particles remains constant is known as .... body.
   (iv) The fundamental quantities are ...., ... and ....
   (v) The basic units in SI system are ....
   (vi) The force with which the body is attracted towards the earth's centre is called ....
   (vii) .... remains constant any where in this universe.
   (viii) The quantity of matter contained in a body is called ....

Ans. (i) rest (ii) kinematics (iii) rigid (iv) mass, length and time (v) metre, kilogram, second, ampere, kalvin and candela. (vi) weight (vii) mass (viii) mass.

2. **Pick up the correct words:-**
   (i) Weight of a body is *constant/variable*.
   (ii) A book lying on the table is in *statics/dynamics*.
   (iii) The unit of velocity is *fundamental/derived unit*.
   (iv) Weight is measured in kg/N in SI unit.
   (v) A quantity having magnitude, direction and sence is called a free *vector/ vector*.

Ans. (i) *variable* (ii) *statics* (iii) derived (iv) N (v) vector.

Chapter 2

Laws of Forces

INTRODUCTION

In practice, an engineer generally comes across such situations where number of forces are acting on a body. For example, an engineer while designing an aeroplane is concerned with the air pressure on the face of plane, weight of the plane etc. Only the full knowledge of these forces and the laws governing them can enable him to make safe and economical design. This chapter is devoted to study the basic concept of a force and analysis of force systems. Once the reader gets familiar with the various types of forces acting on the body, he can easily solve the problems encountered in the engineering field.

2.1. FORCE

A push or pull which produces or tends to produce, stops or tends to stop the motion in a body is called a force.

Example: (i) A cricket ball lying on the ground is at rest but when it is pushed (see Fig. 2.1), it begins to move. This push which has produced motion of the ball is the force.

Similarly, when you apply your hand to a moving ball (see Fig. 2.2), it comes to rest. In this case, the motion of the ball is stopped. This pull which has stopped the motion of the ball is the force.

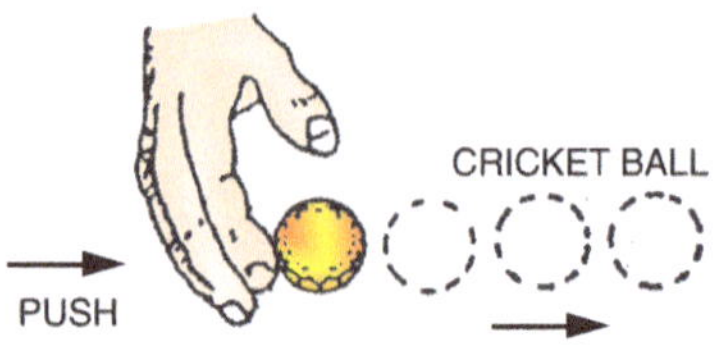

Fig. 2.1. *Hand pushing the ball*

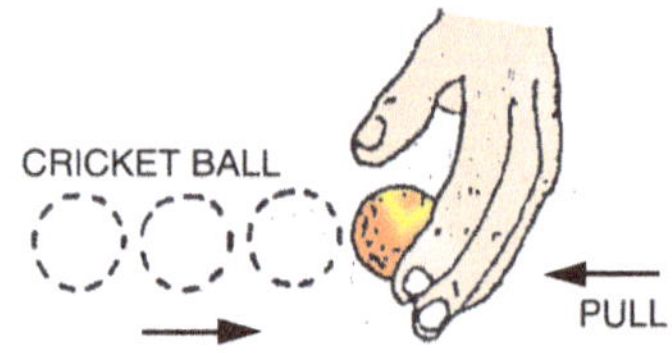

Fig. 2.2. *Hand stopping the ball*

(ii) A single man is pushing the truck to move it as shown in Fig. 2.3. Here the man is unable to produce motion in the truck but he is tending to do so. This tendency to produce motion is the force.

Fig. 2.3. *Man tending to push the truck*

Fig. 2.4. *Man tending to stop the truck*

Similarly, a man trying to stop a moving truck fails to do so, as shown in Fig. 2.4. Here, though the man cannot stop the truck yet he is tending to stop it. This tendency to stop motion is the force.

2.2. EFFECTS OF A FORCE

A force has the following two types of effects on a body:

1. External effect
2. Internal effect

1. External effect: Force may produce the following external effects on a body:

(i) It may produce or stop the motion in a body. For example in Fig. 2.1, the external effect of a force is to produce motion in the ball. Similaly, in Fig. 2.2, the external effect of a force is to stop the motion of the ball.

(ii) It may balance the forces already acting on the body, thus bringing it to rest or in equilibrium.

2. Internal effect: The Internal effect of a force is to produce deformation (change of size or shape) in the body. If the body is rigid, it will not undergo any deformation under the action of forces acting on it.

Note: *In this book the bodies are considered rigid so that external effect of the force system will be dealt with.*

2.3. UNITS OF FORCE

The following are two systems of units of force:-

1. Gravitatioal units
2. Absolute units.

1. Gravitational units: *The gravitational or engineering unit of force is the force with which the earth attracts a unit mass of the body.*

i.e., Gravitational unit of force = Weight of unit mass.

Thus, the gravitational units* of force are lb wt., gm wt, kg-wt in F.P.S., C.G.S and M.K.S. system respectively. The engineers generally use gravitational of force.

It may be added here that giving force in gravitational units, the term 'wt' is generally omitted e.g. force of 5 kg means a force of 5 kg wt.

2. Absolute units: *The absolute or scientific units of force is based on Newton's Second Law of motion.*

i.e., Force = Mass × Acceleration

The absolute units of force are; poundal, dyne, newton in the F.P.S., C.G.S and M.K.S. (or SI) systems respectively.

Poundal: *It is that much force which produces an acceleration of 1 ft/sec² in a mass of 1 pound.*

Dyne: *It is that much force which produces an acceleration of 1 cm/sec² in a mass of 1 gm.*

Newton: *It is that much force which produces an acceleration of 1 m/sec² in a mass of 1 kg.*

$$1 \text{ Newton} = 1 \text{ kg} \times \text{m/sec}^2$$
$$= 1000 \text{ gm} \times 100 \text{ cm/sec}^2 = 10^5 (1 \text{ gm} \times 1 \text{ cm/sec}^2)$$
$$1 \text{ Newton} = 10^5 \text{ dynes}$$

2.4. CONVERSION FROM GRAVITATIONAL SYSTEM TO ABSOLUTE SYSTEM

The unit of force be converted from gravitational system to absolute or vice-versa by the following relation:

Gravitational units of force = 'g' on that system × Absolute unit of force on the system (on any system)

For example, the gravitational unit of force in M.K.S. system is 1 kg wt. The absolute unit in M.K.S. system is 1 newton and value of g is 9.81 m/sec^2.

∴ 1 kg wt = 9.81 times 1 newton, or 1 kg wt = 9.81 newton

The table shows units of force in all systems and their conversion.

S.No.	Systems of units	Gravitational units	Absolute units	Conversion
1.	F.P.S.	1 lb wt	1 poundal	1 bt wt = g poundals = 32.2 poundals
2.	C.G.S.	1 gm wt	1 dyne	1 gm wt = g dynes = 9.81 dynes
3.	M.K.S.	1 kg wt	1 newton	1 kg wt = g newtons = 9.81 newtons
4.	SI	—	newton	—

*Since weight is variable quantity therefore, scientists do not prefer to use gravitational unit of force. But engineers continue it as the concept of weight can be easily sensed.

2.5. CHARACTERISTICS OF A FORCE

The quantities by which a force is fully represented are known as characteristics or elements of a force.

The effects of a force on a body can be fully determined as we know the following quantities called characteristics or elements of a force (see Fig. 2.5):

1. Magnitutde
2. Direction
3. Sense
4. Point of application.

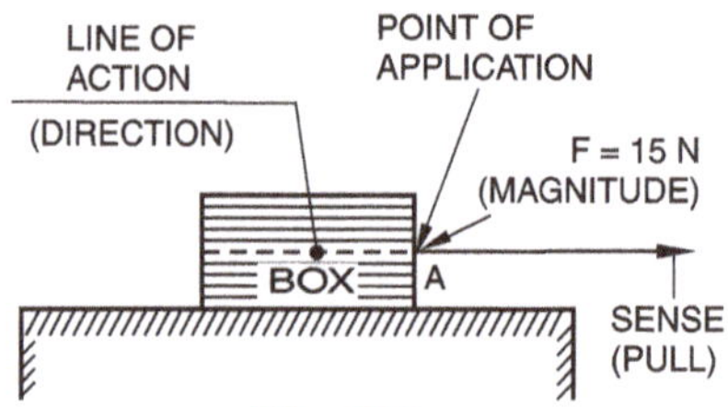

Fig. 2.5.

1. Magnitude: *The quantity of a force is called its magnitude e.g., 15 N.*

2. Direction: *The line along which the force acts is called its direction.*

The direction can be expressed in the following two ways:-

(i) By indicating the angle that the line of action of the force makes with the reference axis (i.e. X-axis or y-axis). For example, the direction of force F (see Fig. 2.6) may be written as, a force making an angle of $\theta°$ with OX′-axis.

(ii) By indicating the angle that the line of action of the force makes with the geographical axis (i.e. North, South, East and West). Thus, in Fig. 2.7 force F_1 can be written as; a force F_1 acting at 45°. North of East (i.e., 45° towards East from North). Similarly force F_2 is acting at 30° North of West.

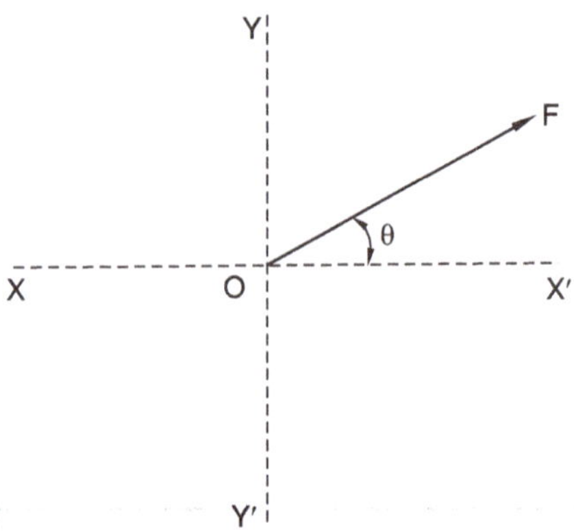

Fig. 2.6.

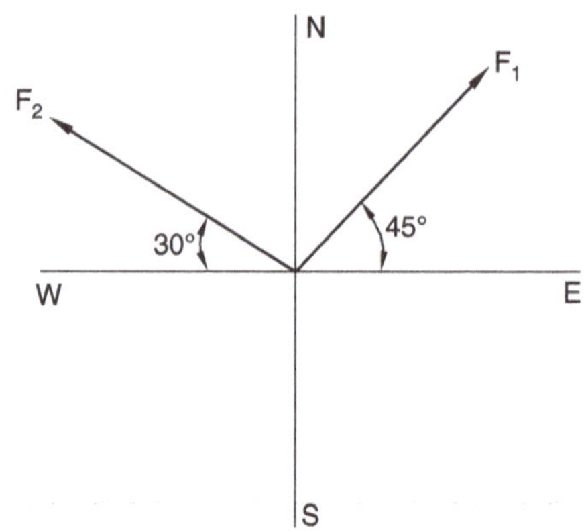

Fig. 2.7.

3. Sense: *The way in which the force acts along the line of action is called the sense of force.*

This is shown by placing an arrow head on the line of action of the force i.e., Pull (see Fig. 2.5).

4. Point of application: *The point at which the force acts on the body is called the point of application.*

For example in Fig. 2.5., point A is the point of application of force on the body.

2.6. MEASUREMENT OF FORCE

A spring balance can be used to measure force by observing the extension produced in the spring. A spring balance is generally callibrated* in vertical position (see Fig. 2.8). The position of the pointer without any load is marked zero. Various division are marked on the spring balance.

The unknown force or weight to be measured is applied to the lower hook of the spring balance. This force produces extension in the spring. The position of the pointer on the scale gives the magnitude of the force applied. The spring balance may be callibrated in grams, kilograms, pounds, newtons etc. according to the requirements.

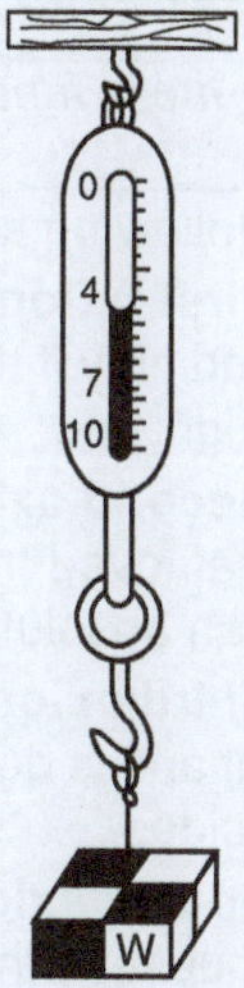

Fig. 2.8. *Spring balance*

2.7. REPRESENTATION OF A FORCE

Force is a vector quantity possessing both magnitude and direction. It can be represented in the following two ways:

1. Vector representation
2. Bow's notation

1. Vector representation: In this method, a force is graphically represented by a straight line (drawn parallel to the line of action of the force), the length of the line represents the magnitude of force and the arrow head indicates the direction of force.

In Fig. 2.9 a force of 10 N acts on the body in the direction AB. This force can be represented by a straight line ab parallel to AB.

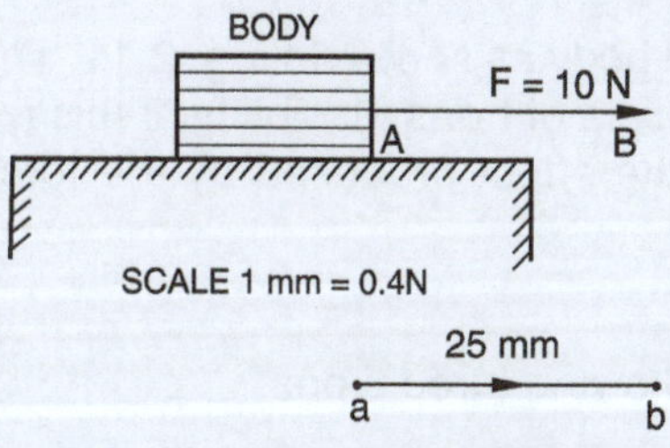

Fig. 2.9. *Vector representation method*

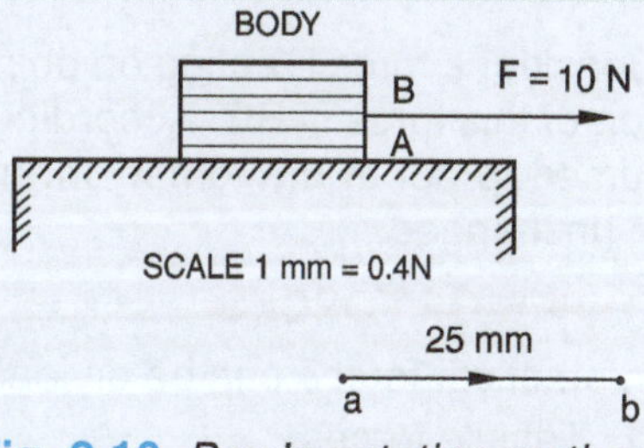

Fig. 2.10. *Bow's notation method*

2. Bow's notation: In this method of representation of a force, capital alphabets are put on either side of the line of action of the force (see Fig. 2.10). The placing of letters is either done clockwise or anti-clockwise.

In Fig. 2.10, a force F is acting on the body in the direction shown. The letter A and B have been put on either side of the line of action of the force. The force F is then called force AB and when the vector for it is drawn, we letter the vector ab (small letter) on some suitable scale.

Bow's notation method is very useful and convenient in solving the problems graphically.

*The spring balance does not indicate the correct force when used in inclined position as the spring balance is callibrated in vertical position.

2.8. AXIOMS IN STATICS

The most fundamental laws in the branch of statics are known as axioms in statics.

The following are the four axioms in statics:

1. First Axiom: An absolutely rigid body is in equilibrium under the action of two forces if and only if the forces are equal in magnitude and opposite in direction along a single straight line.

2. Second axiom: Any balanced system of forces may be added to a system of forces or removed from that system without altering the action of the given system of forces on an absolutely rigid body.

3. Third axiom: The resultant of two forces applied at one point is applied at the same point and is depicted by the diagonal of a parallelogram constructed on the given forces as sides.

4. Fourth axiom: If a nonrigid body is in a state of equilibrium, that equilibrium will not be upset if the body becomes absolutely rigid.

2.9. PRINCIPLE OF TRANSMISSIBILITY

The principle of transmissibility can be defined as under:

"If a force acts at a point in a rigid body, it may be considered to act any other point on the line of action of the force."

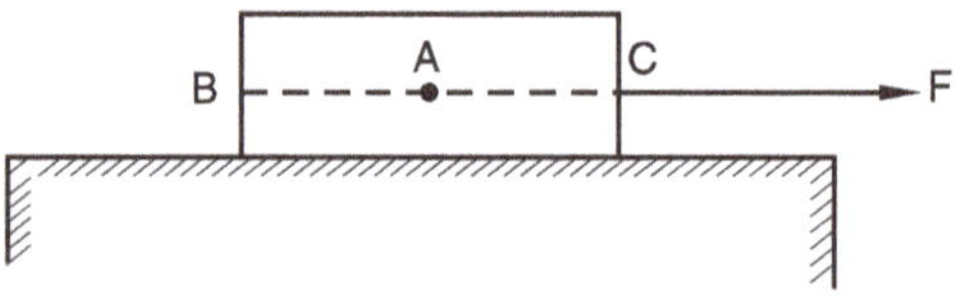

Fig. 2.11. *Principal of transmissibility*

Consider a force F acting on point A in a rigid body as shown In Fig. 2.11. The line of action of this force is BC. According to the principle of transmissibility, if this force F is assumed to act at any other on the line BC, the effect produced by the force will remain unchanged.

2.10. TYPES OF FORCES

In practice, forces can be divided into the following three types:

1. Tensile forces
2. Compressive forces
3. Shear forces

1. Tensile forces: *When two opposite forces of equal magnitude having same line of action, tend to increase the length of the body, the forces applied are called tensile forces.*

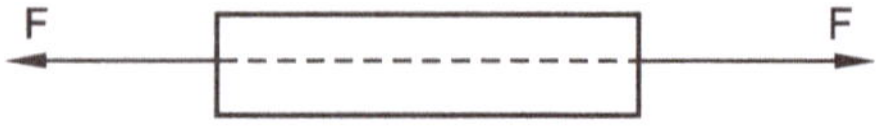

Fig. 2.12. *Tensile forces*

The body under such a condition is said to be in tension. For example, In Fig. 2.12, the bar is subjected to tensile forces and it is in tension

2. Compressive forces: *When two opposite forces of equal magnitude having same line of action, tend to decrease the length of the body, the forces applied are called compressive forces.*

The body under such a condition is said to be in compression. In Fig. 2.13, the forces applied on the bar are compressive forces and the bar is in compression.

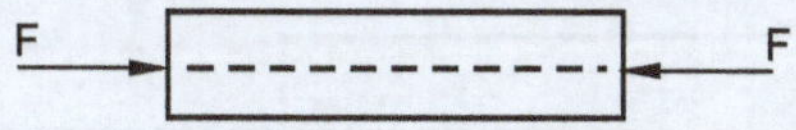

Fig. 2.13. *Compressive forces*

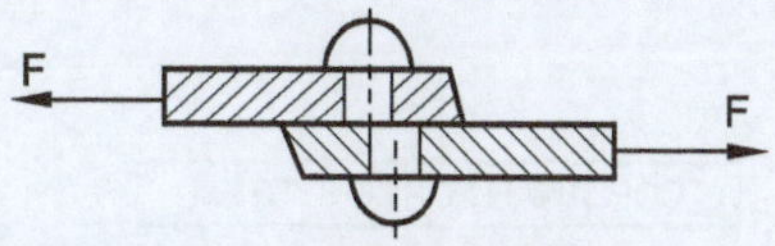

Fig. 2.14. *Shear forces*

3. Shear forces: *When two opposite forces of equal magnitude having different lines of action in such a way that one section of the body tries to slide over the othe after rupture, the applied forces are called shear forces.*

In Fig. 2.14, the rivet is under the action of shear forces. These forces will try to rupture the rivet and the upper portion of the rivet tries to slide over the lower portion.

2.11. ACTION AND REACTION

The force which one body exerts on the second is called the action and the force which the second body exerts on the first is called the reaction.

For example, In Fig. 2.15, a block of weight W is lying on the table. This block is exerting a force W in the downward direction. Therefore, action on the table is the force W and the force R exerted by the table on the block is the reaction. According to newton's third laws of motion, action and reaction are equal and opposite. Therefore, the block is in equilibrium.

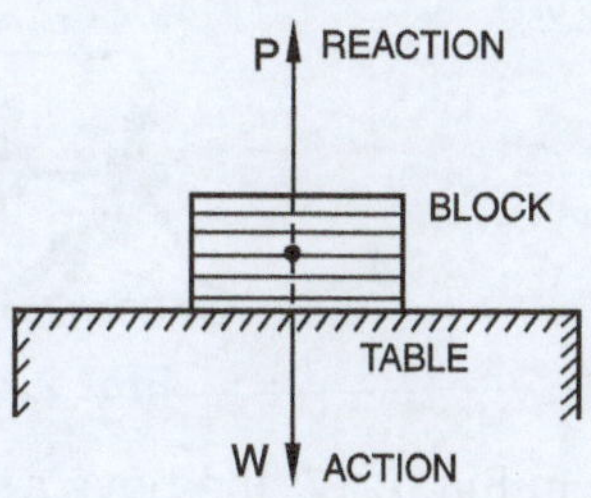

Fig. 2.15. *Action & Reaction*

It should be noted that action and reaction are mutual i.e., one is always called into play by the presence of the other.

2.12. FORCE SYSTEM

A set of forces acting simultaneously on a body is called a force system.

In practice, we come across such siuations where a number of forces act on a body. This is called *force system* or *system of forces*.

It is very essential to analyse the system of forces so that their net effect on a body can be determined.

Illustration: Consider a wooden block resting on a smooth inclined plane, supported by a force P as shown In Fig. 2.16. The force system on the block consists of:

(i) Weight W
(ii) Reaction R of the block on the inclined plane
(iii) Applied force P

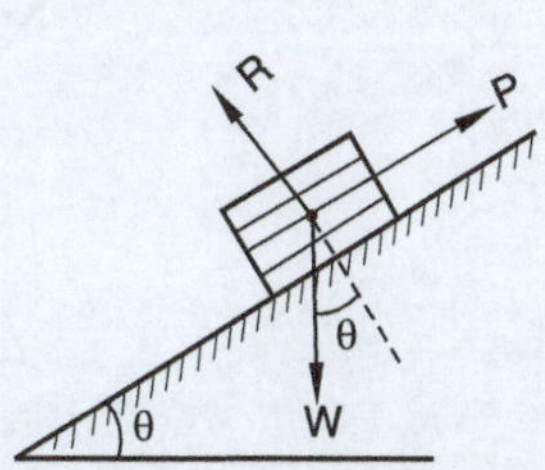

Fig. 2.16. *Force system*

2.13. CLASSIFICATION OF A FORCE SYSTEM

Broadly a force system may be classified as under:-

1. Coplanar force system
2. Non-coplanar force system.

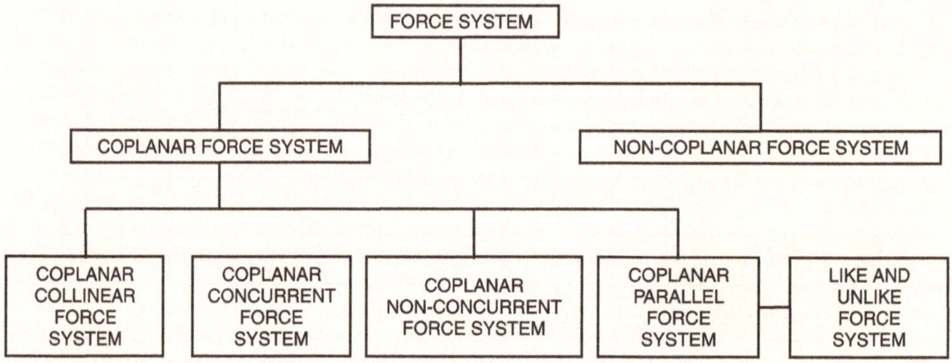

1. Coplanar force system: *The system in which the forces lie in the same plane are called coplanar force system.*

The coplanar force system can be further subdivided into:-

(i) Coplanar collinear force system: *The system in which the forces have common line of action and lie in the same plane are called coplanar collinear force system.*

Fig. 2.17. *Coplanar collinear force system*

In Fig. 2.17, the boys are pulling the rope, where two forces F_1 and F_2 act in the same plane and have the same line of action. Therefore, they are known as co-planar collinear forces.

(ii) Coplanar *concurrent force system: *The system in which the forces meet at a point and lie in the same plane are called coplanar concurrent force system.*

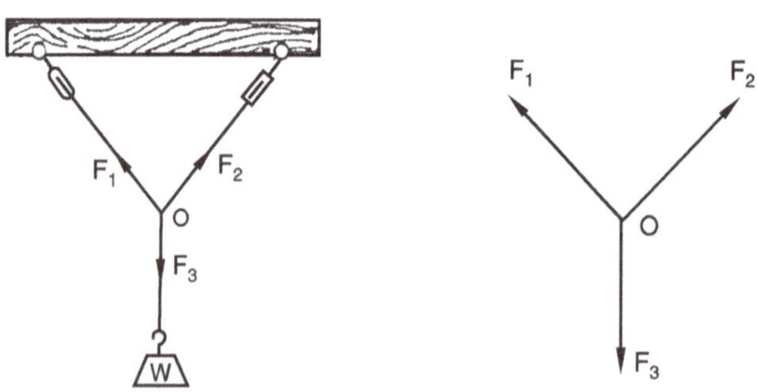

Fig. 2.18. *Coplanar collinear force system*

In Fig. 2.18, a weight W is hung from O where three forces F_1, F_2 and F_3 are acting and they meet at a point O. Therefore, they are coplanar concurrent forces.

(iii) Coplanar non-concurrent force system: *The system in which the forces do not meet at one point but lie in same plane are called coplanar non-concurrent force system.*

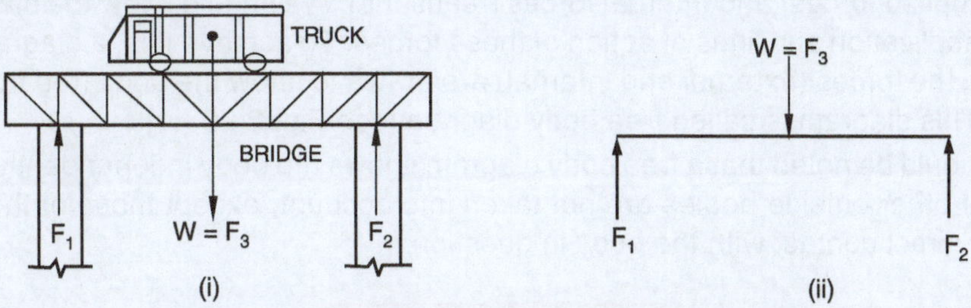

Fig. 2.19. *Coplanar non-current force system*

In Fig. 2.19, the truck on the bridge is shown, where three forces F_1, F_2 and $F_3 = W$ are acting in the same plane but do not meet at a point. Therefore, they are coplanar non-concurrent forces.

(iv) Coplanar parallel force system: *The system in which the forces lie in the same plane having their lines of action parallel to one another are called coplanar parallel force system.*

In Fig. 2.19 (ii), three forces F_1, F_2 and $F_3 = W$ are acting in the same plane having their lines of action parallel to one another. Therefore, they are coplanar parallel forces.

The coplanar parallel parallel forces may be further divided into:-

(a) Coplanar like parallel force system.

(b) Coplanar unlike parallel force system.

When the coplanar parallel forces act in the same direction, they are called coplanar like parallel force system. However, when the coplanar parallel forces act in the opposite direction, they are called coplanar unlike parallel force system. Thus in Fig. 2.19 (ii) F_1 and F_2 are coplanar like parallel forces; F_1 and F_3 or F_2 and F_3 are coplanar unlike parallel forces.

2. Non coplanar force system: *The sytem in which the forces do not lie in the same plane are called non-coplanar force system.*

Non-coplanar forces can be sub-divided into non-coplanar like parallel forces, non coplanar unlike forces, etc.

Note: *All the forces considered in this book are coplanar forces.*

2.14. FREE BODY DIAGRAM

A diagram which shows the forces on the body free of other bodies is called the free body diagram (*see Fig. 2.20*).

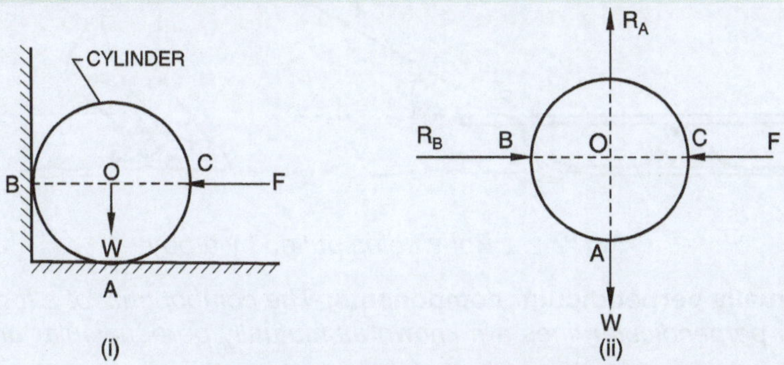

Fig. 2.20. *Free body diagram*

To solve problems In Engineering Mechanics, it is very important to study the various forces acting on a body[see Fig. 2.20 (i)]. The forces involved are external forces (applied forces) and Internal forces (reactions). We should know the magnitude, point of application and lines of action of these forces. To achieve this, a diagram of the body and the forces (external and internal) are drawn to show the body free from other bodies. This diagram is called free body diagram [see Fig. 2.20 (ii)].

It should be noted that a free body diagram shows the body independently and the effects of other outside bodies are not taken into account, except those of the bodies having a direct contact with the body in question.

SELECTED PROBLEM

Problem 1: A cyclinder of weight W = 10 N is lying in a corner as shown in Fig. 2.20 (i) A force F = 15 N is applied horizontal to the cylinder. Draw the free body diagram.

Solution: The weight W of the cylinder acts at the centre O. Due to this weight W, normal reaction R_A acts vertically upwards at point A.

Similarly the horizontal force F produces a horizontal reaction R_B at B [see Fig. 2.20 (i)].

The various forces acting on the cylinder are shown in Fig. 2.20 (ii) and this is the free body diagram.

It may by noted that in order to keep the cylinder in equilibrium.

$$W = R_A = 10 \text{ N} \quad \text{and} \quad F = R_B = 15 \text{ N}$$

2.15. COMPONENTS OF A FORCE

The two forces which can be replaced by a single force acting on a body without causing any change in the effect produced, are called components of the force.

Generally, a force is resolved into two components. These components may be:

1. Mutually perpendicular components
2. Non-perpendicular components

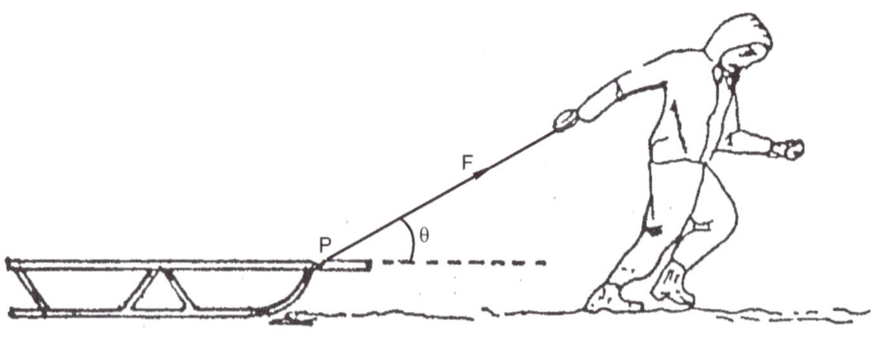

Fig. 2.21. *Body being pulled by a boy*

1. Mutually perpendicular components: *The components of a force along the two mutually perpendicular lines are known as mutually perpendicular or rectangular components.*

Illustration: Consider a boy pulling a body along the ground with a force F. Suppose the rope makes an angle θ with the horizontal as shown in Fig. 2.21.

The force F is resolved into the following two mutually perpendicular components.

(i) Components along OX-axis, $F_x = F\cos\theta$

(ii) Components along OY-axis, $F_y = F\sin\theta$

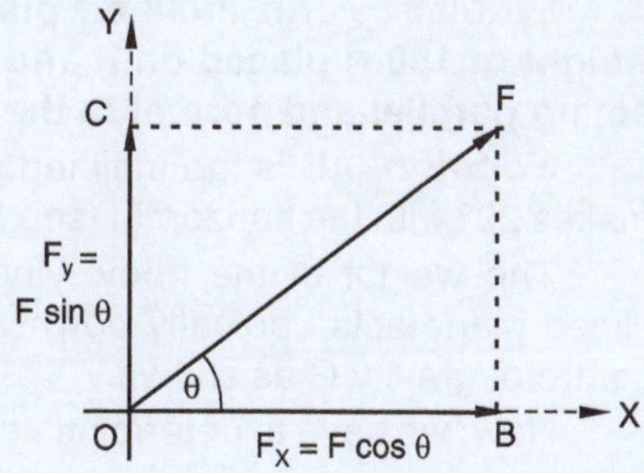

Fig. 2.22.

2. Non-perpendicular components: *The components of a force along two axes which are not mutually perpendicular are known as non-perpendicular components.*

Fig. 2.23 shows a force F acting on a body at an angle α with the horizontal. The force F is resolved into two components (F_1 & F_2).

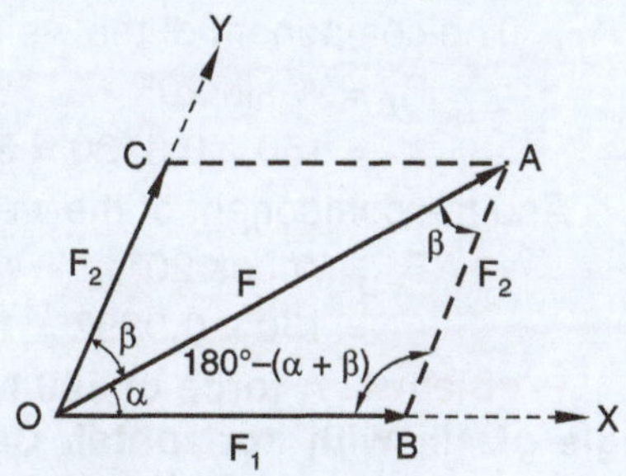

Fig. 2.23.

$$\angle AOB = \alpha \text{ and } \angle COA = \angle OAB = \beta$$

$$\therefore \quad \angle OBA = 180° - (\alpha + \beta)$$

Using sine formula, in Δ AOB

$$\frac{BA}{\sin\alpha} = \frac{OB}{\sin\beta} = \frac{OA}{\sin 180° - (\alpha+\beta)}$$

[$\because$ BA = OC = F_2 and OB = F_1]

$$\therefore \quad \frac{F_2}{\sin\alpha} = \frac{F_1}{\sin\beta} = \frac{F}{\sin(\alpha+\beta)}$$

$$\therefore \quad F_1 = \frac{F\sin\beta}{\sin(\alpha+\beta)} \quad \text{and} \quad F_2 = \frac{F\sin\alpha}{\sin(\alpha+\beta)}$$

SELECTED PROBLEMS

Problem 2: A force of 100 N acts on a body in such a manner that it make an angle of 60° with the vertical axis. Resolve the force into two mutually perpendicular directions.

Solution: The force of 100 N making an agnle of 90° – 60° = 30° with the horizontal is represented by OC in magnitude and direction.

Resolving the force 100 N along OX-axis and OY-axis (see Fig. 2.24).

Component along OX-axis, $F_x = 100 \cos 30°$

$= 100 \times \sqrt{3}/2 =$ **86.6 N [Ans]**

Component along OY-axis, $F_y = 100 \sin 30°$

$= 100 \times 1/2 =$ **50 N [Ans]**

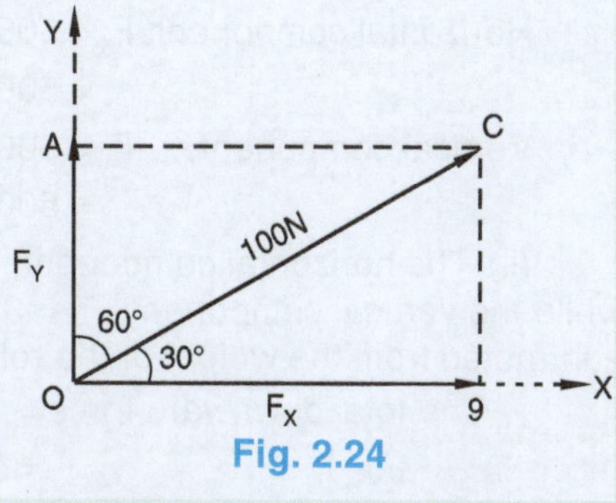

Fig. 2.24

Problem 3: An inclined plane is set an angle of 20° to the horizontal. A weight of 150 N placed on it and is at rest. Calculate components of this weight acting parallel and normal to the plane.

Solution: AB is the inclined plane which makes 20° with the horizontal (see Fig. 2.25)

The weight of the body lying on the inclined plane acts vertically downward through centre of gravity G as shown.

Now, we have to determine its rectangular components i.e., F_x along the plane and F_y normal to the plane.

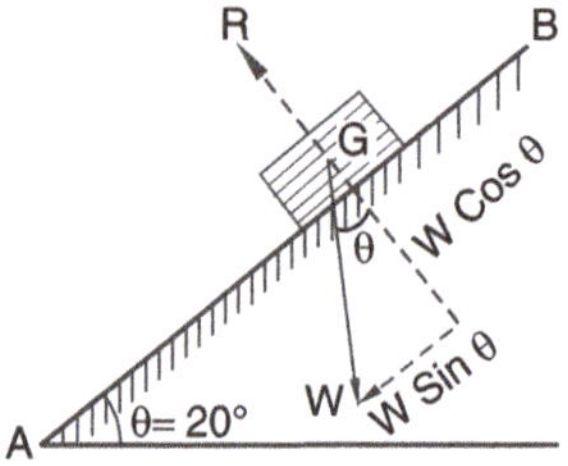

Fig. 2.25.

From geometry of the figure, the angle made by the weight with the normal = 20°

∴ The component of the weight of the body along the plane,

$F_x = W \sin 20°$

$= 150 \times 0.3420 =$ **51.30 N [Ans]**

And the component of the weight of the body normal to the plane,

$F_y = W \cos 20°$

$= 150 \times 0.9397 =$ **140.954 N [Ans]**

Problem 4: A force of 500 N is applied to the handle of a 1500 N roller at an angle of 40° with horizontal. Calculate: (i) The horizontal and vertical components of this force (ii) The force exerted by the roller on the ground, when the roller is (a) pulled and when it is (b) pushed.

Solution : (a) When the roller is pulled [see Fig. 2.26 (i)]

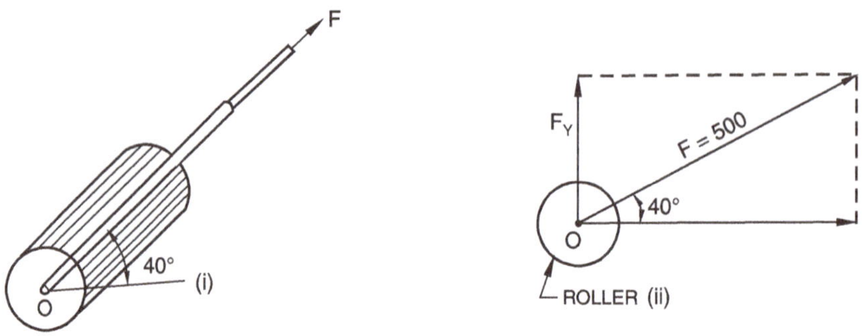

Fig. 2.26.

(i) As shown in Fig. 2.26 (ii) the magnitude of two components F_x and F_y are calculated as below:

Horizontal component, $F_x = 500 \times \cos 40°$

$= 500 \times 0.766 =$ **383 N [Ans]**

Vertical component, $F_y = 500 \times \sin 40°$

$= 500 \times 0.6428 =$ **321.4 N [Ans]**

(ii) The horizontal component, F_x = 383 N is therefore, causing the roller to move, while the vertical component, F_y = 321.4 N acting in the upward direction and must be subtracted from the weight of the roller to find the total downward force on the ground.

∴ The total downward force = 1500 – 321.4 N

= **1178.6 N [Ans]**

(b) When the roller is pushed [see Fig. 2.27 (i)]

(i) As shown in Fig. 2.27 (ii) the magnitude of the two components F_x and F_y are:

$$F_x = 500 \times \cos 40° = 500 \times 0.766 = \mathbf{383\ N\ [Ans]}$$

and $$F_y = 500 \times \sin 40° = 500 \times 0.6428 = \mathbf{321.4\ N\ [Ans]}$$

(ii) In this case too, the horizontal component, F_x = 383 N is causing the roller to move, and the vertical component, F_y = 321.4 N acting in the downward direction and must be added to the weight of roller to find the resultant downward force on the ground.

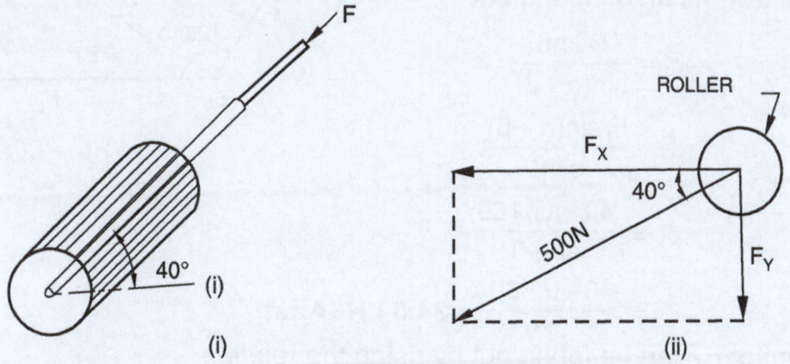

Fig. 2.27

∴ The resultant downward force = 1500 + 321.4

= **1821.4 N [Ans]**

The solution of this problem shows that it is easier to pull a lawn roller than to push it.

Problem 5: A force of 80 N is acting at an angle of 30° to the horizontal. Resolve it along two axis having an angle of 70° in between them. Take one of the axis as horizontal.

Solution : Here F = 80 N, $\alpha = 30°$, $\beta = 40°$

$$\therefore \quad F_1 = F_x \frac{\sin\beta}{\sin(\alpha+\beta)} \quad \text{[see Fig. 2.28]}$$

$$= 80 \times \frac{\sin 40°}{\sin(30°+40°)}$$

$$= 80 \times \frac{0.6428}{0.9397} = \mathbf{54.72\ N\ [Ans]}$$

$$F_2 = F_x \frac{\sin\alpha}{\sin(\alpha+\beta)}$$

$$= 80 \times \frac{\sin 30°}{\sin(30°+40°)}$$

$$= 80 \times \frac{0.5}{0.9397} = \mathbf{42.57\ N\ [Ans]}$$

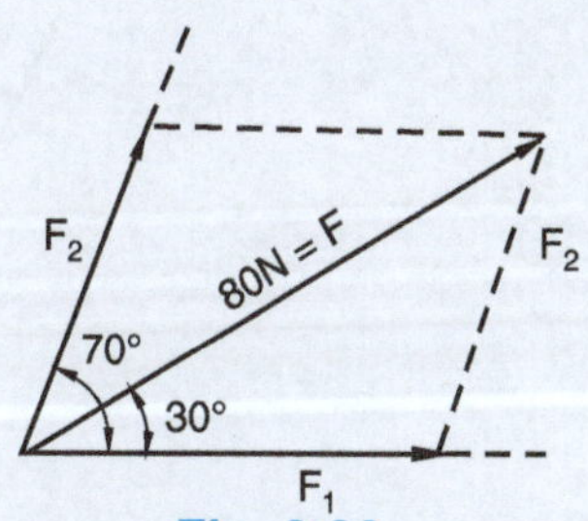

Fig. 2.28.

Problem 6: Two forces act at an angle of 120°. The bigger force is of 40 newtons and their resultant is perpendicular to the smaller one. Find the resultant and the smaller force.

Solution : Let the bigger force be F_1 and and smaller force F_2 acting at an angle of 120° as shown in Fig. 2.29.

Here $F_1 = 40$ N, $\beta = 90°$, $\alpha + \beta = 120°$

$\therefore\ \alpha = 120° - 90° = 30°$

Component F_1 can be found out

$$F_1 = \frac{F \sin\beta}{\sin(\alpha+\beta)}$$

$$\therefore \quad F = \frac{F_1 \sin(\alpha+\beta)}{\sin\beta}$$

$$= \frac{40 \times \sin 120°}{\sin 90°}$$

$$= \frac{40 \times 0.866}{1} = \textbf{34.64 N [Ans]}$$

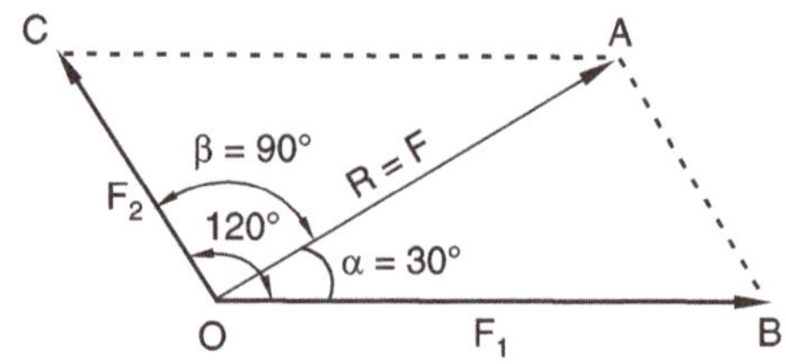

Fig. 2.29.

Component F_2 can be found out by using the relation,

$$F_2 = \frac{F \sin\alpha}{\sin(\alpha+\beta)} = \frac{34.64 \times \sin 30°}{\sin 120°} = \frac{34.64 \times 0.5}{0.866} = \textbf{20 N [Ans]}$$

Problem 7: Two forces of magnitude P + Q and P – Q make an angle 2α with one another and their resultant makes an angle θ with the bisector of the angle between them. Show that P tan θ = Q tan α.

Solution : Let the forces of magnitude P + Q and P – Q act along OA and OB respectively. Let us suppose that the resultant acts along OX.

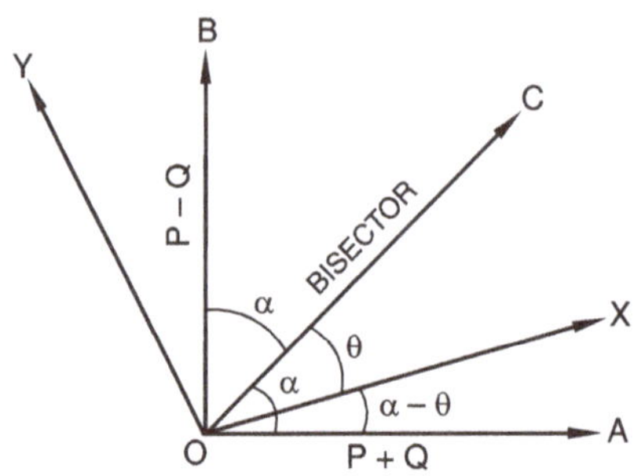

Fig. 2.30.

Then from Fig. 2.30 $\angle AOX = \alpha - \theta$ and $\angle BOX = \alpha + \theta$

Resolving the forces along OY-axis

Resolved part of resultant along OY-axis = Resolved parts of (P + Q) and (P – Q) along OY-axis

$$0 = -(P - Q)\ \sin(\alpha - \theta) + (P - Q)\ \sin(\alpha + \theta)$$

$$= (-P \tan\theta + Q \tan\alpha)\ 2 \cos\alpha \cos\theta$$

$$\therefore \quad -P \tan\theta + Q \tan\alpha = 0 \qquad [\text{Since } \cos\alpha \cos\theta \neq 0]$$

$$\therefore \quad P \tan\theta = Q \tan\alpha$$

2.16. RESULTANT OF CONCURRENT FORCES

If a number of forces are acting simultanously on a body, then a single force which can produce the same effect in the body as is doen by all forces acting together is called resultant force.

The process of finding the resultant of a number of concurrent forces acting on a body is called composition of forces.

It is generally represented by R. In fact, resultant force represents the net effect of various forces acting simultaneously on the body.

2.17. METHODS OF FINDING THE RESULTANT FORCE

The following are the methods used for finding the resultant of number of forces acting on a body:

1. Parallelogram law of forces method
2. Triangle law of forces method
3. Polygon law of forces method.

2.18. PARALLELOGRAM LAW OF FORCES

According parallelogram law of forces,

"If two concurrent forces acting simultaneously on a body are represented in magnitude and direction by the two sides of a parallelogram drawn from a point, then the diagonal of the parallelogram drawn from the same point, represents the resultant force in magnitude and direction."

This method is suitable for finding the resultant of only two concurrent forces.

2.19. RESULTANT OF TWO FORCES

The following are the two cases of resultant of two forces:

1. Resultant of two collinear coplanar forces: In this method the resultant is found by subtracting or adding of two forces (as the case may be). Suppose two forces F_1 and F_2 act as shown in Fig. 2.31 (i). Their resultant will be difference of the two forces [see. Fig. 2.31 (ii)].

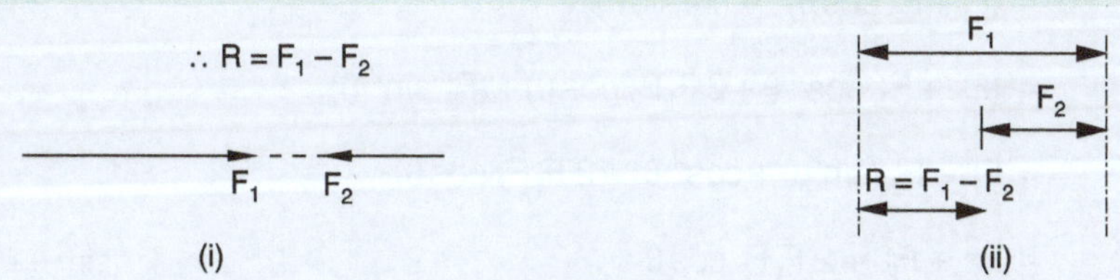

Fig. 2.31.

Now let forces F_1 and F_2 act as shown Fig. 2.32. Their resultant will be sum of the two forces.

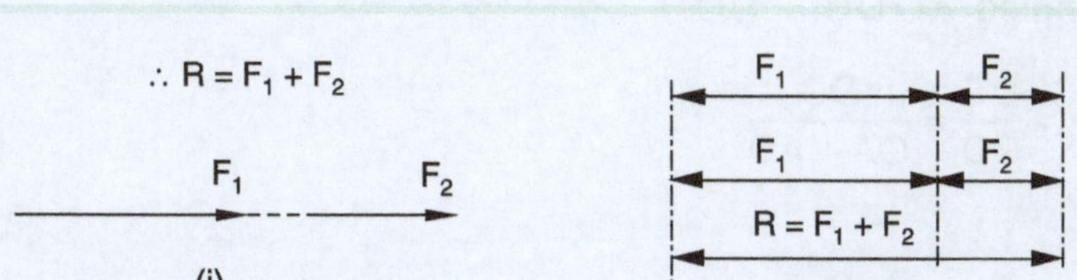

Fig. 2.32.

2. Resultant of two concurrent coplanar forces:

The resultant of two concurrent coplaner coplaner forces can be determined by :

(a) Analytical method

(b) Graphical method

(a) Analytical method:

Magnitude and direction of resultant: Consider two concurrent forces F_1 and F_2 acting on a body at O as shown in Fig. 2.33 (i)

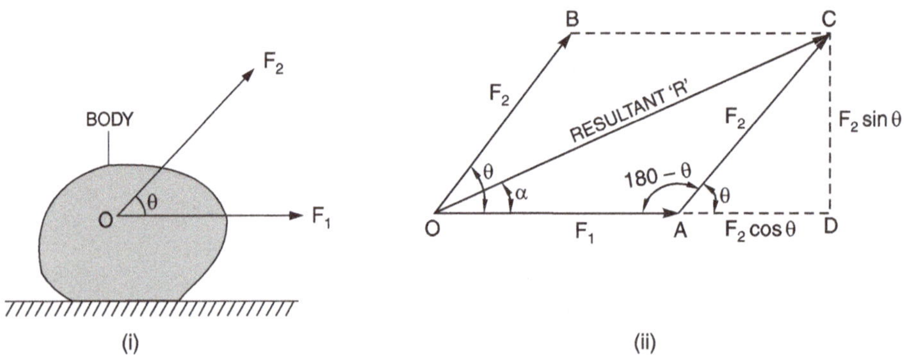

Fig. 2.33.

Let these two forces be represented in magnitude and direction by two sides OA and OB of the parallelogrma OACB. Then, the line joining OC represents the resultant R in magnitude and direction.

Now, we may draw perpendicular CD to meet OA produced as shown in Fig. 2.33 (ii).

Thus, from right-angled ΔODC,

$$OC^2 = OD^2 + DC^2 = (OA + AD)^2 + (DC)^2$$

$$R^2 = (F_1 + F_2\cos\theta)^2 + (F_2\sin\theta)^2 \qquad [\because AD = F_2\cos\theta]$$

$$= (F_1^2 + F_2^2\cos^2\theta + 2F_1F_2\cos\theta) + (F_2^2\sin^2\theta)$$

$$= F_1^2 + F_2^2(\sin^2\theta + \cos^2\theta) + 2F_1F_2\cos\theta)$$

$$= F_1^2 + F_2^2 + 2F_1F_2\cos\theta \qquad [\because (\sin^2\theta + \cos^2\theta) = 1]$$

$$\therefore \quad R = \sqrt{F_1^2 + F_2^2 + 2F_1F_2\cos\theta} \qquad \text{...(i)}$$

Now for finding the direction of the resultant (R), we may proceed as:

In right-angled ΔODC, $\angle COD = \alpha$

$$\tan\alpha = \frac{DC}{OD} = \frac{DC}{OA + AD} \qquad [\because DC = F_2\sin\theta]$$

$$[AD = F_2\cos\theta]$$

$$\therefore \quad \tan\alpha = \frac{F_2\sin\theta}{F_1 + F_2\cos\theta} \qquad \text{...(ii)}$$

Particular cases: *Some special cases relating to the above topic are discussed as:*

Case I: *When the two forces act along the same straight line and in the same direction i.e.,* $\theta = 0°$:-

In this case, by substituting the value of θ in equation (i), we get

$$R = \sqrt{F_1^2 + F_2^2 + 2F_1F_2 \cos 0°} \qquad [\because \cos 0° = 1]$$

$$= \sqrt{F_1^2 + F_2^2 + 2F_1F_2}$$

$$\therefore \quad R = F_1 + F_2$$

$$\tan\alpha = \frac{F_2 \sin\theta}{F_1 + F_2 \cos\theta} = \frac{F_2 \times 0}{F_1 + F_2 \times 1} = 0$$

Case II: *When the two forces act along the same straight line, but opposite in direction i.e., if* $\theta = 180°$

$$R = \sqrt{F_1^2 + F_2^2 + 2F_1F_2 \cos 180°} \qquad [\because \cos 180° = -1]$$

$$= \sqrt{F_1^2 + F_2^2 - 2F_1F_2}$$

$$\therefore \quad R = F_1 - F_2$$

$$\tan\alpha = \frac{F_2 \sin\theta}{F_1 + F_2 \cos\theta} = \frac{F_2 \times 0}{F_1 + F_2 \times -1} = 0$$

Case III: *When the two forces act at right angles i.e., if* $\theta = 90$:-

$$R = \sqrt{F_1^2 + F_2^2 2F_1F_2 \cos 90°}$$

$$= \sqrt{F_1^2 + F_2^2} \qquad [\because \cos 90° = 0]$$

$$\therefore \quad R = \sqrt{F_1^2 + F_2^2}$$

$$\tan\alpha = \frac{F_2 \sin 90°}{F_1 + F_2 \cos 90°} = \frac{F_2}{F_1} \qquad [\because \sin 90° = 1]$$

Case IV: *When the two forces are equal i.e.,* $F_1 = F_2$

$$R = \sqrt{F_1^2 + F_2^2 + 2F_1F_2 \cos\theta} = \sqrt{2F_1^2(1 + \cos\theta)}$$

$$R = \sqrt{4F_1^2 \cos^2\theta/2} \qquad [\because F_1 = F_2]$$

$$\therefore \quad R = 2F_1 \cos\theta/2 \qquad [\because 1 + \cos\theta = 2\cos^2\frac{\theta}{2}]$$

$$\tan\alpha = \frac{F_1 \sin\theta}{F_1 + F_2 \cos\theta} = \frac{\sin\theta}{1 + \cos\theta} \qquad [\because \sin\theta = 2\sin\frac{\theta}{2}\cos\frac{\theta}{2}]$$

$$= \frac{2\sin\theta/2 \cos\theta/2}{2\cos^2\theta/2} = \tan\theta/2 \qquad [\because 1 + \cos\theta = 2\cos^2\frac{\theta}{2}]$$

$$\therefore \quad \alpha = \theta/2$$

(b) Graphical method: The above result can be found out graphically by representing the forces on some suitable scale and constructing the parallelogram. The diagonal of the parallelogram passing through point of application of the forces will represent, the resultant force in magnitude (on same scale) and direction (see Fig. 2.34)

SELECTED PROBLEMS

Problem 8: Two forces of 100 N and 60 N act at a point. If the angle between the lines of action of the two forces is 60°, find graphically the mangitude and direction of their resultant and check the resultant by analytical method.

Solution: $F_1 = 100$ N, $F_2 = 60$ N and $\angle\theta = 60°$ as shown in Fig. 2.34 (a).

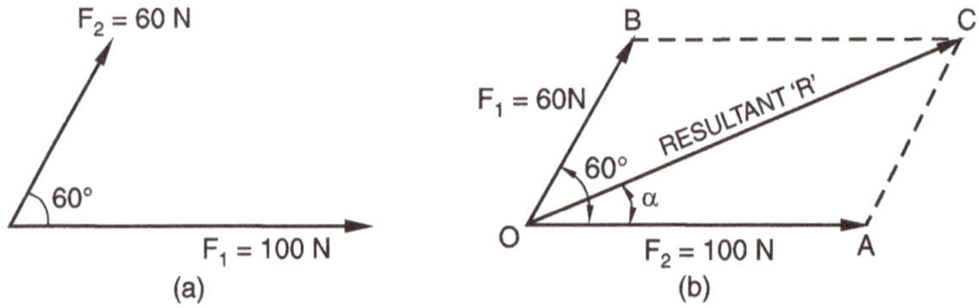

Fig. 2.34.

(i) By graphical method: From any point O, draw OA and OB to a convenient scale (say 3 N = 1 mm) representing the forces $F_1 = 100$ N = 33.33 mm and $F_2 = 60$ N = 20 mm respectively in magnitude and direction such as that the $\angle AOB = 60°$.

Now, if we construct the parallelogram OACB with OA and OB as the adjacent sided as shown in Fig. 2.34 (b), the line joining OC will represent the resultant 'R' in magnitude and direction to the scale already used. On measuring, OC = 46.6 mm and multiply by scale, we get

Magnitude of the resultant = 46.6 × 3 = **140 N approx. [Ans]**

And, the angle α which it makes with OA i.e., with the line of action of force, F_1 = 100 N = **21.78° [Ans]**

(ii) Analytical method: The resultant '*R*' can be found by using the relation,

$$R = \sqrt{F_1^2 + F_2^2 + 2F_1F_2\cos\theta}$$

$$= \sqrt{100^2 + 60^2 + 2\times 100 + 60\times\cos 60°}$$

$$= \sqrt{19600} = \textbf{140 N [Ans]}$$

The direction of the resultant can be found by using the relation,

$$\tan\alpha = \frac{F_2\sin\theta}{F_1 + F_2\cos\theta}$$

$$= \frac{60\sin 60°}{100 + 60\cos 60°} = \frac{60\times 0.866}{100 + 60\times 0.5}$$

$$= \frac{51.96}{130} = 0.3997$$

$$\therefore \qquad \alpha = \tan^{-1} 0.3997 = \textbf{21.78°[Ans]}$$

Since the result obtained *analytically* is almost the same as determined *graphically*, hence solution of the problem is correct.

Problem 9: Two forces of 100 newtons and 80 newtons respectively, having included angle of 135° are acting on a particle. Find the resultant in magnitude and direction, when

(i) both the forces are pull;

(ii) the force of 100 newtons is pull but that of 80 newton is push.

Solution: (i) *When both the forces are pull* (see to Fig. 2.35)

Here, $F_1 = 100$ N, $F_2 = 80$ N and $\theta = 135°$

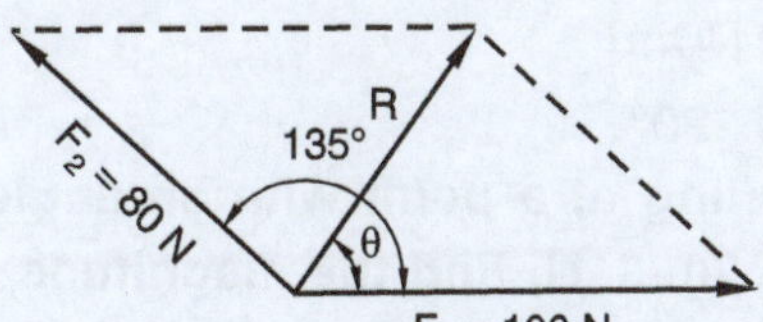

Fig. 2.35.

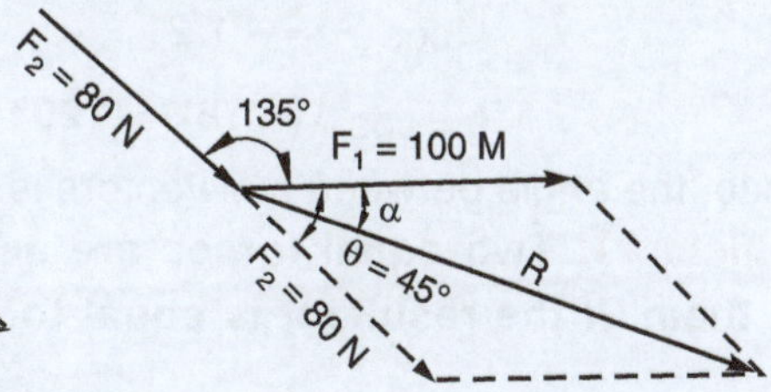

Fig. 2.36.

Now the resultant 'R' can be find out by using the relation.

$$R = \sqrt{F_1^2 + F_2^2 + 2F_1F_2\cos\theta}$$

$$= \sqrt{100^2 + 80^2 + 2 \times 100 \times 80 \times \cos 135°}$$

$$= \sqrt{10000 + 6400 - 11312} = \sqrt{5088}$$

$\therefore$ $R =$ **71.33 N [Ans]**

The angle which this resultant makes with the force of 100 N can be find out by the relation,

$$\tan\alpha = \frac{F_2 \sin\theta}{F_1 + F_2\cos\theta}$$

$$= \frac{80\sin 135°}{100 + 80\cos 135°} = \frac{80 \times 0.707}{100 + 80 \times -7.07} = 1.302$$

$\therefore$ $\alpha = \tan^{-1} 1.302 =$ **52.47° [Ans]**

(ii) *When force of 100* N *is pull but that of 80* N *is push* [Fig. 2.36].

Here, $F_1 = 100$ N (pull) $F_2 = 80$ N (push)

and angle $\theta = 180° - 135° = 45°$

Resultant, $R = \sqrt{100^2 + 80^2 + 100 \times 80 \times \cos 45°}$

$= \sqrt{27712} =$ **166.47 N [Ans]**

Also $$\tan\alpha = \frac{F_2\sin\theta}{F_1 + F_2\cos\theta} = \frac{80 \times \sin 45°}{100 + 80 \times \cos 45°}$$

$$= \frac{56.56}{156.56} = 0.3613$$

$\therefore$ $\alpha = \tan^{-1} 0.3613 =$ **19.86° [Ans]**

Problem 10: Two vector of equal magnitude when added gave a resultant equal to the either vector. What is the angle between them?

Solution: $F_1 = F_2 = R = F$, θ = Angle between vectors

We know, the resultant, $R = \sqrt{F_1^2 + F_2^2 + 2F_1F_2\cos\theta}$

or $F^2 = F^2 + F^2 + 2F \times F \times \cos\theta$

or $-2F^2\cos\theta = F^2$

or $\cos\theta = -1/2$

$\therefore$ $\theta = \cos^{-1}(-1/2)$ = **120° [Ans]**

Hence, the angle between two vectors is 120°

Problem 11: Two equal forces are acting at a point with an angle of 60° between them. If the resultant is equal to $20\sqrt{3}$ N, find the magnitude of each force.

Solution: $\theta = 60°$, $R = 20\sqrt{3}$ N and $F_1 = F_2 = F$

$$R = \sqrt{F_1^2 + F_1^2 + 2F_1F_2\cos\theta}$$

$$20\sqrt{3} = \sqrt{F^2 + F^2 + 2FF\cos 60°}$$

$$20\sqrt{3} = \sqrt{3F^2}$$

Squaring both sides, we get

$$400 \times 3 = 3F^2$$

$\therefore$ $F = \sqrt{400 \times 3/3}$ = **20 N [Ans]**

Problem 12: Find the greatest and the least resultant of two forces whose magnitude are 20 N and 15 N respectively by making the suitable assumptions.

Solution: Here, $F_1 = 20$ N and $F_2 = 15$ N

$$R = \sqrt{F_1^2 + F_2^2 + 2F_1F_2\cos\theta}$$

$$= \sqrt{20^2 + 15^2 + 2 \times 20 \times 15 \times \cos\theta}$$

$$= \sqrt{400 + 225 + 600\cos\theta}$$

Case I: For R to be maximum, $\cos\theta$ should be maximum

i.e., $\theta = 0°$ $[\because \cos 0° = 1]$

Substituting the value of $\theta = 0°$ in equation (i)

$$R_{max} = \sqrt{400 + 225 + 600\cos 0°} = \sqrt{1225}$$

R_{max} = **35 N [Ans]**

Problem 13: The resultant of forces PN and 30 N is 40 N Inclined at 60° to the 30 N force. Find the magnitude and direction of P.

Solution: In ΔOAB

$$AB^2 = OA^2 + OB^2 - 2\,OB \times OA\cos\angle BOA$$

$$P^2 = (30)^2 + (40)^2 - 2 \times 30 \times 40 \cos 60^\circ$$

$$P = \sqrt{900 + 1600 - 2400 \times 0.5}$$

$$\therefore \quad P = \mathbf{36.06\ N\ [Ans]}$$

Applying sine rule for ΔOAB

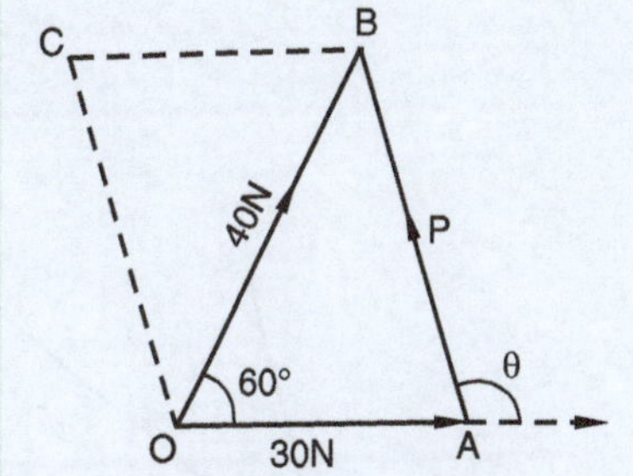

Fig. 2.37.

$$\frac{P}{\sin 60^\circ} = \frac{40}{\sin(180^\circ - \theta)}$$

$$\frac{36.06}{0.866} = \frac{40}{\sin(180^\circ - \theta)}$$

$$\therefore \sin(180^\circ - \theta) = \frac{40 \times 0.866}{36.06} = 0.96$$

$$180^\circ - \theta = 73.74^\circ$$

$$\therefore \theta = \mathbf{106.26^\circ\ [Ans]}$$

Problem 14: The resultant of two equal forces acting at an angle of 2α, is twice as great as when they act at angle of 2β. Show that $\cos\alpha = 2\cos\beta$.

Solution: $F_1 = F_2 = F$

Case I: When $\theta = 2\alpha$

$$R = \sqrt{F_1^2 + F_2^2 + 2F_1F_2\cos\theta}$$

$$= \sqrt{F^2 + F^2 + 2FF\cos 2\alpha}$$

$$= \sqrt{2F^2 + 2F^2\cos 2\alpha}$$

$$= \sqrt{2F^2(1 + \cos 2\alpha)}$$

$$= \sqrt{2F^2(1 + 2\cos^2\alpha - 1)} \qquad [\because \cos 2\alpha = 2\cos^2\alpha - 1]$$

$$\therefore \quad R = 2F\cos\alpha$$

Case II: When $\theta = 2\beta$

$$R_1 = \sqrt{F_1^2 + F_2^2 + 2F_1F_2\cos\theta}$$

$$= \sqrt{F^2 + F^2 + 2FF\cos 2\beta}$$

$$R_1 = \sqrt{2F^2 + 2F^2\cos 2\beta} = \sqrt{2F^2(1 + \cos 2\beta)}$$

$$= \sqrt{2F^2(1 + 2\cos^2\beta - 1)} \qquad [\because \cos 2\beta = 2\cos^2\beta - 1]$$

$$\therefore \quad R_1 = 2F\cos\beta$$

But $R = 2R_1$ (Given)

$$\therefore 2F\cos\alpha = 2(2F\cos\beta)$$

$$2F\cos\alpha = 4F\cos\beta$$

$$\therefore \quad \mathbf{\cos\alpha = 2\cos\beta\ [Ans]}$$

Problem 15: Two forces acting at a point are such that if the direction of one is reversed, the direction of the resultant is turned through a right angle. Prove that the forces must be equal in magnitude.

Solution: F_1 and F_2 be the given forces acting at angle θ to each other as shown in Fig. 2.38.

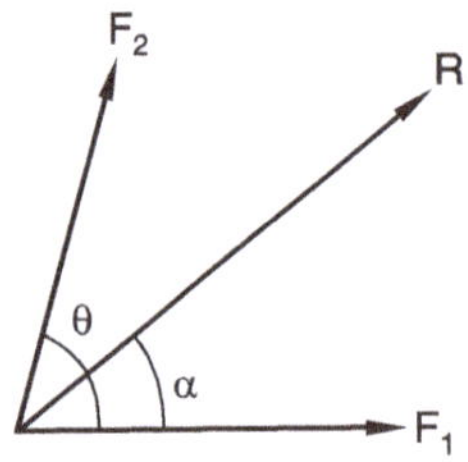

Fig. 2.38.

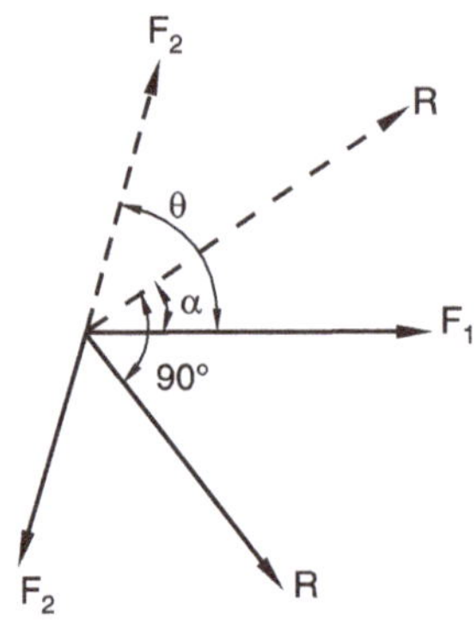

Fig. 2.39.

We know $\tan\alpha = \dfrac{F_2 \sin\theta}{F_1 + F_2 \cos\theta}$ ...(i)

Let the direction of F_2 is reversed as shown in Fig. 2.39.

Now the resultant R makes an angle of $(90° - \alpha)$ with F_1

$$\text{Thus, } \tan(90° - \alpha) = \frac{F_2 \sin(180 - \theta)}{F_1 + F_2 \cos(180° - \theta)}$$

or $$\cot\alpha = \frac{F_2 \sin\theta}{F_1 - F_2 \cos\theta}$$ $[\because \tan(90° - \alpha) = \cot\alpha]$

$[\because \sin(180° - \theta) = \sin\theta]$

or $$\frac{1}{\cot\alpha} = \frac{F_1 - F_2 \cos\theta}{F_2 \sin\theta}$$ $[\because \cos(180° - \theta) = -\cos\theta]$

$$\tan\alpha = \frac{F_1 - F_2 \cos\theta}{F_2 \sin\theta}$$ ...(ii)

Equating equation (i) and (ii), we get

$$\frac{F_2 \sin\theta}{F_1 + F_2 \cos\theta} = \frac{F_1 - F_2 \cos\theta}{F_2 \sin\theta}$$

$$(F_2 \sin\theta)(F_2 \sin\theta) = (F_1 + F_2 \cos\theta)(F_1 - F_2 \cos\theta)$$

$$F_2^2 \sin^2\theta = F_1^2 - F_2^2 \cos^2\theta$$ $[\because a^2 - b^2 = (a+b)(a-b)]$

or $$F_2^2 \sin^2\theta + F_2^2 \cos^2\theta = F_1^2$$

$$F_2^2(\sin^2\theta + \cos^2\theta) = F_1^2$$

$$F_2^2 = F_1^2$$

$\therefore$ $\mathbf{F_1 = F_2}$ **[Proved]**

Problem 16: **Find the magnitude of two forces such that if they act at right angles, their resultant is $\sqrt{10}$ newtons. But if they act an angle of 60°, their resultant is $\sqrt{13}$ newtons.**

Solution: Let the two forces be F_1 and F_2.

Case I: (i) When the two forces act at right angles:

$$R = \sqrt{10}\ N \quad \text{and} \quad \theta = 90°.$$

Now, the resultant, $R = \sqrt{F_1^2 + F_2^2 + 2F_1F_2\cos\theta}$

or $R^2 = F_1^2 + F_2^2 + 2F_1F_2\cos\theta$

or $(\sqrt{10})^2 = F_1^2 + F_2^2 + 2F_1F_2\cos 90°$

or $10 = F_1^2 + F_2^2$ $\quad [\because \cos 90° = 0]$

$\therefore \quad F_1^2 + F_2^2 = 10 \quad$ ...(i)

Case II: (ii) When the two forces act an an angle of 60°

Here, $R = \sqrt{13}\ N \quad \text{and} \quad \theta = 60°$

Now the resultant $R = \sqrt{F_1^2 + F_2^2 + 2F_1F_2\cos\theta}$

or $R^2 = F_1^2 + F_2^2 + 2F_1F_2\cos\theta$

or $(\sqrt{13})^2 = F_1^2 + F_2^2 + 2F_1F_2\cos 60° \quad$ ...(ii)

But from (i), $F_1^2 + F_2^2 = 10$ and $\cos 60° = 0.5$

Therefore, by putting the value in equation (ii), we get

$\therefore \quad 13 = 10 + 2F_1F_2 \times 0.5$

or $F_1F_2 = 13 - 10 = 3 \quad$ ...(iii)

We know $(F_1 + F_2)^2 = F_1^2 + F_2^2 + 2F_1F_2$

or $(F_1 + F_2)^2 = 10 + 2 \times 3 = 16 \quad [\because F_1 \times F_2 = 3 \text{ as above}]$

$\therefore \quad F_1 + F_2 = 4 \quad$ ...(iv)

and also $(F_1 - F_2)^2 = F_1^2 + F_2^2 - 2F_1F_2$

$(F_1 - F_2)^2 = 10 - 2 \times 3 = 4$

$\therefore \quad F_1 - F_2 = 2 \quad$ ...(v)

Now adding equation (iv) and (v), we get

$2F_1 = 6$

$\therefore$ $F_1 = 3$ **N [Ans]**

But $F_1 + F_2 = 4$

$F_2 = 4 - 3 = 1$ **N [Ans]**

Problem 17: **The resultant of two forces P and Q is R, If Q be doubled R is doubled, if Q be reversed, R is again doubled, show that P : Q : R = $\sqrt{2} : \sqrt{3} : \sqrt{2}$**

Solution: Let θ be the angle between forces P and Q

$$\text{Then } R^2 = P^2 + Q^2 + 2PQ\cos\theta \quad \text{...(i)}$$

(i) When Q be doubled, R becomes 2R

$$\therefore \quad (2R)^2 = P^2 + (2Q)^2 + 2P \times 2Q\cos\theta$$

$$4R^2 = P^2 + 4Q^2 + 4PQ\cos\theta \quad \text{...(ii)}$$

(ii) When Q becomes – Q, R again becomes 2R

$$\therefore \quad (2R)^2 = P^2 + (-Q)^2 + 2P(-Q\cos\theta$$

$$4R^2 = P^2 + Q^2 - 2PQ\cos\theta \quad \text{...(iii)}$$

Adding equations (i) and (ii) $5R^2 = 2P^2 + 2Q^2$ ...(iv)

Multiplying (iii) by 2 and adding to (ii), we get

$$12R^2 = 3P^2 + 6Q^2 \quad \text{[Dividing by 3]}$$

$$4R^2 = P^2 + 2Q^2 \quad \text{(v)}$$

Subtracting (v) from (iv), we get $R^2 = P^2$

$$\therefore \quad R = P$$

Substituting the value of R in (v), we get

$$4P^2 = P^2 + 2Q^2$$

$$3P^2 = 2Q^2$$

$$3/2\ P^2 = Q$$

$$\sqrt{3/2}P = Q$$

$$\text{Now} \quad P : Q : R = P : \sqrt{3/2}\ P : P \quad [\because R = P]$$

$$= 1 : \sqrt{3/2} : 1$$

$$\therefore \quad \mathbf{P : Q : R = \sqrt{2} : \sqrt{3} : \sqrt{2}} \quad \textbf{[Proved]}$$

Problem 18: The resultant of two forces F_1 and F_2 acting at an angle θ is equal to (2m + 1) $\sqrt{F_1^2 + F_2^2}$. When they act at an angle of $(\pi/2 - \theta)$, the resultant is equal to (2m – 1) $\sqrt{F_1^2 + F_2^2}$, show that $\tan\theta = \dfrac{m-1}{m+1}$.

Solution: The given two forces are F_1 and F_2.

(a) When the forces act at an angle θ

$$\text{The resultant, } R = (2m + 1)\sqrt{F_1^2 + F_2^2} \quad \text{...(i)}$$

$$\text{But we know, } \quad R = \sqrt{F_1^2 + F_1^2 + 2F_1F_2\cos\theta} \quad \text{...(ii)}$$

By equating equation (I) and (II), we get

$$(2m + 1)\sqrt{F_1^2 + F_2^2} = \sqrt{F_1^2 + F_2^2 + 2F_1F_2\cos\theta}$$

$$\text{or} \quad (2m + 1)^2 (F_1^2 + F_2^2) = F_1^2 + F_2^2 + 2F_1F_2\cos\theta$$

$$[(2m + 1)^2 - 1][F_1^2 + F_2^2] = 2F_1F_2\cos\theta$$

$$\text{or} \quad (4m^2 + 4m)(F_1^2 + F_2^2) = 2F_1F_2\cos\theta$$

$$\text{or} \quad 4m(m + 1)(F_1^2 + F_2^2) = 2F_1F_2\cos\theta \quad \text{...(iii)}$$

(b) When the forces act an angle of $(\pi/2-\theta)$

$$\text{The resultant,} \quad R = (2m-1)\sqrt{F_1^2+F_2^2} \qquad \text{... (iv)}$$

$$\text{But,} \quad R = \sqrt{F_1^2+F_2^2+2F_1F_2\cos(\pi/2-\theta)}$$

$$\text{or} \quad R = \sqrt{F_1^2+F_2^2+2F_1F_2\sin\theta} \qquad \text{...(v)}$$

$[\because \cos(\pi/2-\theta) \neq \sin\theta]$

Equating equation (iv) amd (v), we get

$$(2m-1)\sqrt{F_1^2+F_2^2} = \sqrt{F_1^2+F_2^2+2F_1F_2\sin\theta}$$

$$\text{or} \quad (2m-1)(F_1^2+F_2^2) = F_1^2+F_2^2+2F_1F_2\sin\theta$$

$$\text{or} \quad [(2m-1)^2-1][F_1^2+F_2^2] = 2F_1F_2\sin\theta$$

$$\text{or} \quad (4m^2-4m)(F_1^2+F_2^2) = 2F_1F_2\sin\theta$$

$$\text{or} \quad 4m(m-1)(F_1^2+F_2^2) = 2F_1F_2\sin\theta \qquad \text{... (vi)}$$

Now dividing equations (vi) by (iii), we get

$$\frac{4m(m-1)(F_1^2+F_2^2)}{4m(m+1)(F_1^2+F_2^2)} = \frac{2F_1F_2\sin\theta}{2F_1F_2\cos\theta}$$

$$\frac{m-1}{m+1} = \frac{\sin\theta}{\cos\theta} = \tan\theta$$

$$\mathbf{\tan\theta = \frac{m-1}{m+1}} \quad \textbf{[Proved]}$$

2.20. TRIANGLE LAW OF FORCES

According to triangle law of forces,

"If two concurrent forces acting simultaneously on a body are represented in magnitude and direction by two sides of a triangle taken in order, then the third side taken in opposite order represents the resultant force in magnitude and direction".

This method is used to find resultant of two forces graphically.

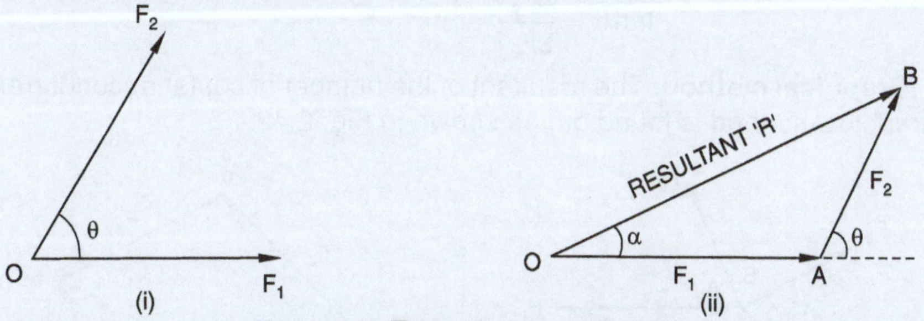

Fig. 2.40. *Triangle law of Forces*

Consider two concurrent forces F_1 and F_2 acting simultaneously on a body at an angle θ [see Fig. 2.40 (i)]. Represent the forces F_1 and F_2 in magnitude (on selected scale) and direction by the vector OA and AB in order, as shown in Fig. 2.40 (ii). The third side OB taken in opposite order completely represents the resultant force R in magnitude (on same scale) and direction.

2.21. POLYGON LAW OF FORCES

According to polygon law of forces,

> *"If a number of concurrent forcs acting simultaneously on a body are represented in magnitude and direction by the sides of a polygon taken in order, then the closing side of the polygon taken in the opposite order represents the resultant force in magnitude and direction".*

This method is used to find the resultant of more than two forces by (i) analytical method and (ii) graphical method as under:

1. Analytical method: The resultant of four forces can be determined analytically by using the method of resolution of forces (see Fig. 2.41).

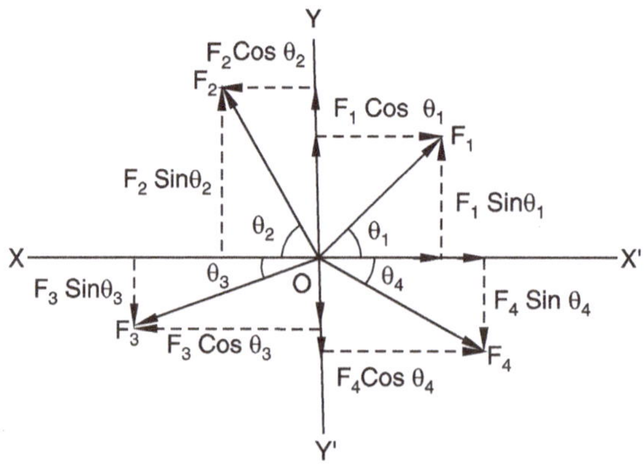

Fig. 2.41.

Now, resolve all the force horizontally and vertically, we get

$\Sigma F_x = F_1 \cos\theta_1 - F_2 \cos\theta_2 - F_3 \cos\theta_3 + F_4 \cos\theta_4$ and

$\Sigma F_y = F_1 \sin\theta_1 + F_2 \sin\theta_2 - F_3 \sin\theta_3 + F_4 \sin\theta_4$

Then, the resultant 'R' is given by the relation,

$$R = \sqrt{(\Sigma F_x)^2 + (\Sigma F_y)^2}$$

Now, the resultant R makes an angle α with X-axis, then

$$\tan\alpha = \frac{\Sigma F_y}{\Sigma F_x}$$

2. Graphical method: The resultant of the number of coplanar concurrent forces by the graphical method is found out as shown in Fig. 2.42.

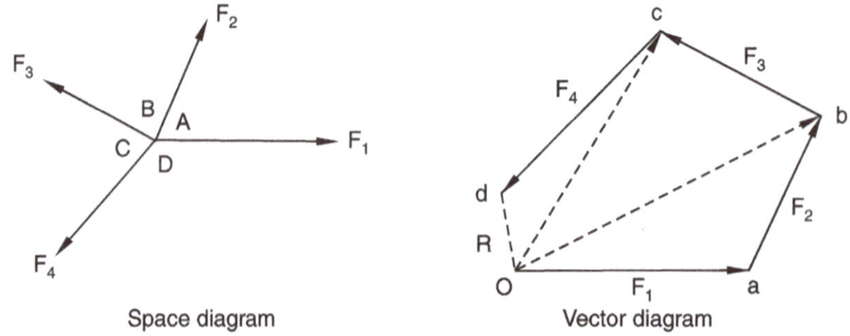

Fig. 2.42. *Polygon Law of Forces*

(I) First of all draw the space diagram for the given forces.

(II) Select a suitable point and go on adding the vector oa = F_1, ab = F_2, bc = F_3, cd = F_4, representing the forces to a convenient scale.

(III) Then, the closing line R joining the first point in the vector diagam, gives the resultant od of the given force system in magnitude as well as in direction.

SELECTED PROBLEMS

Problem 19: A practicle is acted upon by the following four forces:

(i) 20 N inclined 30° to the North of East;

(ii) 25 N towards North;

(iii) 30 N towards North West;

(iv) 35 N inclined 40° to South of West.

Find the magnitude and direction of the resultant force by analytical method and graphical method.

Solution : The various forces acting on the particle are shown in Fig. 2.43(i).

1. Analytical method:

Now resolving all the forces horizontally, we get

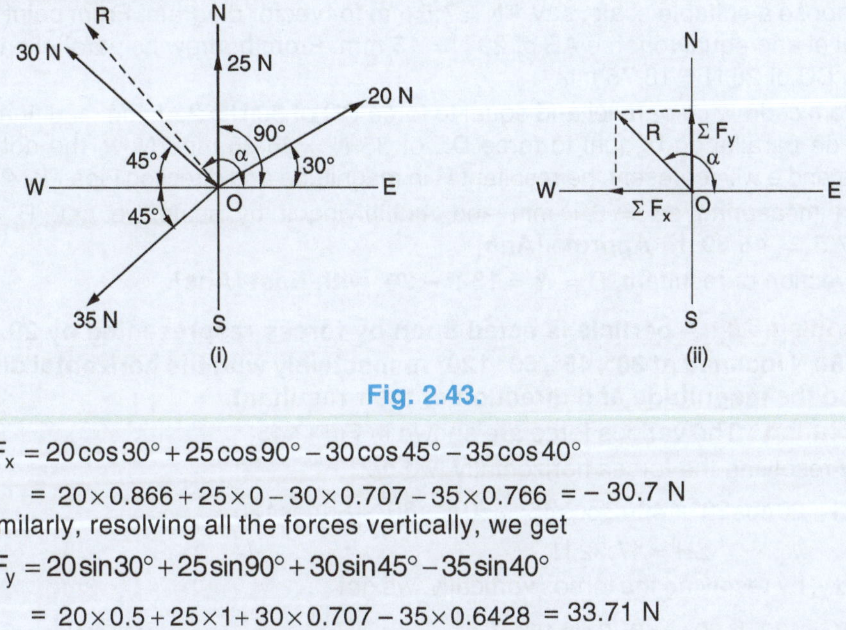

Fig. 2.43.

$$\Sigma F_x = 20\cos 30° + 25\cos 90° - 30\cos 45° - 35\cos 40°$$

$$= 20 \times 0.866 + 25 \times 0 - 30 \times 0.707 - 35 \times 0.766 = -30.7 \text{ N}$$

Similarly, resolving all the forces vertically, we get

$$\Sigma F_y = 20\sin 30° + 25\sin 90° + 30\sin 45° - 35\sin 40°$$

$$= 20 \times 0.5 + 25 \times 1 + 30 \times 0.707 - 35 \times 0.6428 = 33.71 \text{ N}$$

For finding the resultant 'R', we may use the relation,

$$R = \sqrt{(\Sigma F_x^2 + (\Sigma F_y)^2}$$

$$= \sqrt{(-30.7)^2 + (33.71)^2} = \sqrt{2058.85} = \mathbf{45.59 \text{ N [Ans]}}$$

For finding the direction of resultant, α

$$\tan\alpha = \frac{\Sigma F_y}{\Sigma F_x} = \frac{33.71}{30.7} = 1.098$$

$$\alpha = \tan^{-1} 1.098 = 47° - 40 \text{ with West}$$

$$= \mathbf{132° - 20' \text{ with East [Ans].}}$$

2. Graphical Method:

The various forces acting on a particle are shown in the space diagram according to bow's notation [see Fig. 2.44 (i)].

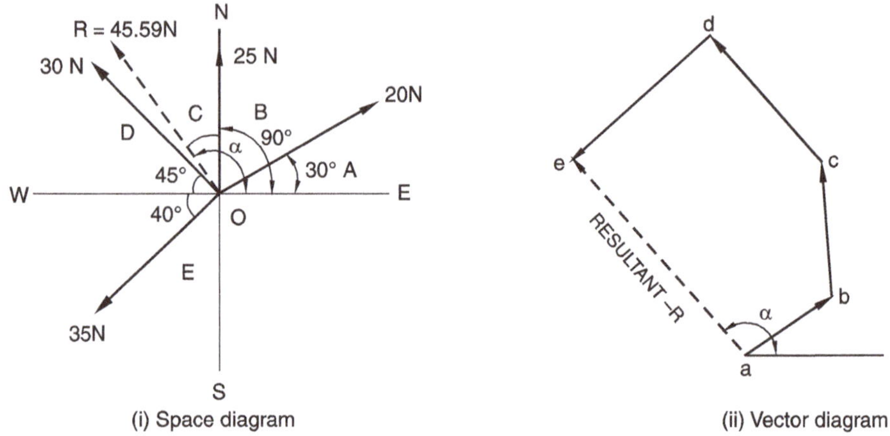

Fig. 2.44.

Choose a suitable scale, say 1 N = 7.5 mm for vector diagram. From point a, draw ab parallel and equal to force AB of 20 N = 15 mm. From b, draw bc parallel and equal to force BC of 25 N = 18.75 mm

From c, draw cd parallel and equal to force CD of 30 N = 22.5 mm and finally from d draw de parallel and equal to force DE of 35 N = 26.25 mm. Now, the dotted line joining a and e will represent the resultant R in magnitude and direction [see Fig. 2.44 (ii)].

On measuring ea = 6.1 mm and multiplying it by scale we get. R = ea = 6.1 × 7.5 = **45.59 N Approx [Ans]**

Direction of resultant, R = α = **132° – 20′ with East [Ans].**

Problem 20 : A particle is acted upon by forces respresented by 20, $40\sqrt{2}$, 60 and 80 N inclined at 30°, 45°, 60° 120° respectively with the horizontal direction OX. Find the magnitude and direction of their resultant.

Solution : The various force are shown in Fig. 2.45.

By resolving the forces horizontally, we get

$$\Sigma H = 20\cos 30^\circ + 40\sqrt{2}\cos 45^\circ + 60\cos 60^\circ + 80\cos 120^\circ$$

$$\therefore \qquad \Sigma H = 47.32 \text{ N}$$

Now, by resolving the forces vertically, we get

$$\Sigma V = 20\sin 30^\circ + 40\sqrt{2}\sin 45^\circ + 60\sin 60^\circ + 80\sin 120^\circ$$

$$\therefore \qquad \Sigma V = 171.24 \text{ N}$$

$$\therefore \text{ Resultant,} \quad R = \sqrt{\Sigma H^2 + \Sigma V^2}$$

$$= \mathbf{177.65\ N\ [Ans]}$$

$\therefore$ Direction of resultant,

$$\tan\alpha = \frac{\Sigma V}{\Sigma H} = \frac{171.24}{47.32}$$

or $\qquad \alpha$ = **74° – 33′** with OX axis.

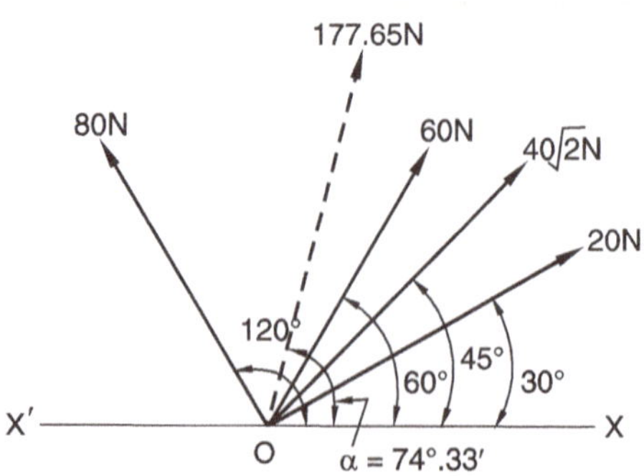

Fig. 2.45.

Problem 21 : The resultant of the four forces of x, 12, 15 and 20 N making angle of y°, 70°, 120°–25', 155° respectively with a fixed line is 39.5 N making angle 111°–47' with the fixed line. Find the force x and its angle y° with the fixed line.

Solution : The various forces and their resultant act are shown in Fig. 2.46.

Now resolving all the forces horizontally, we get

$$\Sigma F_x = x\cos y° + 12\cos 70° + 15\cos 120° - 25' + 20\cos 155°$$

$$= x\cos y° + 12\times 0.3420 + 15(-0.5037) + 20\times(-0.9063)$$

or $\Sigma F_x = x\cos y° - 21.58$

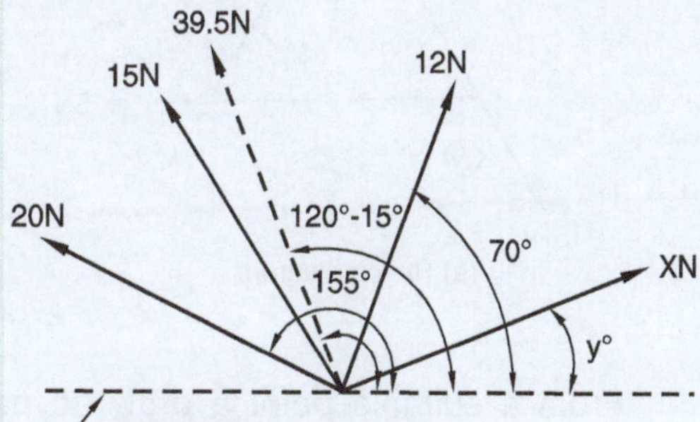

Fig. 2.46.

But ΣF_x must by equal to the horizontal component of the resultant which is 39.5 cos 111°–47′

$= 39.5 \times (-0.366) = -14.46$

$\therefore \quad x\cos y° - 21.58 = -14.46$

$$x\cos y° = 7.122 \quad \text{... (i)}$$

Similarly, resolving all the forces vertically, we get

$$\Sigma F_y = x\sin y° + 12\sin 70° + 15\sin 120° - 15 + 20\sin 155°$$

$$= x\sin y° + 12\times 0.9397 + 15\times 0.8628 + 20\times 0.4226$$

or $\Sigma F_y = x\sin y° + 32.69$

But ΣF_y must be equal to the vertical component of the resultant

$= 39.5 \sin 111°–47' = 39.5 \times 0.9336 = 36.69$

$\therefore \quad x\sin y° + 32.69 = 36.69$

or $x\sin y° = 4$ ... (ii)

Now dividing equation (ii) by (i), we get

$$\frac{x\sin y°}{x\cos y°} = \frac{4}{7.122}$$

$$\tan y° = 0.562$$

$\therefore$ Direction, $y° = \tan^{-1} 0.562 =$ **29-33° [Ans]**

From equation (ii) $x\sin 29.33° = 4$

$\therefore$ Force, $x = \dfrac{4}{\sin 29.33°} = \dfrac{4}{0.4895}$ = **8.166 N [Ans]**

Problem 22: A particle is acted upon by three forces, equal to 5 N, 10 N, 13 N along the three sides of an equiateral triangle, taken in order. Find graphically or otherwise the magnitude and direction of the resultant force.

Solution : The various forces acting on the particle are shown in the space diagram by bow's notation [see Fig. 2.47 (a)]

Draw the vector diagram for the force system as discussed below [see Fig. 2.47 (b)]:

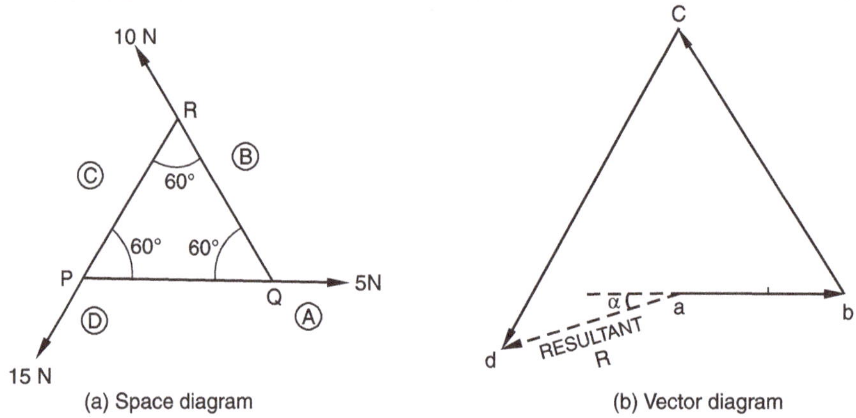

Fig. 2.47.

From a suitable point a draw ab parallel and equal to the force of 5 N to a convenient scale. From b, draw bc parallel and equal to the force of 10 N and cd parallel and equal to the force of 13 N as shown in Fig. 2.47 (b). Then, the line joining ad will represent the resultant 'R' in magnitude (on the same scale) as well as in direction.

On measuring, the magnitude of the resultant, R = ad = **7 N [Ans]**.

The direction which this resultant makes with force of 5 N

$$\alpha = \mathbf{20°\text{–}47'} \text{ [Ans]}$$

Problem 23: Force of 2, 3, 5, $\sqrt{3}$ and 2 newtons respectively act at one of the angular points of a regular hexagon towards the other five angular points taken in order. Find the magnitude and direction of the resultant force.

Solution : The various forces acting on the angular points of a regular hexagon are shown in Fig. 2.48.

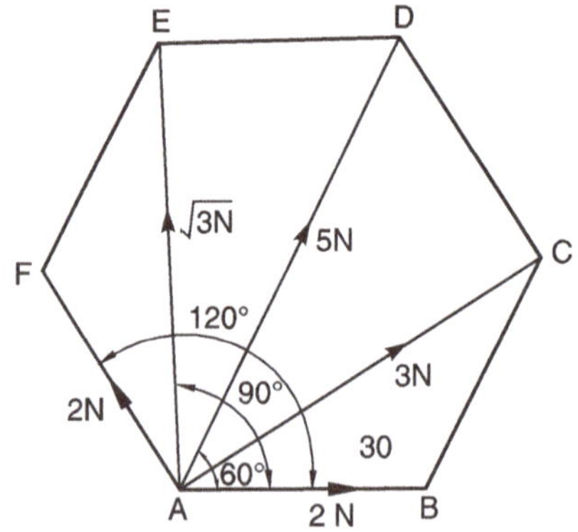

Fig. 2.48.

Now resolving all the forces horizontally, we get

$$\Sigma F_x = 2\cos 0° + 3\cos 30° + 5\cos 60° + \sqrt{3}\cos 90° + 2\cos 120°$$

$$= 2\times 1 + 3\times 0.866 + 5\times 0.5 + \sqrt{3}\times 0 + 2\times -0.5 = 6.1\,\text{N}$$

Similarly resolving all the forces vertically, we get

$$\Sigma F_y = 2\sin 0° + 3\sin 30° + 5\sin 60° + \sqrt{3}\sin 90° + 2\sin 120°$$

$$= 2\times 0 + 3\times 0.5 + 5\times 0.866 + \sqrt{3}\times 1 + 2\times 0.866 = 9.29 \text{ N}$$

For finding the value of R, use the relation,

$$R = \sqrt{F_x + F_y} = \sqrt{(6.1)^2 + (9.29)^2} = \mathbf{11.11\ N\ [Ans]}$$

$$\tan\alpha = \frac{\Sigma F_y}{\Sigma F_x} = \frac{9.29}{6.1} = 1.52$$

$$\therefore \quad \alpha = \tan^{-1} 1.52 = \mathbf{56°\text{-}66\ [Ans]}$$

2.22. EQUILIBRIUM OF FORCES AND EQUILIBRANT

When the resultant of a number of force acting on a body is zero and the body is at rest, the body is said to be in equilibrium.

In practice, we come across a number of situations where many forces acting on the body. In order to keep the body in equilibrium, a force equal in magnitude but opposite in direction to the resultant.

Thus, if two forces F_1 and F_2 acting at point O on a body (see Fig. 2.49) have resultant R, then in order to keep the body in euilibrium, a force R_e at O is the equilibrant i.e., the force which keeps the body in equilibrium.

Thus, a single force which balances the effect of a number of forces acting on a body is called the equilibrant.

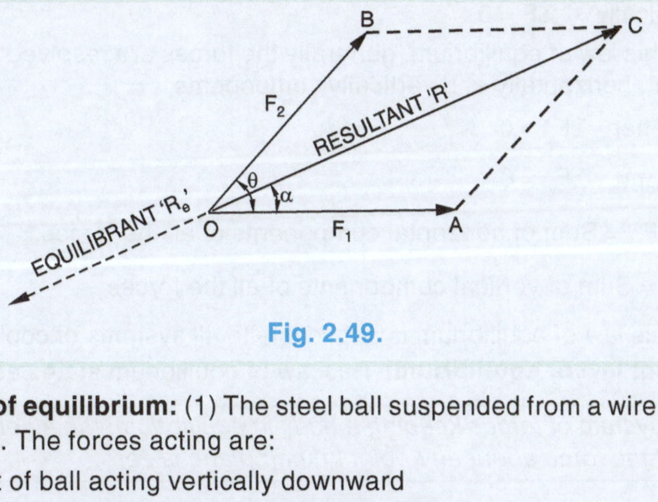

Fig. 2.49.

Example of equilibrium: (1) The steel ball suspended from a wire (see Fig. 2.50) is in equilibrium. The forces acting are:

(i) Weight of ball acting vertically downward

(ii) Tension in the wire acting vertically upward.

These two concurrent forces are equal in magnitude and opposite in direction so that resultant is zero. Hence, the ball is in equilibrium.

2. In Fig. 2.51, two men are trying to start a bulky horse. But the horse exerts just the right amount of force to make the resultant zero. Hence, the system is in equilibrium.

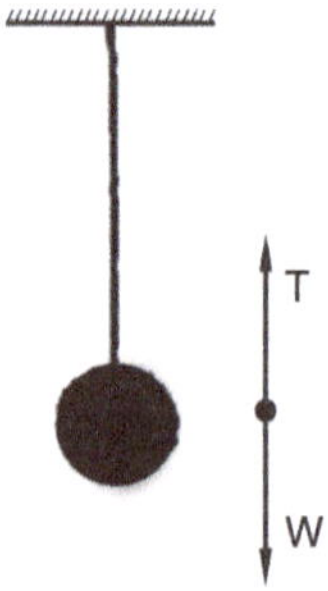

Fig. 2.49. *Ball hanging from support*

Fig. 2.49. *Two men pulling the horse*

Fig. 2.52. *Men pulling out the blanket*

3. In Fig. 2.52, a number of persons are pulling the blanket outward in many directions so that the resultant is zero. Hence, the blanket is in equilibrium

2.23. LAWS OF EQUILIBRIUM

The following are the laws of equilibrium, depending upon the forces acting on a rigid body and keeping it in equilibrium:-

1. The force law of equilibrium;
2. The moment law of equilibrium.

1. Force law of equilibrium: This law of equilibrium states as under:-

> *"For any system of forces keeping a body in equilibrium, the algebraic sum of forces in any direction is zero."*

Mathematically $\Sigma F = 0$

To apply this law of equilibrium, generally the forces are resolved into rectangular components i.e., horizontally and vertically components,

$$\text{Then } \Sigma F_x = 0$$

$$\text{and, } \Sigma F_y = 0$$

Where, ΣF_x = Sum of horizontal components of all the forces,

and, ΣF_y = Sum of vertical components of all the forces.

Utility: This law of equilibrium is applicable to all systems of coplanar forces.

2. Moment law of equilibrium: This law of equilibrium states as under:-

> *"For any system of forces keeping a body in equilibrium, the algebraic sum of the moments of all the force about any point in their plane is zero."*

Mathematically $\Sigma M = 0$

or $\Sigma F \times d = 0$

Where, ΣM = Algebric sum of moments of all the forces acting on the body about any point in the plane of forces.

SELECTED PROBLEMS

Problem 24: A pole rests vertically with its base on the ground and is held in position by five ropes all in the horizontal plane and drawn tight. From the pole the first rope runs due North, the second 60° West of North, the third 30° South of West and fourth 30° East of South. The tension in these four ropes are 40 N, 25 N, 30 N and 35 N respectively. Find the direction of the fifth rope and its tension.

Solution : The various forces acting on pole are shown in Fig. 2.53

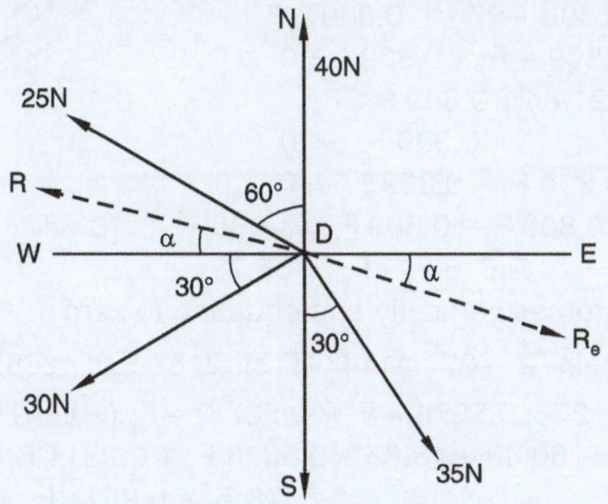

Fig. 2.53.

Now resolving all the forces horizontally, we get

$$\Sigma H = -25 \sin 60° - 30 \cos 30° + 35 \sin 30°$$

$$= -25 \times 0.866 - 30 \times 0.866 + 35 \times 0.5$$

$$\therefore \quad \Sigma H = -30.13 \text{ N}$$

Similarly, resolving all the forces vertically, we get

$$\Sigma V = 40 + 25 \cos 60° - 30 \sin 30° - 35 \cos 30°$$

$$= 40 + 25 \times 0.5 - 30 \times 0.5 - 35 \times 0.866$$

$$\therefore \quad \Sigma V = 7.19 \text{ N}$$

Resultant, $R = \sqrt{\Sigma H^2 + \Sigma V^2}$

$$= \sqrt{(-30.13)^2 + (7.19)^2} = 31.13 \text{ N}$$

$$\tan \alpha = \frac{\Sigma V}{\Sigma H} = \frac{7.19}{30.13} = 0.2386$$

$$\alpha = \tan^{-1} 0.2386 = 13.42°$$

Resultant, R acts 13.42° North of West (see Fig. 2.53)

$\therefore$ Tension in the fifth rope will be given by equilibrant (R_e) which is equal but opposite to R. Thus, tension in fifth rope is **31.13 N** acts at **13.42° South of East [Ans]**

Problem 25: A wheel has five pokes spaced equality from one another. If magnitude of forces in three consecutive pokes is 23 N, 32 N, and 27 N respectively. Find magnitude of forces in other two pokes.

Solution: Let the wheel be having 5 pokes OA, OB, OC, OD and OE equally spaced (see Fig. 2.54)

Let the forces in pokes OD and OE as F_1 and F_2.

Considering the equilibrium of the wheel under five pushes in the pokes and resolving all the forces horizontally and equating to zero.

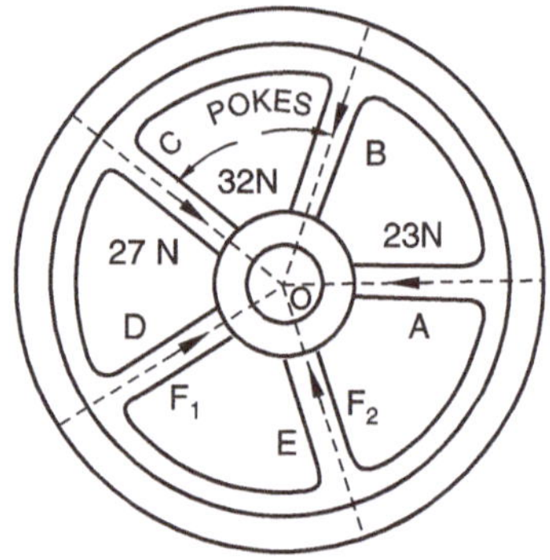

Fig. 2.54.

$$-23\cos 0° - 32\cos 72° - 27\cos 144° - F_1\cos 216° - F_2\cos 288° = 0$$

$$-23 \times 1 - 32 \times 0.309 - 27 \times -0.809 - F_1 \times 0.809 - F_2 \times 0.309 = 0$$

$$\text{or } -23 - 9.89 + 21.84 + 0.809\, F_1 - 0.309\, F_2 = 0$$

$$\text{or } -11.05 + 0.809\, F_1 - 0.309\, F_2 = 0$$

$$\text{or } 0.809\, F_1 - 0.309\, F_2 = 11.05 \quad \text{[Dividing both the sides by 0.309]}$$

$$\text{or } 2.618\, F_1 - F_2 = 35.76 \quad \text{...(i)}$$

Resolving all the forces vertically and equating to zero

$$-23\sin 0° - 32\sin 72° - 27\sin 144° - F_1\sin 216 - F_2\sin 288° = 0$$

$$0 - 32 \times 0.9511 - 27 \times 0.5878 - F_1\,(-0.5878) - F_2\,(-0.9511) = 0$$

$$0 - 30.43 - 15.87 + 0.5878\, F_1 + 0.9511\, F_2 = 0$$

$$\text{or } 0.5878\, F_1 + 0.9511\, F_2 = 46.3$$

[Dividing both the sides by 0.9511]

$$\text{or } 0.618\, F_1 + F_2 = 48.68 \quad \text{...(ii)}$$

Now adding (i) and (ii), we get $3.236\, F_1 = 84.44$

$\therefore$ $F_1 = 84.44/3.236 =$ **26.1 N [Ans]**

and, by substituting the value of F_1 in (ii), we get

$$0.618 \times 26.1 + F_2 = 48.68$$

$\therefore$ $F_2 = 48.68 - 16.13 =$ **32.55 N [Ans]**

Problem 26: Find the magnitude and direction of equilibrium of a force system as shown in Fig. 2.55 (i).

Solution : Consider the given force system as shown in Fig. 2.55.

To find the magnitude and direction of the resultant force of the given force system, we may take the forces as shown in Fig. 2.55 (ii).

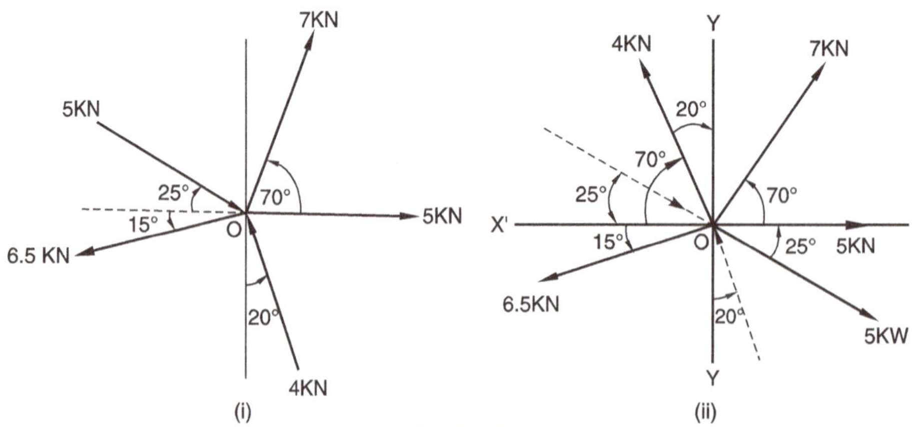

Fig. 2.55.

Resolving all the forces horizontally and vertically, we get

$$\Sigma F_x = 5\cos 0° + 7\cos 70° - 4\sin 20° - 6.5\cos 15° + 5\cos 25°$$
$$= 5 \times 1 + 7 \times 0.3420 - 4 \times 0.3420 - 6.5 \times 0.9659 + 5 \times 0.9063$$
$$= 4.279 \text{ kN} \quad \text{and}$$

$$\Sigma F_y = 5\sin 0° + 7\sin 70° + 4\cos 20° - 6.5\sin 15° + 5\sin 25°$$
$$= 5 \times 0 + 7 \times 0.9397 + 4 \times 0.9397 - 6.5 \times 0.2588 - 5 \times 0.4226$$
$$= 6.542 \text{ kN}$$

The magnitude of the resultant 'R' can be found by using the realtion:

$$R = \sqrt{(\Sigma F_x)^2 + (\Sigma F_y)^2} = \sqrt{(4.279)^2 + (6.542)^2}$$
$$= \sqrt{75.37} = 8.68 \text{ kN [Ans.]}$$

$$\text{and } \tan\alpha = \frac{\Sigma F_y}{\Sigma F_x} = \frac{6.542}{4.279} = 1.529$$

$$\therefore \quad \alpha = \tan^{-1} 1.529 = 56.81° \text{ with the direction OX}$$

Hence the equilibrant of the given force system will by **8.68 kN** acting at an angle of **56.81 with OX axis [Ans.]**

Problem 27: A block of stone weighting 5 tonnes is to be raised by a crane. The angle between the load cable and the crane boom is 30°. The boom is held in place by a tie cable, which forms an angle of 20° with the boom as shown in Fig. 2-55. Determine graphically or otherwise the magnitude of the thrust force of the boom and of the tie force of the cable.

Solution: The crane shown in Fig. 2.56 (i) is in equilibrium under the following forces:-

(i) Load, W = 5 tonnes, (ii) Thrust force of the boom, F

(iii) Tie force of the cable, T

Since all these force keep the crane in equilibrium, the force diagram must close

To draw force or vector diagram, we may use the bow's notation as shown in space diagram. [see fig 2-56 (ii)].

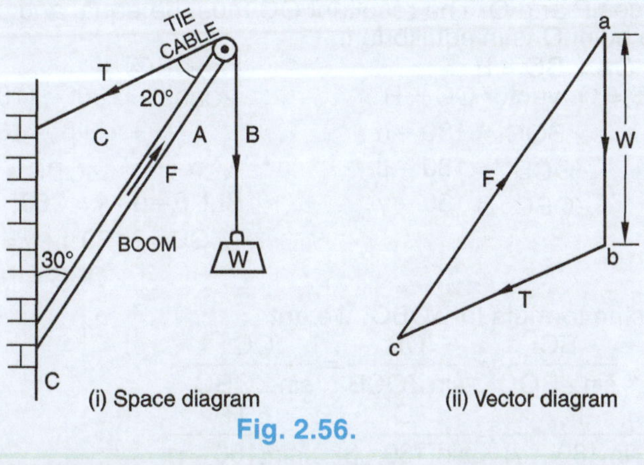

(i) Space diagram (ii) Vector diagram

Fig. 2.56.

Now, draw ab parallel and equal to force AB = 5 tonnes to a convenient scale (say 5 mm = 1 tonne). From b, draw bc parallel to force BC i.e., T and from a, ac parallel to CA. i.e., F to meet bc in c.

On measuring (to same scale), we get

Tie force of the cable, T = ac = **7.3** t **[Ans.]** and

Thrust force of the boom, F = bc = **11.21 [Ans.]**

2.24. LAMI'S THEOREM

According to Lami's theorem,

> *"If three concurrent forces acting on a body, keep it in equilibrium, then each force is proportional to sine of the angle between the other two."*

Thus, if the three forces P, Q and R acting at point O are in equilibrium (see Fig. 2.57), then according to lami's theorem.

$$P : Q : R :: \sin\alpha : \sin\beta : \sin\gamma$$

or
$$\frac{P}{\sin\alpha} = \frac{Q}{\sin\beta} = \frac{R}{\sin\gamma}$$

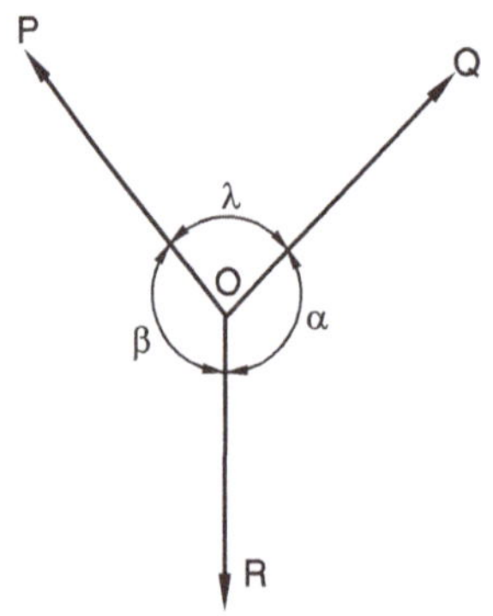

Fig. 2.57.

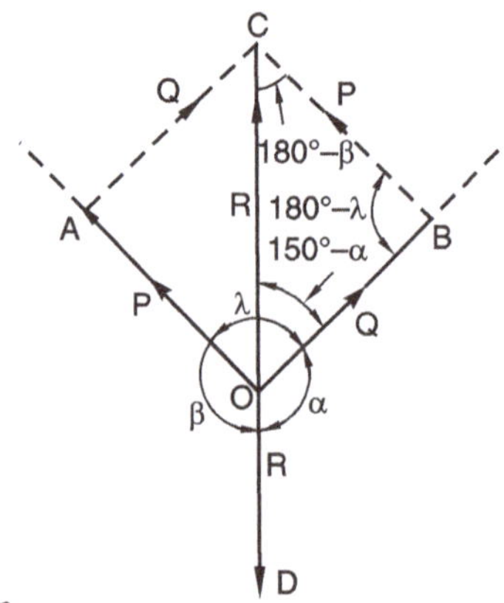

Lami's Theorem

Fig. 2.58.

Proof: Consider three forces P, Q and R acting at point O, keep it in. equilibrium. Let the angle between their lines of actions be α, β, and γ as shown in Fig. 2.58. These three forces are represented completely in magnitude and direction by OA, OB and OD respectively. Complete the parallelogram OACB. The diagonal OC represents the resultant of forces P and Q. The resultant OC must be equal and opposite to the third force R. As the point O is in equilibrium.

∴ Vector OA = BC = P

Vector OB = Q, vector OC = R

In ΔOBC, $\angle BOC = 180 - \alpha$

$\angle OCB = 180 - \beta$

$\angle OBC = 180 - \gamma$

In ΔOBC,

$\angle OBCZ = 180° - (180° - \propto) - (180 - \beta)$

$= \alpha + \beta - 180°$

$\therefore \; \alpha + \beta = \angle OBC + 180°$

But $\alpha + \beta + \gamma = 360°$

$(\angle OBC + 180°) + \gamma = 360°$

$\therefore \; \angle OBC = 180 - \gamma$

Applying sine formula to ΔOBC, we get

$$\frac{BC}{\sin\angle BOC} = \frac{OB}{\sin\angle OCB} = \frac{OC}{\sin\angle OBC}$$

or
$$\frac{P}{\sin(180-\alpha)} = \frac{Q}{\sin(180-\beta)} = \frac{R}{\sin(180-\gamma)}$$

$$\therefore \quad \mathbf{\frac{P}{\sin\alpha} = \frac{Q}{Sin\beta} = \frac{R}{\sin\gamma}}$$

SELECTED PROBLEMS

Problem 28: A smooth sphere of weight 'W' is supported in contact with a smooth vertical wall by a string fastened to a point on its surface, the other end being attached to a point on the wall. If the length of string is equal to the radius of sphere, find tension in the string and reaction of the wall.

Solution : Consider the sphere be attached to the point A on the wall by mean of a string AD such that the length AD of the string is equal to the radius say 'R' of the sphere as shown in fig 2-59.

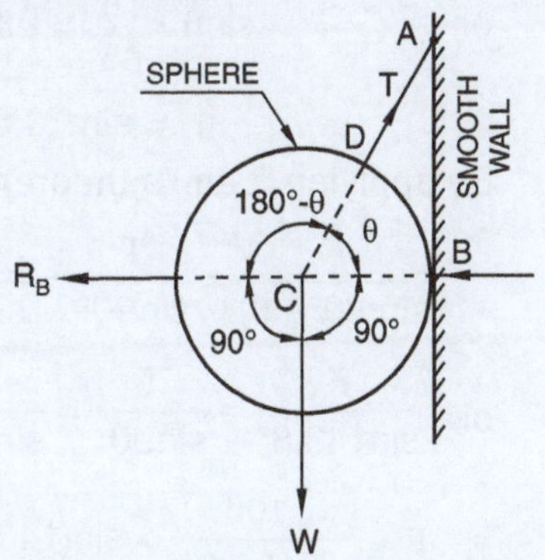

Fig. 2.59.

Thus, the sphere is in equilibrium under the following forces:-

(i) Weight 'W' acting vertically downwards.

(ii) Tesnsion 'T' in the string AD.

(iii) Reaction 'R_B' of the wall at the point of contact B of the sphere.

Now in rt. ∠d Δ ABC

$$BC = R, \ AC = CD + DA = 2R \quad \text{and} \quad \angle ACB = \theta$$

Thus $\cos\theta = BC/AC$ or $\cos\theta = R/2R = 1/2$

$\therefore \qquad \theta = \cos^{-1} 1/2 = 60°$

By applying "Lami's theorem at C, we get,

$$\frac{R_B}{\sin(90°+\theta)} = \frac{T}{\sin 90°} = \frac{W}{\sin(180°-\theta)}$$

or
$$\frac{R_B}{\sin 150°} = \frac{T}{\sin 90°} = \frac{W}{\sin 120°} \qquad [\text{since } \theta = 60°]$$

∴ Reaction of the wall.

$$R_B = \frac{W}{\sin 120°} \times \sin 150° = \frac{W \times 1/2}{\sqrt{3}/2} = \frac{\mathbf{W}}{\sqrt{\mathbf{3}}} \ \textbf{[Ans]}$$

$$\text{Tension in string, } T = \frac{W}{\sin 120°} \times \sin 90° = \frac{W \times 1}{\sqrt{3}/2} = \frac{\mathbf{2W}}{\sqrt{\mathbf{3}}} \ \textbf{[Ans]}$$

Problem 29: A ball weighing 100 newtons is suspended vertically by a string 5 m long. Find the magnitude of the force which should be applied horizontally to hold the ball one metre above the lowest point. Also determine tensions in the string.

Solution : Let the ball of weight W = 100 N be suspended vertically from hook 'A' by means of a string AC = 5 m as shown in Fig. 2.60.

Now consider the force 'F' applied horizontally to hold the ball 1 m above the lowest point.

Thus the ball is in equilibrium under the following forces:-

(i) Weight 'W' = 100 N acting vertically downwards.

(ii) Force 'F' applied horizontally on the ball.

(iii) Tension 'T' in the string.

Now In right-angled ΔABC

$$AB = 4m, AC = 5m, \angle ACB = \theta$$

$$\text{Then} \quad \sin\theta = \frac{AB}{AC}$$

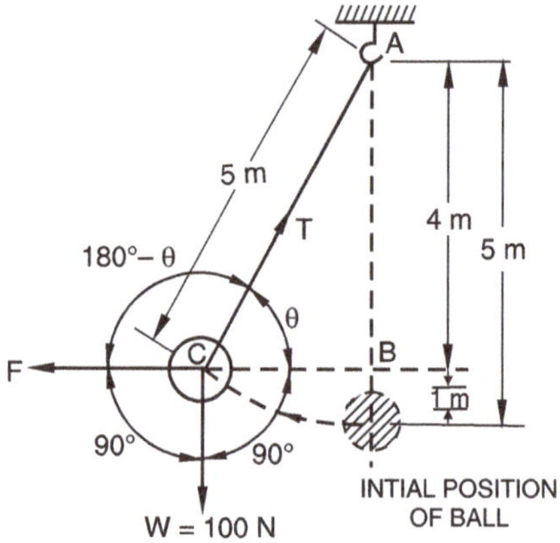

Fig. 2.60.

or $$\sin\theta = \frac{4}{5} = 0.8$$

$$\therefore \quad \theta = \sin^{-1} 0.8 = 53.13°$$

By applying "Lami's theorem" at C, we get

$$\frac{F}{\sin(90° + \theta)} = \frac{T}{\sin 90°} = \frac{W}{\sin(180° - \theta)}$$

or $$\frac{F}{\sin 143.8°} = \frac{T}{\sin 90°} = \frac{W}{\sin(180° - 53.13°)}$$

$$\therefore \quad F = \frac{100}{\sin 126.87°} \times \sin(143.8°) = \frac{100 \times 0.6}{0.8} = \mathbf{75\ N\ [Ans]}$$

$$\therefore \quad T = \frac{100}{\sin 126.87°} \times \sin 90° = \frac{100 \times 1}{0.8} = \mathbf{125\ N\ [\ Ans]}$$

Problem 30: A body of weight 1000 N is supported by two string of 4m and 3 m in length, attached at the same horizontal level 5 m apart. Find the tension in the strings.

Solution : The weight W = 1000 N is suspended by string OA = 3 m and OB = 4 m as shown in Fig. 2.61.

In Δ AOB $5^2 = 4^2 + 3^2$

$$\therefore \quad \angle AOB = 90°$$

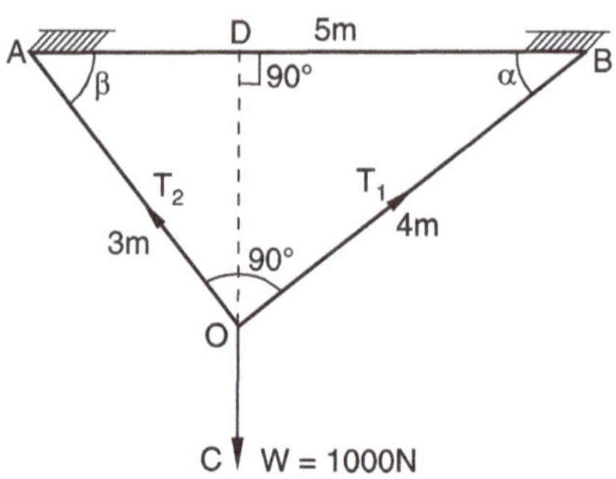

Fig. 2.61.

In rt. $\angle$d ΔAOB $\tan\beta = \frac{4}{3}$

$$\therefore \quad \beta = 53.6°$$

$$\therefore \quad \angle AOD = 180° - (90° + 53.6°) = 36.4°$$

$$\angle AOC = 180° - \angle AOD = 180° - 36.4°$$

In rt. angled Δ AOB, $\tan\alpha = \frac{3}{4}$ $\therefore$ $\alpha = 36.4°$

$$\therefore \quad \angle BOD = 180° - (90° + 36.4°) = 53.6°$$

$$\therefore \quad \angle BOC = 180° - \angle BOD = 180° - 53.6°$$

Since the body is equilibrium under the action of three forces T_1, T_2 and 1000 N, therefore according to "Lami's Theorem".

$$\frac{T_1}{\sin\angle AOC} = \frac{T_2}{\sin\angle BOC} = \frac{1000}{\sin 90°}$$

$$\frac{T_1}{\sin(180-36.4^\circ)}=\frac{T_2}{\sin(180-53.6^\circ)}=\frac{1000}{\sin 90^\circ}$$

$$\frac{T_1}{\sin 36.4^\circ}=\frac{T_2}{\sin 53.6^\circ}=\frac{1000}{\sin 90^\circ}$$

$\therefore$ $T_1 = 1000 \times \sin 36.4^\circ = 1000 \times 0.5934 =$ **593.4 N [Ans]**

and $T_2 = 1000 \times \sin 53.6^\circ = 1000 \times 0.8048 =$ **804.8 N [Ans]**

Problem 31: A machine weighing 3 tonnes is supported by two chains attached to some point on the machine. One of these chains goes to the eye bolt in the wall and is inclined 30° to the horizontal and the other goes to the hook in ceiling and is inclined at 45° to the horizontal. Find the tensions in the two chains.

Solution : Let the machine weight, W = 3 tonnes be suspended as shown in Fig. 2.62.

Thus the machine is in equilibrium under the following three forces:

(i) Weight 'W' acting vertically downwards.

(ii) Tension 'T_1' in chain OA.

(iii) Tension 'T_2' in chain OB.

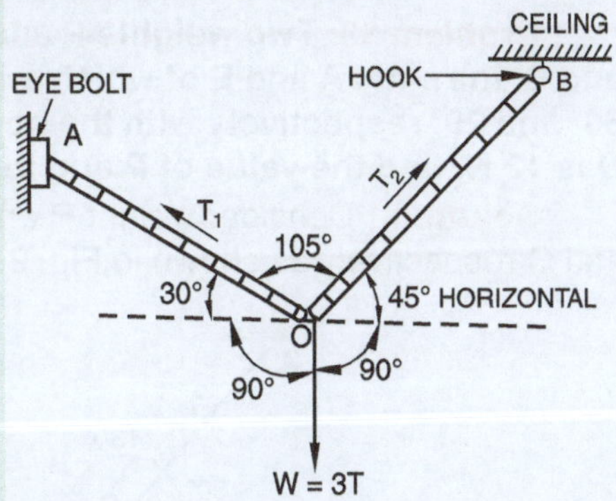

Fig. 2.62.

Now applying Lami's theorem at O, we get

$$\frac{T_1}{\sin 135^\circ}=\frac{T_2}{\sin 120^\circ}=\frac{W}{\sin 105^\circ} \quad [\because W = 3\text{ t}]$$

$$T_1 = \frac{3}{\sin 105^\circ}\times \sin 135^\circ = \frac{3\times 0.707}{0.9659} = \mathbf{2.19\ t\ \ [Ans]}$$

$$\text{and } T_2 = \frac{3}{\sin 105^\circ}\times \sin 120^\circ = \frac{3\times 0.866}{0.9659} = \mathbf{2.689\ t\ \ [Ans]}$$

Problem 32: A body weighing 260 N is supported by two strings 5 and 12 metres long. Their outer ends being fastened to the extremities of a rod 13 metres long. If the rod is so placed that the body lies vertically below the middle of the rod. Find the tension in the strings.

Solution : The body weighing 260 N be suspended from the extremities of the rod AB by means of two string AO = 5 m and BO = 12 m such that the body at O lies vertically below middle point C of rod AB as shown in Fig. 2.63.

Thus the body is in equilibrium under the following forces:

(i) Weight 'W' acting downwards at O.

(ii) Tension T_1 in string OA, and

(iii) Tension 'T_2' in string OB.

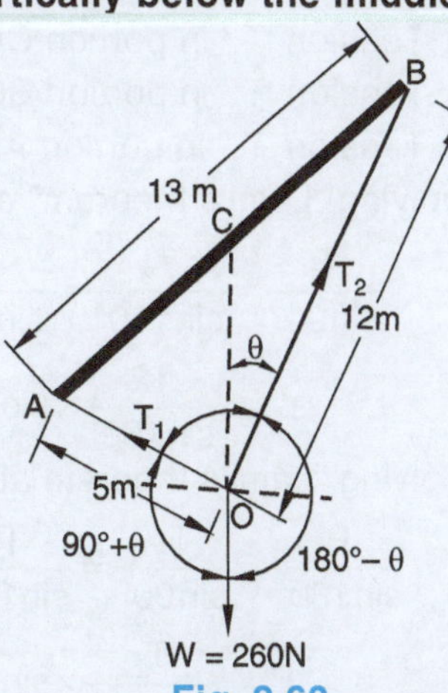

Fig. 2.63.

Now in ΔAOB, AB = 13 m, OA = 5 m

OB = 12 m, $\angle ABO = \theta$, and

$\therefore \angle AOB = 90^\circ$ $[\because AB^2 = OA^2 + OB^2]$

Then $\sin\theta = OA/AB$ and $\cos\theta = OB/AB$

$\therefore$ $\sin\theta = 5/13$, and $\cos\theta = 12/13$

Applying "Lami's theorem" at O, we get

$$\frac{T_1}{\sin(180^\circ - \theta)} = \frac{T_2}{\sin(90^\circ + \theta)} = \frac{W}{\sin 90^\circ}$$

or $$\frac{T_1}{\sin\theta} = \frac{T_2}{\cos\theta} = \frac{260}{1}$$

$\therefore$ $T_1 = (260/1)\times\sin\theta = (260/1)\times(5/13) =$ **100 N [Ans]**

and $T_2 = (260/1)\times\cos\theta = (260/1)\times(12/13) =$ **240 N [Ans]**

Problem 33: Two weights P and Q are fixed to a string ABCD at the points B and C, the ends A and D of which are fixed. Portions AB and CD make angle ends 50° and 20° respectively with the horizontal and portion BC remains horizontal. If Q is 12 N, find the value of P and tension in portion BC of the string.

Solution : Consider weight P and Q be fixed to the string ABCD at the points B and C respectively as shown in Fig. 2.64.

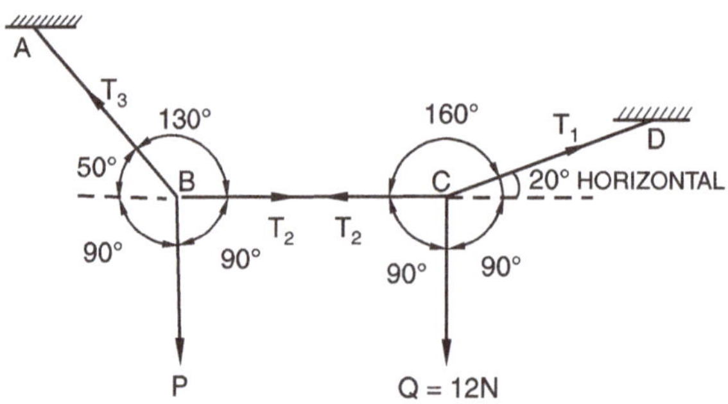

Fig. 2.64.

The string ABCD is in equilibrium under the following forces:

(i) Weight 'P' acting at B.

(ii) Weight 'Q' acting at C.

(iii) Tension 'T_1' in portion CD of the string.

(iv) Tension 'T_2' in portion BC of the string.

(v) Tension 'T_3" in portion AB of the string.

Applying "Lami's theorem" at point C, we get

$$\frac{T_1}{\sin 90^\circ} = \frac{T_2}{\sin 110^\circ} = \frac{Q}{\sin 160^\circ}$$

$\therefore$ $$T_2 = \frac{12}{\sin 160^\circ}\times\sin 110^\circ = \frac{12\times 0.9397}{0.3420} = \textbf{32.97 N [Ans]}$$

Applying "Lami's theorem at point B, we get

$$\frac{P}{\sin 130^\circ} = \frac{T_3}{\sin 90^\circ} = \frac{T_2}{\sin 140^\circ}$$

$\therefore$ $$P = \frac{T_2}{\sin 140^\circ}\times\sin 130^\circ = \frac{32.97\times 0.766}{0.6428} = \textbf{39.29 N [Ans]}$$

Problem 34: In a jib crane, the length of the crane post is 2 m, the tie rod makes 45° angle with the crane post at A while the jib makes 30° angle with the crane post at B. The crane is to lift a load of 2000 N, find out the forces in the tie and jib.

Solution: Let F_1 and F_2 be the forces in the tie and jib respectively. The various forces acting on the crane are F_1, F_2 and 2000 N weight.

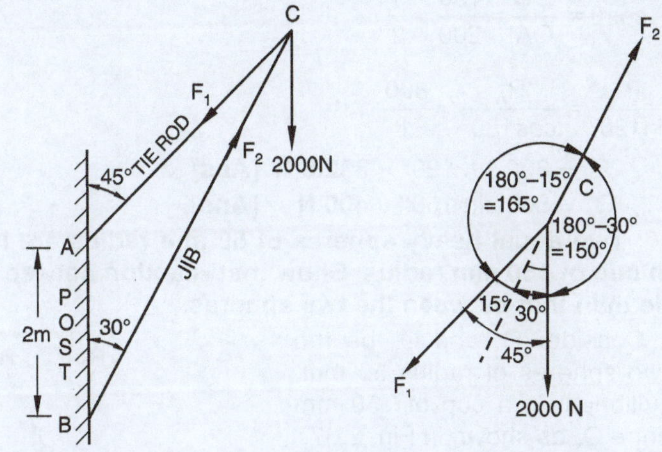

Fig. 2.65.

Consider the equilibrium of point C and apply Lami's theorem

$$\frac{F_1}{\sin 150^\circ} = \frac{F_2}{\sin 45^\circ} = \frac{2000}{\sin 165^\circ}$$

$$\frac{F_1}{0.5} = \frac{F_2}{0.707} = \frac{2000}{0.259}$$

$$F_1 = \frac{0.5 \times 2000}{0.259} = \textbf{3861 N (Tension) [Ans]}$$

$$F_2 = \frac{0.707 \times 2000}{0.259} = \textbf{5459.46 (compression) [Ans]}$$

Problem 35: A uniform wheel 400 mm in diameter rests against a rigid rectangular block 100 mm thick as shown in fig, 2.66. Find the least pull through the centre of the wheel to just turn it over the corner of the block. If all the surfaces are smooth, find also the reaction of the block. The weight of the wheel is 800 N.

Solution : Consider the equilibrium of wheel when it is just to turnover the block. In this position the following forces are acting on the wheel :

1. The pull, P.
2. Weight of wheel, 800 N
3. Reaction, R_A

In this condition, there will be no contact of wheel with floor so $R_m = 0$

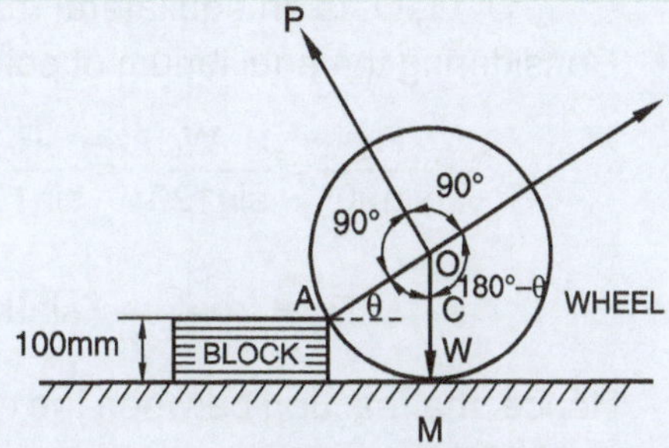

Fig. 2.66.

Now, applying "Lami theorem", we get

$$\frac{P}{\sin(180° - \theta)} = \frac{R_A}{\sin(90° + \theta)} = \frac{800}{\sin 90°}$$

In right angle ΔAOC

$$\cos\theta = \frac{OC}{OA} = \frac{100}{200} = \frac{1}{2} \qquad \therefore\ \theta = 60°$$

$$\frac{P}{\sin 120°} = \frac{R_A}{\cos 150°} = \frac{800}{1}$$

$\therefore$ $P = 800 \sin 120° =$ **692.8 N [Ans]**

and, $R_A = 800 \sin 150° =$ **400 N [Ans]**

Problem 36: Two equal heavy spheres of 50 mm radius are in equilibrium within a smooth cup of 150 mm radius. Show that reaction between the cup and sphere is double than that between the two spheres.

Solution : Consider O_1 and O_2 be the centres of the two spheres of radius 50 mm, which are in equilibrium in a cup of 150 mm radius having centre O_3 as shown in Fig. 2.67.

Thus the spheres are in equilibrium under the following forces:

(i) Weight 'W' of each sphere.

(ii) Reaction R_A and R_B, between the spheres and the cup acting normally at the points of contact A and B respectively.

(iii) Reaction 'R_C' between the two spheres as shown.

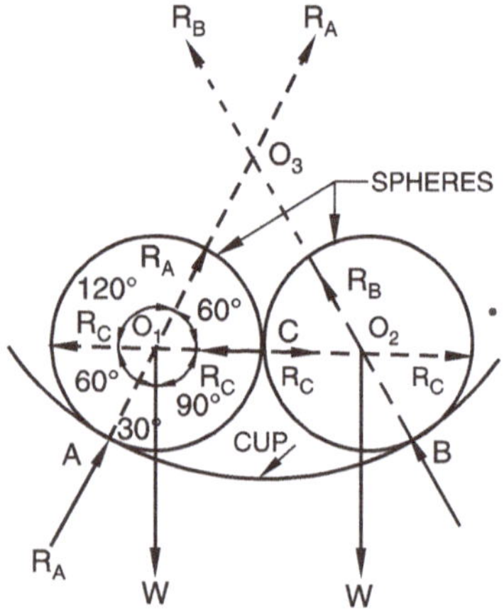

Fig. 2.67.

Now, from geometry of the figures, we find that

$$O_3A = O_3B = 150 \text{ mm},$$
$$O_1A = O_2B = 50 \text{ mm}$$
$$\therefore \quad O_1O_3 = O_3A - O_1A = 150 - 50 = 100 \text{ mm}$$

and $O_2O_3 = O_3B - O_1B = 150 - 50 = 100$ mm

also $O_1O_2 = O_1C + O_2C = 50 + 50 = 100$ mm

$\therefore$ $\Delta\, O_1 O_2 O_3$ is an equilateral triangle having each of its angle = 60°

Considering the equilibrium of point C and applying "Lami's theorem" to it, we get

$$\frac{R_A}{\sin 90°} = \frac{W}{\sin 120°} = \frac{R_C}{\sin 150°}$$

$$\therefore \quad R_A = \frac{R_C}{\sin 150°} \times \sin 90° = \frac{R_C \times 1}{1/2} = \mathbf{2\ R_C}\ \textbf{[Ans]}$$

Hence, the reaction between the cup and one sphere is double than that between the two spheres.

2.25. BEAM

A structural member which carries loads at right angles to its axis is called a beam.

2.26. TYPES OF BEAMS

The following are the different types of beams used in engineering practice:-

1. Cantilever beam
2. Simply supported beam
3. Overhanging beam
4. Fixed beam
5. Continuous beam

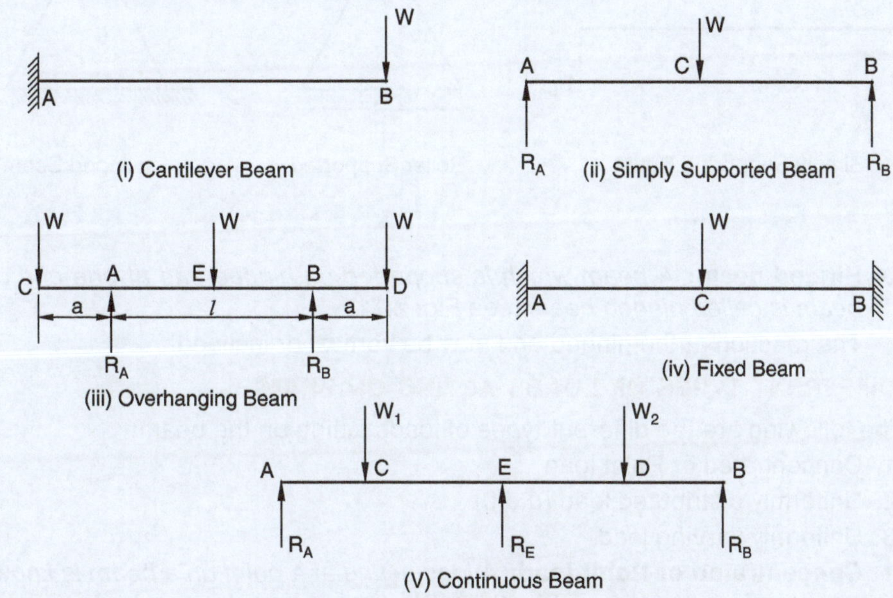

Fig. 2.68. *Different types of beams*

1. Cantilever beam. *A beam which is fixed at one end and free at the other end is known as cantilever beam* [see Fig. 2.68 (i)]

2. Simply supported beam: *A beam supported freely on the supports at its both ends is known as simply supported beam* [see Fig. 2.68 (ii)]

3. Overhanging beam: *A beam in which its end portion is extended beyond the support is known as overhanging beam* [see Fig. 2.68 (iii)]

4. Fixed beam: *A beam whose both ends are fixed in walls is known as fixed beam* [see Fig. 2.68 (iv)]

5. Continuous beam: *A beam supported on more than two supports is known as continous beam* [see Fig. 2.68 (v)]

2.27. END SUPPORTS OF BEAMS

The following are the different end supports for the beams:-

1. Simply supported beams
2. Roller supported beams
3. Hinged beams

1. Simply supported beam: *A beam which is supported at its both ends is called simply supported beam or feely supported beam.*

The supports may be in the form of walls or column. The reactions at the support are always vertical upwards as shown in Fig. 2.69

2. Roller supported beam: *A beam which is supported on rollers at one end of the beam in order to permit free movement in horizontal direction.*

The reaction at the roller end is vertical as shown in Fig. 2.71

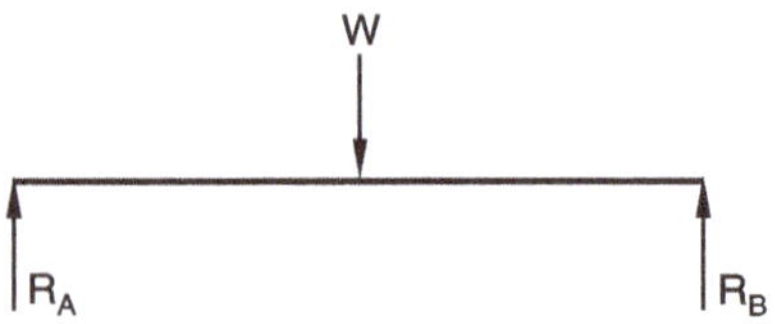

Simply Supported Beam

Fig. 2.69.

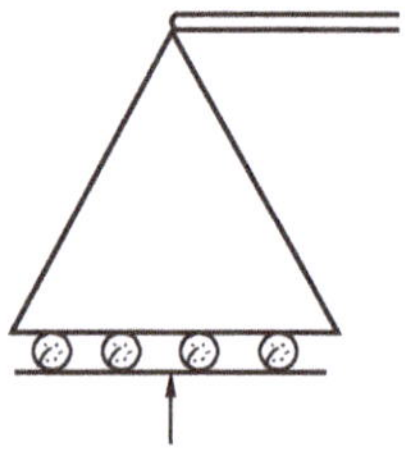

Roller Supported

Fig. 2.70.

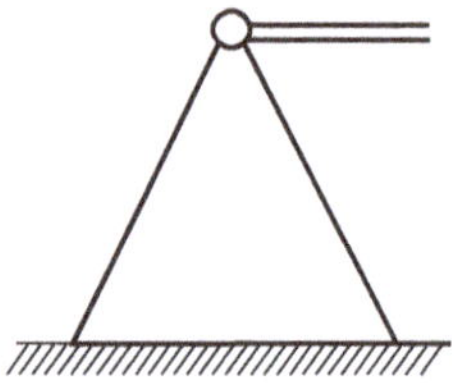

Hinged Beam

Fig. 2.71.

3. **Hinged beam:** *A beam which is supported on hinged end at one end of the beam is called hinged beam* (see Fig. 2.71)

   The reaction at the hinged end may be vertical or inclined.

2.28. DIFFERENT TYPES OF LOADS ACTING ON BEAM

The following are the different types of loads acting on the beam:-

1. Concentrated or Point load
2. Uniformly distributed load (u.d.l.)
3. Uniformly varying load

1. Concentrated or Point load: *A load acting at a point on a beam is known as concentrated or a point load* [see Fig. 2.72 (i)]

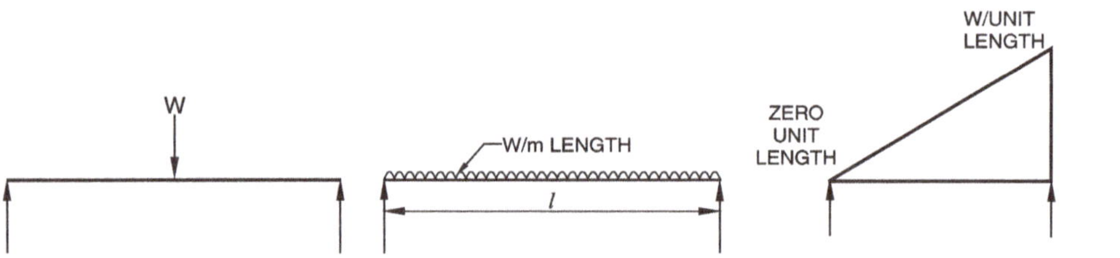

(i) Concentrated load (ii) Uniformly distributed load (iii) Uniformly varying load

Fig. 2.72. *Types of loads*

2. Uniformly distributed load (u.d.l.): A load which is spread uniformly over the *entire or on small portion of the beam is known as uniformly distributed load* [see Fig. 2.5 (ii)]

3. Uniformly varying load: *A load which is spread in such* a way that rate of loading varies from one point to another is known as uniformly varying load [see Fig. 2.72 (iii)]

SELECTED PROBLEMS

Problem 37: A simply supported beam 5 m long is subjected to three point loads of 3 kN, 4kN and 2 kN each at distances of 2 m 3 m and 4 m from the left end. Find the reactions at the supports.

Solution: Let R_A and R_B be the reactions at supports A and B [see Fig. 2.73]

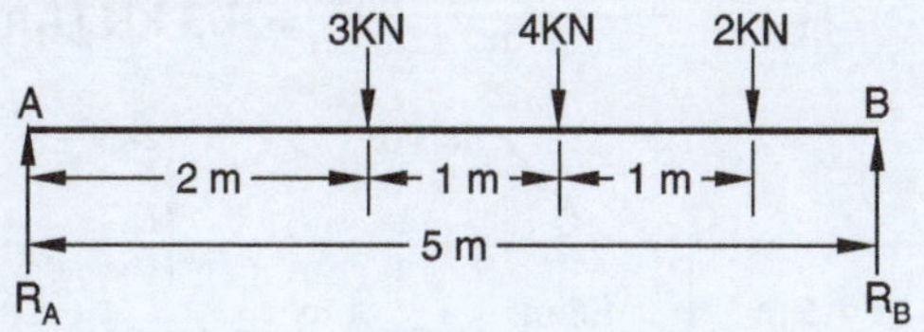

Fig. 2.73.

To find the reaction R_B, take moments about A

$$R_B \times 5 - 3 \times 2 - 4 \times 3 - 2 \times 4 = 0$$

$$R_B \times 5 = 6 + 12 + 8$$

$$\therefore \quad R_B = \mathbf{5.2\ kN\ [Ans]}$$

But $\quad R_A + R_B = \text{Total load on beam}$

$$R_A + 5.2 = 3 + 4 + 2$$

$$\therefore R_A = 9 - 5.2 = \mathbf{3.8\ kN\ [Ans]}$$

Problem 38: A simply supported beam 6 m long is carrying uniformly distributed load of 2 kN/m over a length of 3m from its left end. Find the reactions at the supports.

Solution: Let R_A and R_B the reactions at supports A and B [see Fig. 2.72]

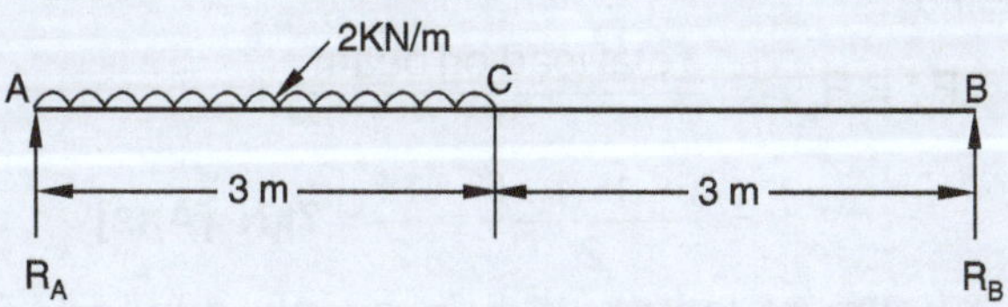

Fig. 2.74.

To find reaction at B, take moments about A:

$$R_B \times 6 - 2 \times 3 \times \frac{3}{2} = 0$$

$$R_B = \mathbf{1.5\ kN\ [Ans.]}$$

$$\text{But } R_A + R_B = \text{Total load on the beam} = 2 \times 3 = 6$$

$$R_A = 6 - 1.5 = \mathbf{4.5\ kN\ [Ans]}$$

Problem 39: Find the reactions of the beam of length 5m. It carries a u.d.l. of 4 kN/m on a span of 1.5 m at a distance of 2 m from right end support and a concentrated load of 2 kN at 1m from right end.

Solution: To find the reactions at B, take moment about A;

$$R_B \times 5 - 4 \times 1.5 \times (1.5 + \frac{1.5}{2}) - 2 \times 4 = 0$$

or $$R_B = \frac{4 \times 1.5 \times 2.25 + 8}{5} = \textbf{4.3 kN [Ans]}$$

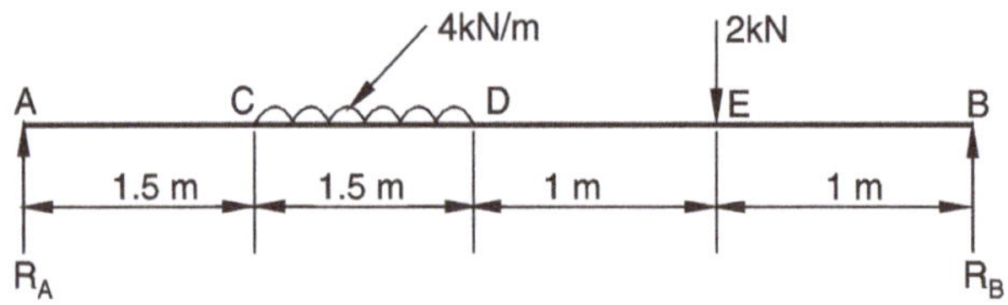

Fig. 2.75.

But $R_A + R_B$ = Total load on the beam

$R_A + 4.3 = 4 \times 1.5 + 2$

$\therefore$ $R_A = 6 + 2 - 4.3 = $ **3.7 kN [Ans]**

Problem 40: A beam AB 10m long has supports at its ends A and B. It carries a point load 5 kN at 3m from B, a load of 5kN – 3m from A and also a U.d.l. of 1 kN/m between the point loads. Find the reactions.

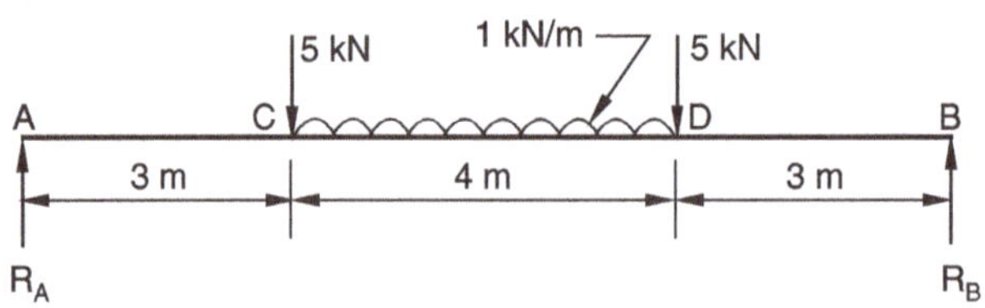

Fig. 2.76.

Solution: As the load on the beam is symmetrical therefore reaction at each support will be equal i.e.

$$R_A = R_B = \frac{\text{Total load on beam}}{2}$$

$$= \frac{5 + 1 \times 4 + 5}{2} = \frac{14}{2} = \textbf{7kN [Ans]}$$

Problem 41: A beam of length 10m is simply supported at two supports which are 6m apart, with an overhang of 2m on each side. The beam carries a concentrated load of 500 N and 1000 N at left and right ends of beam. Find the reactions of beam at the supports.

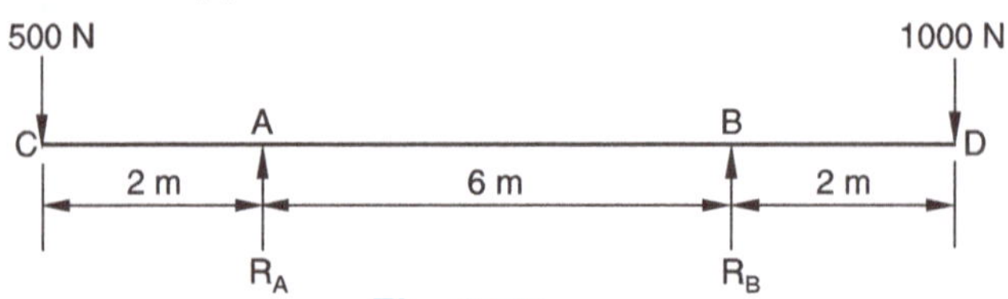

Fig. 2.77.

Solution: Let R_A and R_B be the reactions at supports A and B [see Fig. 2.77].

To find the reaction R_A, take moments about B

$$R_A \times 6 - R_B \times 0 + 1000 \times 2 - 500 \times 8 = 0$$

$$\therefore \quad R_A = \frac{2000}{6} = \mathbf{333.34\ N\ [Ans]}$$

But
$$R_A + R_B = 1000 + 500$$
$$333.34 + R_B = 1500$$

$$\therefore \quad R_B = 1500 - 333.34 = \mathbf{1166.66\ [Ans]}$$

MODEL TEST QUESTIONS

1. Define force and give units in which it is measured.
2. What do you understand by 1 newton force? State how force is expressed in gravitational units.
3. Define the term force and state clearly the effects of a force.
4. What are the various characteristics of a force?
5. What do you mean by a force system. Explain the various force systems.
6. Define components of a force. Derive the expession for the rectangular and non-rectangular components of a force.
7. What do you understand by term resultant of a force system?
8. Define the parallelogram law of forces. What is the use of this law?
9. Show that the resultant $R = \sqrt{P^2 + Q^2}$, when the two component forces P and Q are at right angles to each other. What will be the value of 'R' if the angle between the forces is zero?
10. State and explain triangle law of forces and polygon law of forces.
11. Explain the principle of transmissibility of forces.
12. Explain the equilibrium of forces.
13. What are the general conditions of equilibrium for a body under the action of coplanar concurrent forces?
14. States and prove Lami's Theorem.
15. Differentiate between the following:-
    (a) Vector representation and bow's notation method;
    (b) Action and reaction;
    (c) Tensile and compressive forces;
    (d) Coplanar and non-coplanar forces;
    (e) Coplanar concurrent and coplanar non-concurrent forces;
    (f) Equilibrant and resultant forces.
16. Define a beam. What are the different types of beams?

QUESTIONS FOR SELF EXAMINATION

1. Fill in the blanks with appropriate words and rewrite:-
    (i) The action in the form of ..... or ...... which produces or tends to produce, stops or tends to stop motion of a body is called ....
    (ii) A force may produce deformation in a .... body.
    (iii) The unit of force in SI system is ....
    (iv) One newton is that much force which produces an acceleration of 1 m/sec^2 in a mass of ....

(v) The sense of a force is indicated by ....

(vi) A force can be represented by vector representation method and by .... method.

(vii) The force which tend to increase the length of body is called .... force.

(viii) The "Parallelogram law of forces" is used for determining the resultant of .... forces.

(ix) The gravitational force at moon is .... times than on earth.

(x) The process of splitting up the force into a number of its coponents without changing its effect on the body is called .... of forces.

(xi) The process of finding the .... of a number of forces is composition of forces.

(xii) A load which is spread uniformly over a beam is called ....

Ans. 1 (*i*) push or pull, force (*ii*) flexible; (*iii*) newton; (*iv*) 1 kg; (*v*) arrow head; (*vi*) bow's notation; (*vii*) tensile; (*viii*) two concurrent; (*ix*) 1/6; (*x*) resolution; (*xi*) resultant (*xii*) u.d.l.

2. State whether the following statements are true and false:-

(a) A force can be represented by magnitude and direction only.

(b) A force which one body exerts on the second is called action.

(c) Non coplanar forces are those which belong to the same plane.

(d) Coplanar concurrent forces are those which do not meet.

(e) With the help of "Parallelogram Law of force" we can determine the resultant of two forces.

(f) Law of polygon of forces does not include law of triangle of forces.

(g) The components of a forces along two mutually perpendicular lines are called rectangular components.

(h) A structural member which carries load at right angle to axis is called a beam.

Ans. 2 (a) False; (b) True; (c) False; (d) False; (e) True; (f) False; (g) True; (h) True

PROBLEMS FOR PRACTICE

1. A boy pulls a roller with a force of 50 N. The handle of the roller makes an angle of 30° with the ground. What are the magnitude of the horizontal and vertical components. [**Ans.** 43.3 N, 25 N]

2. What force must be exerted to prevent a truck weighting 15000 N from rolling down on an inclined plane which makes an angle of 4° with the horizontal. [**Ans.** 1046.3 N upward]

3. Determine graphically the resultant of three coplanar collinear forces 12, 15 and 20 kN acting horizontally, when
   (a) the sense of action of all the forces is same;
   (b) the sense of action of 20 kN force is opposite.
   [**Ans.** (a) 47 kN in the direction of all the forces. (b) 7 kN in direction of 12 and 15 kN forces]

4. Two forces P and Q are of the same magnitude and inclined to each other at angle α. Show that their resultant has magnitude 2P cos $\alpha/2$ and is directed along the bisector of the angle between them.

5. At what angle must the forces P + Q and P – Q act in order their resultant may be $\sqrt{P^2 + 3Q^2}$ **[Ans.** 120°]

6. Find the magnitude of two forces such that if they act at right angles, their resultant is $\sqrt{34}$ while if they act an angle of 60°, their resultant is 7 N.

   [Ans. 5 N, 3 N]

7. The the greatest and the least resultant of two forces whose magnitude are 12 newtons and 10 newtons respectively by making the suitable assumption.

   [Ans. R_{max} = 22 N, R_{min} = 2 N]

8. The resultant of two forces, one of them being double the other, is 13 N. If one force is reversed and the other is kept unaltered, the resultant becomes 9 N. Find the magnitude of forces and the conise of the angle at which the forces are inclined in the first case.

   [Ans. F_1 = 10 N, F_2 = 5 N and $\cos\alpha$ = 0.44]

9. Find graphically or otherwise the magnitude and direction of the resultnt of the four concurrent forces 80 N, 120 N, 150, N and 120 N making angles of 30°, 70°, 120°-15′ and 155° respectively with a fixed line.

   [Ans. 395 N acting at 111°-28′ with the fixed line]

10. Determine analytically or graphically the resultant of the following four coplanar forces:-
    (i) 52 N East 10° North;
    (ii) 78 N North 24° West;
    (iv) 126 N West 3° South;
    (v) 114 N South 9° West.

    [Ans. 130 1 N East 17°-24′ North]

11. Six coplanar forces are acting on a particle as detailed below. Find the magnitude and direction of the resultant force.
    (i) A push of 50 N, 10° East of North;
    (ii) A pull of 60 N, 40° East of North;
    (iii) A pull of 40 N due East;
    (iv) A push of 80 N, 45° West of South.
    (v) A pull of 50 N due West;
    (vi) A push of 30 N due 60° West of North.

    [Ans. 109.2 N, N 69°-30′ E]

12. Four forces act a point. First of 200 N actng due North, second of 100 N acting due South. The third of 500 N acting due East and fourth of 300 N acting West. Find magnitude and direction of the resultant force.

    [Ans. 223.6 N at 26°-34′ with East]

13. Forces 20, 30 and 40 N act from the verticles along the sides of an equiateral triangle, taken in order. Find magnitude and direction of the resultant of all the forces. **[Ans.** 17.5 N at 232°-58′ with the force of 20 N]

14. ABC is an isosceles triangle, right angles at C, three equal forces P act along AB, BC, CA. Find the resultant completely .

    [Ans. 0.424 P acting horizontally]

15. Forces of 70, 10, 10 and 30 N respectively act at one of the angular points of a regular pentagon towards the four other points taken in order. Find the magnitude and direction of their resultant force.

[**Ans**. 83.68 N making an angle of 31° -55′]

16. A pendulum weights 50 N. It is pulled aside by a horizontal pull of 20 N. Calculate angle of the rope to the vertical and tension in the string.

[**Ans**. 21°-48′, 53.9 N]

17. A weight of 100 newtons is suspended by a knot from two strings attached to a horizontal beam. These strings make angle of 45° and 30° respectively with horizontal. Find tension in the strings.

[**Ans**. 89.65 N, 73.2 N]

18. A string ABCD is suspended from two fixed points A and D. It carries two weights of 100 N and and 'W' at C and B respectively. The inclination of portion DC is 30° and that of portion AB is 60° to the vertical. If $\angle$ABC = 120°, find tensions in the different portion of the string and magnitude of 'W'.

[**Ans**. AB = 100 N, BC = 100 N, CD = 173.2 and weight 'W' = 100 N]

19. A smooth circular cylinder of radius 1 m is lying in a triangular groove, one side of which maked 15° with the horizontal and the other 40° with the horizontal. Find the reaction at the surfaces of the contact if there is no friction and weight of the cylinder is 100 N. [**Ans**. 78.47 N, 31.59 N]

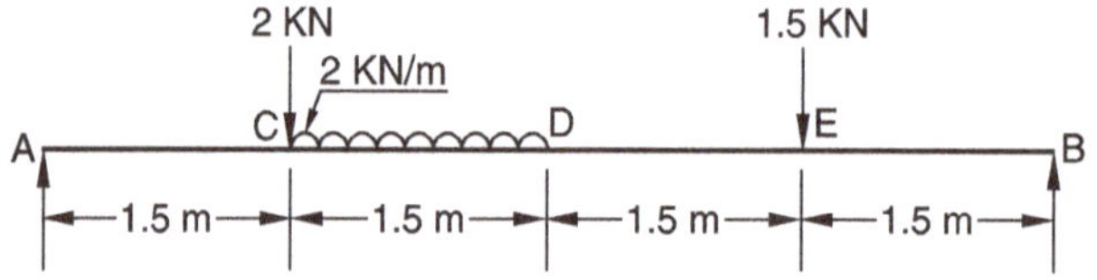

Fig. 2.78.

20. Find the reactions R_A and R_B of a loaded beam as shown in Fig. 2.76.

[**Ans**. 5.25kN & 1.25 kN]

Chapter 3

Moments

INTRODUCTION

In practice when a force is applied to a body without any fixed point in it, the motion produced in the body is linear one. On the other hand, if one point in the body is hinged (in case of crane) and a force is applied at any other point, the body tends to rotate i.e., turning moment is produced about the hinged point. This phenomenon finds wide applications in the engineering field. e.g. a cyclist applying force on the paddle produces a turing effect Therefore, a clear concept of moment is very essential for the solution of engineering problems in the subject of Engineering Mechanics.

3.1. MOMENT

The turning effect of a force on a body about a hinged point is called moment of the force.

Consider a gate hinged at point A and B as shown in Fig. 3.1. When a man applies a force F at a point C (point C is other than the line joining hinged points A and B) the force produces turning effect and the gate is opened. The ability of the force to cause rotation about axis AB is the moment (turning effect) of the force.

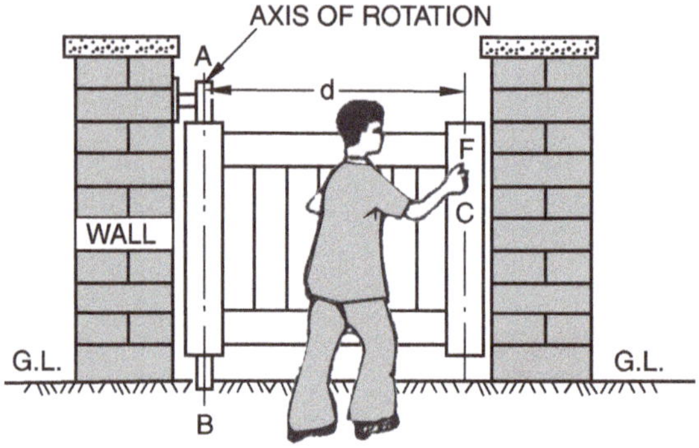

Fig. 3.1. *Man opening the gate*

3.2. MEASUREMENT OF MOMENT

The moment of a force depends upon the following two factors:

(i) Magnitude of force

(ii) Perpendicular distance between this line of action of force and the axis of rotation (called lever arm).

It is measured by the product of a force about a given point (or axis) and the perpendicular distance between line of action of the force and axis of rotaion, i.e.

$$\text{Moment of force} = F \times d$$

Where F = Magnitude of force

d = Perpendicular distance between the line of action of force and the axis of rotation.

Examples: (i) Consider a body free to rotate about point O.

Let F be the force acting on it as shown in Fig. 3.2. Draw OM perpendicular to the line of action of F.

Then, moment of F about O,

$$= F \times OM$$

$$= F \times d$$

(ii) In Fig. 3.1. the force F is applied to the gate at a distance d, perpendicular to the axis of rotaion.

$$\therefore \text{ Moment of force } F \times d$$

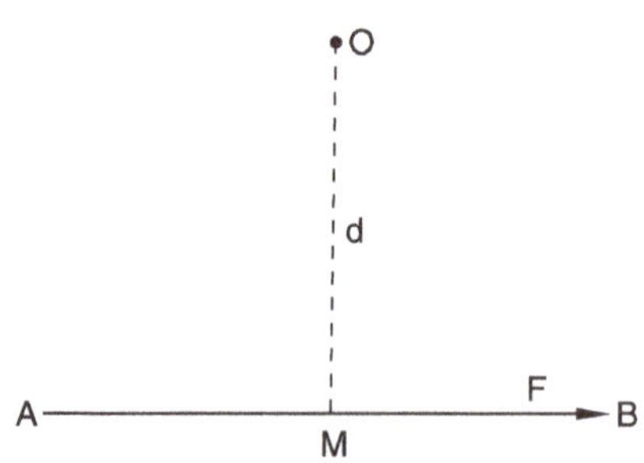

Fig. 3.2

Note: *The moment of force will be zero, it F = 0 or d = 0) Thus, if the force is passing through the axis of rotaion (i.e., d = 0), the moment will be zero. This shows that a force can only produce turning effect if its line of action does not pass through axis of rotation.*

Unit of moment: As we know,

$$\text{Moment} = \text{Force} \times \text{distance}$$

Force is measured in newton (N) and distance in metres (m).

$$\therefore \text{ Unit of moment} = \text{Unit of force} \times \text{Unit of distance}$$
$$= N \times m = Nm$$

Therefore, unit of moment is N-m in S.I. units.

3.3. SIGNS OF MOMENTS

A force can cause the body to rotate in anti-clockwise or closkwise direction depending upon the manner in which the force is applied. The following signs are used in moments:

(i) When a force tends to rotate a body in anti-clocikwise direction, its moment is said to be positive (see Fig. 3.3).

(ii) When a froce tends to rotate a body in clockwise direction, its moment is said to be negative (see Fig. 3.4).

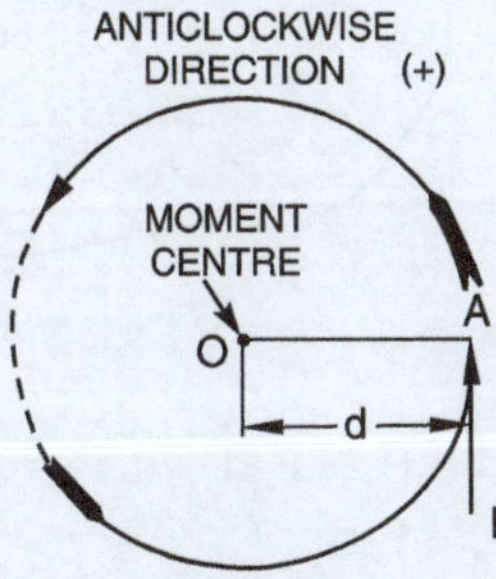

Fig. 3.3

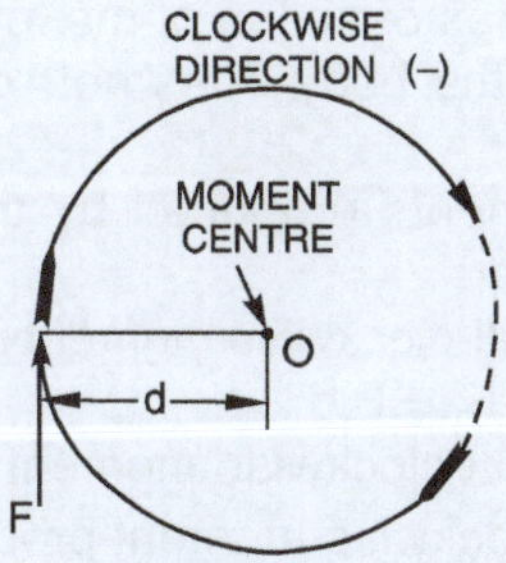

Fig. 3.4

3.4. RESULTANT TURNING MOMENT

When a number of forces are acting simultaneously on a body, the total turning effort on the body about a point is equal to the algebraic sum of moments of all the forces about that point.*

Consider three co-planar foces F_1, F_2 and F_3 acting simultaneously on the body as shown in Fig. 3.5. Let O be the point in the body which is fixed so that the body is capable of turning about this point. Let d_1, d_2 and d_3 be the perpendicular distances from point O to the lines of action of forces F_1, F_2 and F_3 respectively.

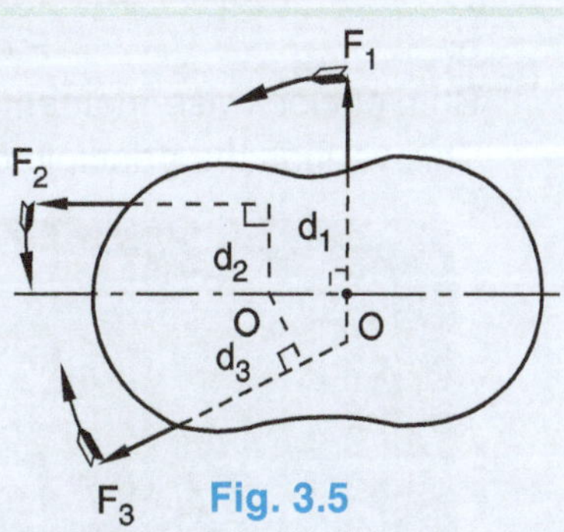

Fig. 3.5

Taking moments of all the foces about point O,

Moment of $F_1 = F_1 \times d_1$ (anti-clockwise)

Moment of $F_2 = F_2 \times d_2$ (anti-clockwise)

Moment of $F_3 = F_3 \times d_3$ (clockwise)

$\therefore$ Resultant turning moment about point O

= Algebraic sum of the moments of all forces about point O

$= (F_1d_1) + (F_2d_2) + (-F_3d_3)$

If the resultant moment is positive, then the body will rotate in anti-clockwise direction. However, if the resultant moment is negative, then the body rotates in clockwise direction.

3.5. PRINCIPLE OF MOMENTS

The principle of moments tells regarding the rotational equilibrium i.e., When the body does not rotate in any direction, it is said to be in rotational equilibrium.

According to this principle:

"If a body is in equilibrium (rotational equilibrium) under the action of a number of forces acting simultaneously on the body, the sum of clockwise moments is equal to sum of anti-clockwise moments about the axis of rotation."

i.e., Under rotational equiliburium,

Sum of clockwise moments = Sum of anti-clockwise moments.

Verification: Consider four forces F_1, F_2, F_3 and F_4 acting on a body as shown in Fig. 3.6.

Let O be the point in the body about which it is capable to rotate. Let d_1, d_2, d_3 are d_4 be the respective perpendicular distances between point O and lines of action of the forces. Let us suppose that the body does not rotate in any direction.

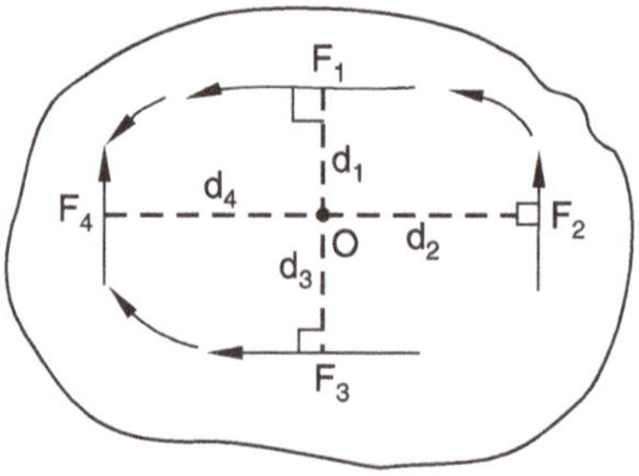

Fig. 3.6

The moments produced by these forces will be as under:

(i) An anti-clockwise moment produced by force $F_1 = F_1d_1$

(ii) An anti-clockwise moment produced by force $F_2 = F_2d_2$

(iii) An clockwise moment produced by force $F_3 = F_3d_3$

(iv) An clockwise moment produced by force $F_4 = F_4d_4$

The resultant of these moments will be equal to the algebraic sum of all the moments about O.

$\therefore$ Resultant moment $= F_1d_1 + F_2d_2 - F_3d_3 - F_4d_4$

Since the body is in rotational equilibrium i.e., it does not rotate in any direction, therefore, the resultant moment is zero.

$$\therefore \quad F_1d_1 + F_2d_2 - F_3d_3 - F_4d_4 = 0$$

$$\text{or} \quad F_3d_3 + F_4d_4 = F_1d_1 + F_2d_2$$

Sum of clockwise moments = Sum of anti-clockwise moments.

This verifies the principle of moments.

Designing of suspension bridge to find out the turning moment of forces at points within the structure

3.6. GENERAL CONDITIONS OF EQUILIBRIUM OF BODIES UNDER COPLANEAR FORCES

The three general conditions of equilibrium are:

1. The algebraic sum of the horizontal components of all the forces is zero, i.e. $\Sigma H = 0$
2. The algebraic sum of the vertical components of all the force is zero, i.e. $\Sigma V = 0$
3. The algebraic sum of the moments of all the forces about any point is zero, i.e. $\Sigma M = 0$

SELECTED PROBLEMS

Problem 1: A force of 120 N acts at a point A as shown in Fig. 3.7. Determine its moment about hinged point O.

Solution: Resolving the force 120 N into two components at A as shown in Fig 3.7.

$\therefore \quad F_H = 120 \cos 60°$ and

$F_V = 120 \sin 60°$

Taking moments about O

Moment due to $F_H = 120 \cos 60° \times 0 = 0$

[$\because$ perpendicular distance from O to the line action of $F_H = 0$]

Moment due to $F_V = 120 \sin 60° \times 10$ (clockwise)

$= 120 \times 0.866 \times 10 = 1039.2$ Nm

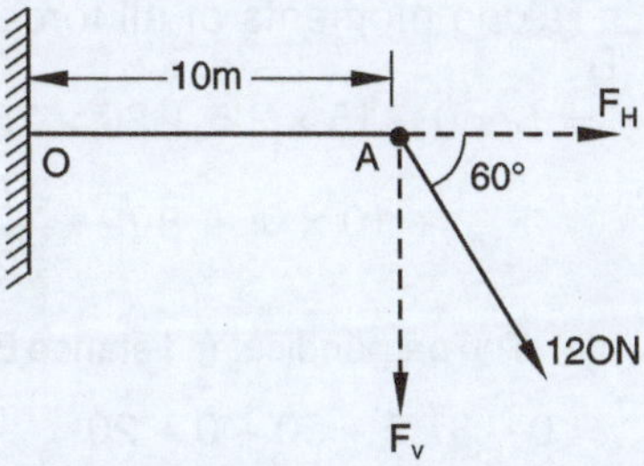

Fig. 3.7

$\therefore$ Moment due to force of 120 N about O = **1039.2 Nm [Ans]**

Problem 2: Determine the moment due to a force of 100 N with respect to the given point O as shown Fig. 3.8.

Solution: Resolving the force 100 N into two components

$$F_H = 100 \cos\theta \quad \text{and} \quad F_V = 100 \sin\theta$$

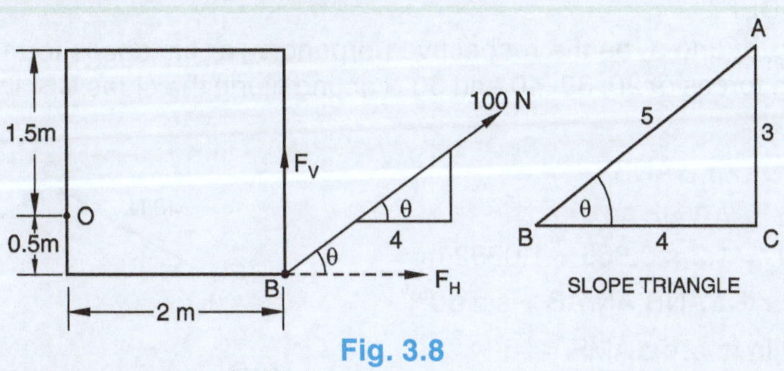

Fig. 3.8

In traingle ABC

$$AB = \sqrt{4^2 + 3^2} = \sqrt{16+9} = 5$$

$$\therefore \quad \cos\theta = \frac{4}{5} \quad \text{and} \quad \sin\theta = \frac{3}{5}$$

Consider clockwise moments as negative and anti-clockwise moments as positive.

Taking moments about $O = F_H \times 0.5 + F_V \times 2$

$$= 100\cos\theta \times 0.5 + 100\sin\theta \times 2$$

$$= 100 \times \frac{4}{5} \times 0.5 + 100\frac{3}{5} \times 2$$

$$= 40 + 20 = 160 \text{ Nm}$$

∴ Moment due to 100 N force with respect to A = **160 Nm (anti-clockwise) [Ans]**

Problem 3: Forces act along the sides BC, BA, DC and DA of a square, equal to 5, 10, 15, 20N respectively. Also a foce of $8\sqrt{2}$ N acts along with the diagonal CA. Calculate the resultant moment of all forces about the corner B when each side of the square is 2.5 metres.

Solution: Fig. 3.9 shows forces acting along the sides of a square ABCD.

Taking moments of all forces abount corner B.

$$= 5 \times 0 - 15 \times 2.5 + 20 \times 2.5$$

$$+ 10 \times 0 + 8\sqrt{2} \times \frac{2.5}{\sqrt{2}}$$

[∵ The perpendicular distance BE = $2.5/\sqrt{2}$ m]

$= 0 - 37.5 + 50 + 0 + 20$

$=$ **32.5 Nm [Ans]**

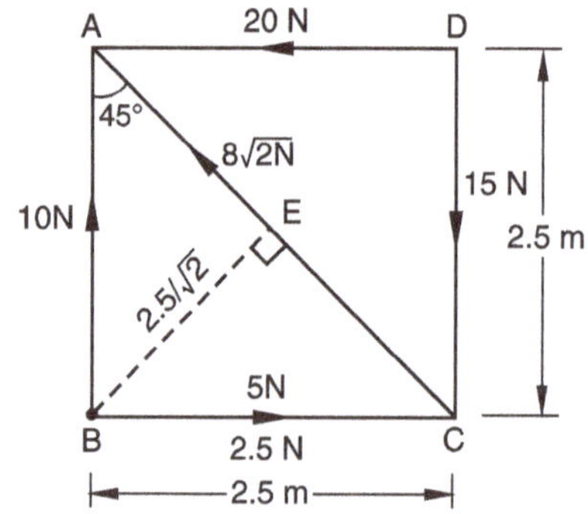

Fig. 3.9

The positive sign shows that the resultant moment of all forces about corner B is in the anti-clockwise direcction.

Problem 4: Forces acting along the sides AB, CB, DC, ED, EF and FA of a regular hexagon ABCDEF are 10, 20, 30, 40, 50 and 50N respectively. Find the resultant moment about the corner A. Take each side of the hexagon equal to 1.2 m.

Solution: Fig. 3.10 shows various forces acting along the sides of a regular hexagon ABCDEF.

Let d_1, d_2, d_3 and d_4 be the respective perpendicular distances form A to the lines of action of the forces of 20, 30, 40 and 50 N acting along the sides CB, DC, DE and EF respectively.

Now in rt. ∠d ΔANB,

$AN = d_1 = AB \sin 60°$

∴ $d_1 = 1.2 \times 0.866 = 1.0392$ m

[∵ In rt. ∠d ΔANB AN/AB = sin 60°]

Similarly in rt. ∠d ΔAMF,

$AM = d_4 = 1.0392$ m and in ΔABC,

AC i.e., [using cosine law]

$$d_2 = \sqrt{AB^2 + BC^2 - 2AC \cdot BC\cos 120°}$$

$$= \sqrt{(1.2)^2 + (1.2)^2 - 2 \times 1.2 \times 1.2 \times (-0.5)}$$

$= 2.0784$ m

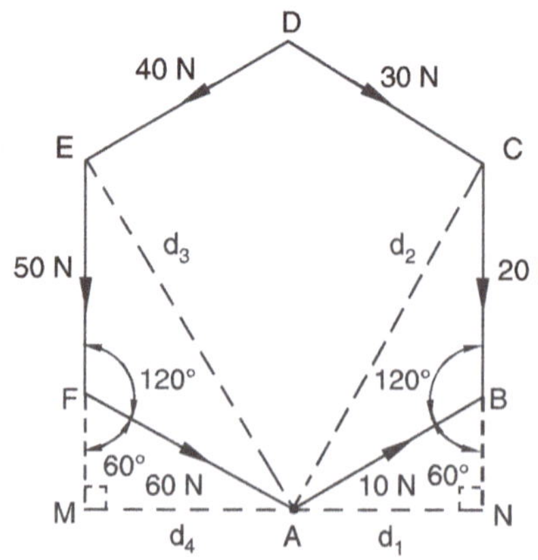

Fig. 3.10

Also from ΔAFE, $AE = d_3 = 2.0784$ m

Now taking moments of all the force about A, we get

Resultant moment, ΣM

$= 10 \times 0 - (20 \times d_1) - (30 \times d_2) + (40 \times d_3) + (50 \times d_4) + 60 \times 0$

$= (20 \times 1.0392) - (30 \times 2.0784) + (40 \times 2.0784) + (50 \times 1.0392)$

$= - 20.784 - 62.352 + 83.136 + 51.96 =$ **51.96 Nm[Ans]**

Hence the resultant moment about A is 51. 96 Nm in the anticlockwise direction.

Problem 5: The connecting rod of a steam engine, shown in Fig. 3.11, is 2 m long and the lenght of crank is 0.5. The thrust in the connecting rod is 1200 N. If the rod and the crank are in position as shown, find turning moment about crank shaft 0.

Solution: The thrust in the connecting rod AB = 1200 N

Let OC = Perpendicular distance of the line of action of thrust in the connecting rod from O and $\angle OAC = \theta$.

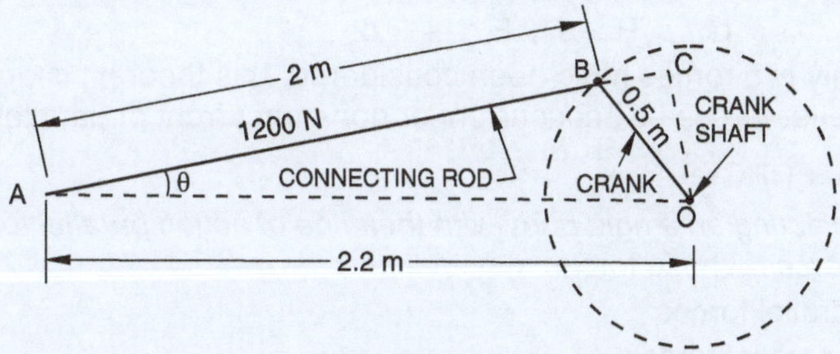

Fig. 3.11.

In ΔAOB, AB = 2 m,

OB = 0.5 m and AO = 2.2 m, $\angle OAB = \angle OAC = \theta$

Using cosine formula,

$\therefore \quad OB^2 = AB^2 + AO^2 - 2AB \times AO \times \cos\theta$.

$$\cos\theta = \frac{AB^2 + AO^2 - OB^2}{2 \times AB \times AO} = \frac{2^2 + 2.2^2 - 0.5^2}{2 \times 2 \times 2.2} = \frac{8.59}{8.8} = 0.9761$$

$\therefore \quad \theta = \cos^{-1} 0.9761 = 12° \, 33'$

$\therefore \; OC/\,OA = \sin\theta$

or $\quad OC = OA \sin 12° \, 33' = 2.2 \times 0.2173 = 0.478$ m

$\therefore$ Turning moment about O = 1200 × 0.478

= 573.7 N m (anticlockwise) [Ans]

3.7. VARIGNON'S THEOREM

This theorem deals with the moments of copanar forces* and their resultant about any point or axis. According to this theorem.

> *"If a number of coplanar foces are acting on a rigid body, then the algebraic sum of their moments about any point or axis is equal to the moment of their resultant about the same point or axis."*

*Concurrent or Non-concurrent or Parallel forces.

Suppose two forces F_1 and F_2 are acting at a point A as shown in fig. 3.12. Their resultant R will be given by the diagonal AD of the parallelogram ABDC.

Let O be any point in the plane of the forces about which moments are to be the perpendicular distances of F_1, F_2 and R respectively from the point O.

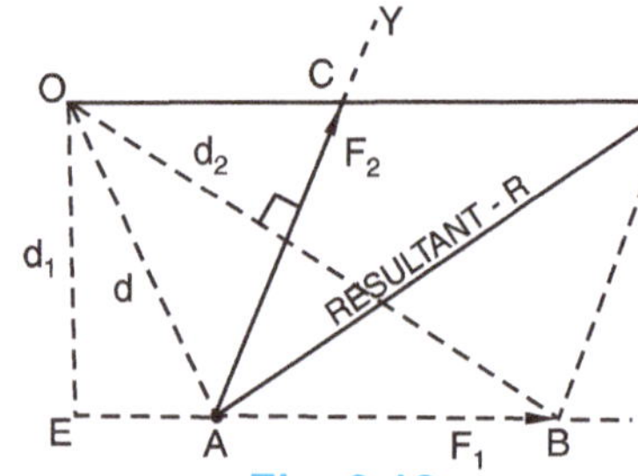

Fig. 3.12

Then, Moment of F_1 about O $= + F_1 \times d_1$ (anticlockwise)

Moment of F_2 about O $= + F_2 \times d_2$ (anticlockwise)

Moment of R about O $= + R \times d$ (anticlockwise)

According to Varignon's theorem,

Moment of R about O = Algebraic sum of moment of F_1 an F_2 about O.

$$\text{or} \quad R \times d = F_1 d_1 + F_2 d_2$$

Though only two forces have been considered, this theorem is applicable to any number of coplanar forces i.e., concurrent or non-concurrent or parallel force.

3.8. PARALLEL FORCES

The forces acting on a rigid body with their line of action parallel to each other are called parallel forces. These forces may be:

(i) Like parallel forces

(ii) Unlike parallel forces

(i) **Resultant of like parallel forces:** *When the parallel forces act in the same direction, they are called like parallel forces.*

Consider two parallel forces F_1 and F_2 acting at points A and B as shown in Fig. 3.13. At A and B introduce two equal and opposite forces F and F so that the equilibrium of the system is not affected.

At point A, obtain R_1 as the resultant of F_1 and F by completing the respective rectangle of forces. Similarly obtain the resultant R_2 of forces F_2 and F at B. The two resultants R_1 and R_2 are produced to meet at point O.

At point O, resolve R_1 into F and F_1; and R_2 into F and F_2 as shown.

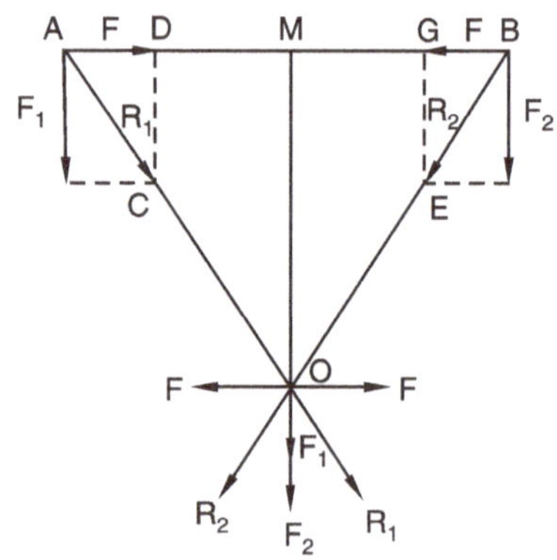

Fig. 3.13. *Like parallel forces*

Now, the equal and opposite forces F and F at O cancel each other while the resultant of F_1 and F_2 is obtained as the the force $F_1 + F_2$, since both the force are acting in the same direction and along the same line of action.

Thus resultant of two like parallel forces is equal to their sum i.e, $(F_1 + F_2)$.

In order to find to point of application of the resultant on AB, produced as under:

ΔADC and ΔAMO are similar.

$$\therefore \qquad \frac{DC}{AD} = \frac{MO}{AM}$$

$$\text{or} \qquad \frac{F_1}{F} = \frac{MO}{AM} \qquad \text{...(i)}$$

$$\text{Similarly} \qquad \frac{EG}{GB} = \frac{MO}{BM}$$

$$\text{or} \qquad \frac{F_2}{F} = \frac{MO}{BM} \qquad \text{...(ii)}$$

Dividing equation (i) by equation (ii), we get

$$\frac{F_1}{F_2} = \frac{BM}{AM}$$

$$\therefore \qquad F_1 \times AM = F_2 \times BM$$

Thus, it can be said that the resultant of two like parallel forces is equal to their sum and acts at such a point which divides the internal distance between the like parallel forces in the inverse ratio of their magnitudes.

(ii) **Resultant of unlike parallel forces:** *When the parallel forces act in the opposite direction, they are unlike parallel forces.*

Consider two unlike parallel forces F_1 and F_2 acting at points A and B as shown in fig. 3.14. At A and B introduce two equal and opposite forces F and F so that the equilibrium of the system is not affected.

At point A, obtain R_1 as the resultant of F_1 and F by completing the respective rectangle of forces. The two resultants when produced meet at point O.

At point O, resolve R_1 into F and F_1; and R_2 into F and F_2 as shown. Now, the equal and opposite forces F and F at O cancel each other while the resultant of F_1 and F_2 obtained as force $(F_2 - F_1)$, since the forces are acting in the opposite direction and along the opposite line of action.

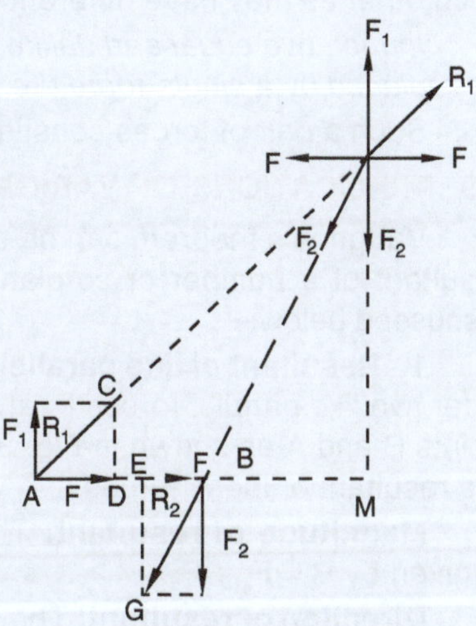

Fig. 3.14. *Unlike parallel forces*

Thus, resultant of two unlike parallel forces is equal to their difference i.e. $(F_2 - F_1)$.

In order to find the point of application of the resultant on AB, proceed as under:

Produced the line of action of $(F_2 - F_1)$ as to meet AB produced at point M as shown in Fig. 3.14.

ΔADC and ΔAMO are similar

$$\therefore \qquad \frac{DC}{AD} = \frac{MO}{AM}$$

$$\text{or} \qquad \frac{F_1}{F} = \frac{MO}{AM} \qquad \text{...(i)}$$

Similarly ΔBEG and ΔBMO are similar

$$\therefore \qquad \frac{EB}{EG} = \frac{MO}{BM}$$

$$\text{or} \qquad \frac{F_2}{F} = \frac{MO}{BM} \qquad \text{...(ii)}$$

Dividing equation (i) by equation (ii), we get,

$$\frac{F_1}{F_2} = \frac{BM}{AM}$$

$$\text{or} \qquad F_1 \times AM = F_2 = BM$$

Thus, it can be said that the resultant of two unlike parallel forces is equal to their difference and acts at such a point, which divides the internal distance between the unlike parallel fores in the inverse ratio of their magnitudes.

Case of Failure: It is intersting to note that there is one particular case in which the above construction fails. If the two unlike parallel forces are equal i.e., $F_1 = F_2$, then the lines AC and GB will be parallel to one another and, therefore, they will not intersect at all. In this case, we cannot find a single resultant force. These forces cannot cancel each other as they have different lines of action.

In fact, two equal and unlike parallel forces having different lines of action cannot be replaced by a single force (i.e., resultant).

Such a pair of forces constitutes a couple, which will be discussed later.

3.9. APPLICATIONS OF VARIGNON'S THEOREM

Varignon's theorem can be used to determine the position of line of action of the resultant of a number of co-planar forces acting on a body. A few such cases are discussed below:–

1. Resultant of like parallel forces: Consider two like parallel forces F_1 and F_2 acting at points B and A as shown in Fig. 3.15. Let R be the resultant of these forces.

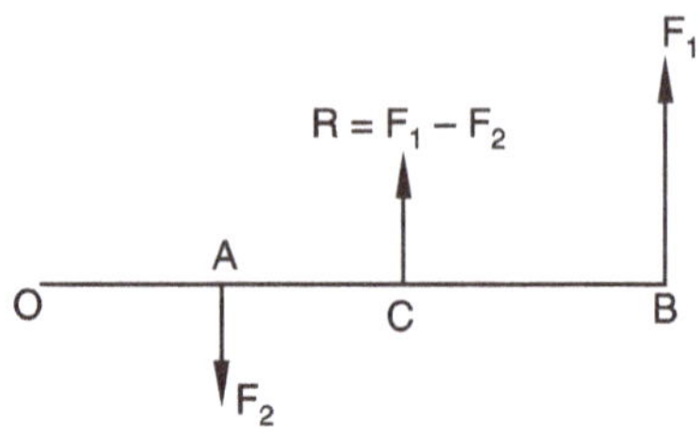

Fig. 3.15

Magnitude of resultant: The magnitude is given by $R = F_1 + F_2$.

Direction of resultant: The resultant acts in the direction of the foces F_1 and F_2.

Line of action: In order to determine the line of action of the resultant, select any point O in the plane of the forces. Through O, draw a straight line perpendicular ot the forces F_1, F_2 and R cutting their lines of action at A, B, and C respectively (see Fig. 3.15).

By Varignon's theorem,

Moment of R about O = Algebraic sum of moments of F_1 and F_2 about O.

$$\text{or} \quad R \times OC = F_1 \times OB - F_2 \times OA$$

$$\text{or} \qquad OC = \frac{F_1 \times OB - F_2 \times OA}{R}$$

Thus OC and hence point C where the resultant acts can be located. The same method can be extended to any number of the parallel foces.

2. Resultant of unlike parallel foces: Consider two unlike parallel force F_1 and F_2 ($F_1 > F_2$) acting at points B and A as shown in Fig. 3.16. Let R be the resultant of these forces.

Magnitude of resultant: The magnitude is given by $R = F_1 - F_2$

Direction of resultant : The resultant acts in the direction of bigger force i.e. F_1.

Line of action: In order to determine the line of action on the resultant, select any point O in the plane of these forces. Through O, draw a straight line perpendicular to force F_1, F_2 and R cutting their lines of action at A, B and C respectivley (see Fig. 3.16).

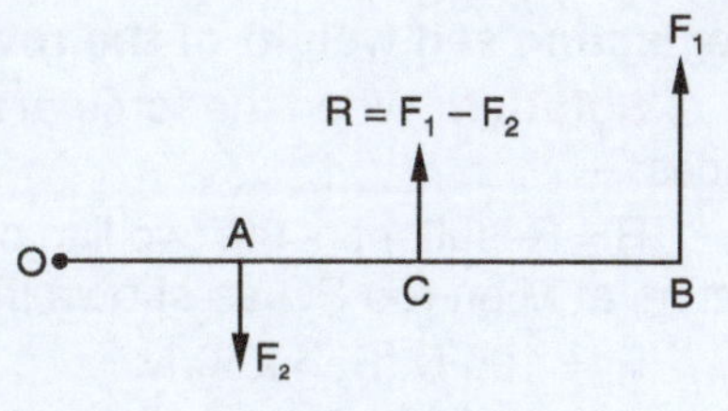

Fig. 3.16

By Varignon's theorem,

Moment of R about O

= Algebraic sum of Moments of F_1 and F_2 about O.

$$R \times OC = F_1 \times OB + F_2 \times OA$$

$$\therefore \quad OC = \frac{F_1 \times OB + F_2 \times OA}{R}$$

Thus OC and hence point C where the resultant acts can be located.

The same method can be extended to any number of unlike parallel forces.

3. Resultant of non-concurrent, non parallel force: *When the lines of action of a number of coplanar forces acting on a body are non-parallel and they do not meet one point, they are called non-concurrent, non-parallel forces.*

Consider three non-parallel and non-concurrent forces F_1, F_2 and F_3 acting on a body and making angle θ_1, θ_2 and θ_3 with horizontal respectively as shown in Fig. 3.17. Let R be the resultant of these forces.

Magnitude of resultant: Resolving forces horizorally and vertically and taking their respectively sum

$$\Sigma F_x = F_1 \cos\theta_1 - F_2 \cos\theta_2 - F_3 \cos\theta_3$$

$$\Sigma F_y = F_1 \sin\theta_1 + F_2 \sin\theta_2 - F_3 \sin\theta_3$$

$$\therefore \; R = \sqrt{\Sigma F_{x^2} + \Sigma F_{y^2}}$$

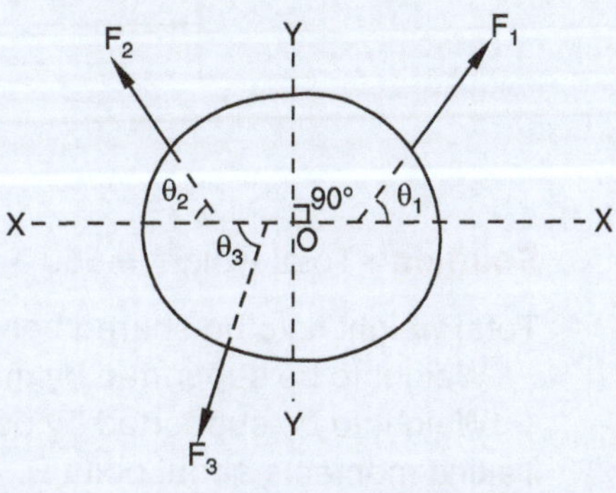

Fig. 3.17

Direction of Resultant: Let the angle β which the resultant R makes with horizontal is given by,

$$\tan\beta = \frac{\Sigma F_y}{\Sigma F_x}$$

Line of action: To determine the line of action of resultant, select any point O in the plane of the forces.

Then by Varignon's theorem,

Moment of R about O = Algebraic sum of moments of all forces (or their component) about O.

Thus the point of application of the resultant can be found out.

SELECTED PROBLEMS

Problem 6: Two like parallel forces of 100 N and 250 N act on the ends of a rod 3.5 m long. Find the resultant force and the point where it acts on the rod, neglecting self weight of the rod.

Solution: When the force are like parallel foces:—

R = Resultant of the two like parallel forces acting at M on rod PQ as shown in Fig. 3.18

$F_1 = 100$ N, $F_2 = 250$ N

PQ = 3.5 m

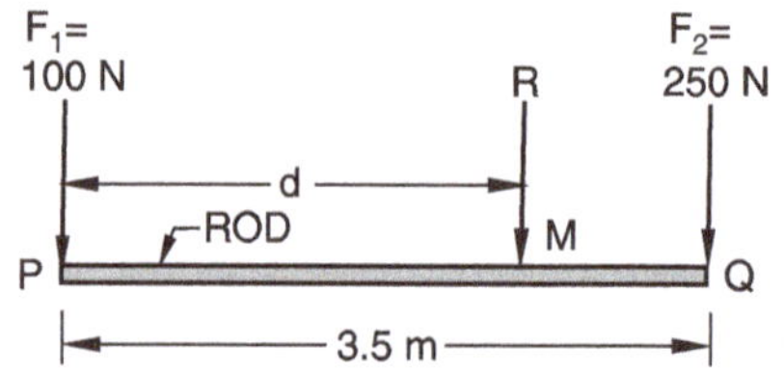

Fig. 3.18

Since the forces are like parallel forces, the resultant 'R' can be determined by using the relation,

$$R = F_1 + F_2 = 100 + 250$$

= 350 Nm in the direction of forces [Ans]

To find the position of the resultant, use the relation

$$F_1 \times PM = F_2 \times QM$$

or $$100 \times PM = 250\ (PQ - PM) = 250\ (3.5 - PM)$$

or $$100\ PM = 875 - 250\ PM$$

or $$350\ PM = 875$$

$\therefore$ PM = d = 875/350 = **2.5 m [Ans]**

Problem 7: A man and a boy carry a weight of 550 N between them by means of a uniform rod 5.5 metre long, weighing 200 N. Where must the weight be placed so that the man may bear twice as much of the weight as the boy?

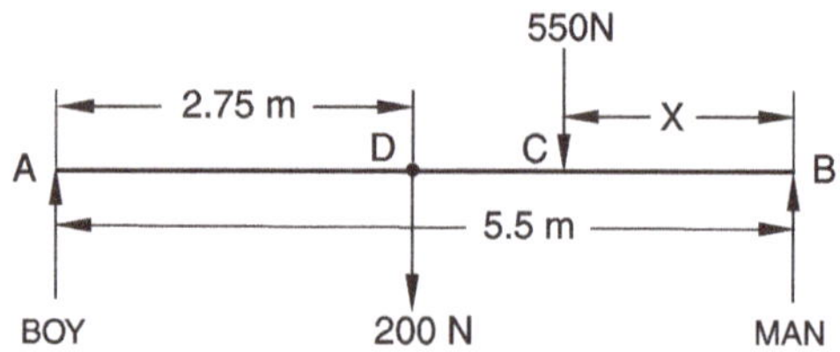

Fig. 3.19

Solution : Total weight = 550 + 200 = 750 N

Total weight is to be shared between man and boy in ratio 2 : 1

$\therefore$ Weight to be supported by man = 750 × 2/3 = 500N

Weight to be supported by boy = 750 × 1/3 = 250 N

Taking moments about point B.

Clockwise moments = Weight supported by boy × AB

= 250 × 5.5 = 1375 N m

Anti-clockwise moments = Wt of beam × CB + 550 × x

200 × 2.75 + 550 × x = 550 + 550 x

By applying principle of moments,

Anti-clockwise moments = clockwise moments

$$550 + 550\,x = 1375$$

$$\therefore \quad x = \frac{1375 - 550}{550} = 1.5 \text{ m} \quad [\text{Ans}]$$

Problem 8: Four coplanar foces of magnitude 10 N, 20 N, 30 N and 40 N are acting respectively along the four sides of a square ABCD as shown Fig. 3.20. Dtermine the magnitude, direction and position of the single force with respect to corner A, which will keep the body in equilbrium.

Solution: Algebraic sum of forces in the horizontal direction,

$$\Sigma H = 10 - 30 = -20 N$$

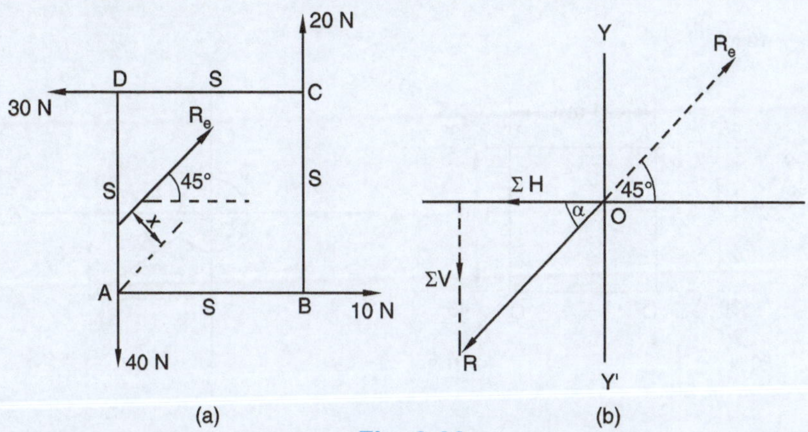

Fig. 3.20

Similarly, the algebraic sum of forces in the vertical direction,

$$\Sigma V = 20 - 40 = -20 N$$

$$\text{Resultant,} \quad R = \sqrt{(\Sigma H)^2 + (\Sigma V)^2}$$

$$= \sqrt{(-20)^2 + (-20)^2} = 28.28 \text{ N}$$

Let the resultant act an angle α with the horizontal

$$\tan\alpha = \frac{\Sigma V}{\Sigma H} = \frac{-20}{-20} = 1$$

$$\therefore \quad \alpha = \tan^{-1}_{1} = 45°$$

Since ΣH and ΣV both are – ve, the resultant R lies in the third quantum.

$\therefore$ To keep the body in equlibrium, the force R_e (equilibrant) = 28.28 N must act at angle of 45° with horizontal as shown in Fig. 3.20. (b).

Let the resultant R act at a perpendicular distance x from point A.

Considering clockwise moments as negative and anti-clockwise moments as positive.

Taking moments of all the forces about A and equatingg them to zero i.e., $\Sigma M_A = 0$, we get

$$30 \times s + 20 \times s - R_e \times x = 0$$

$$R_e \times x = 30 \times s + 20 \times s \qquad [\because s = \text{side of square}]$$

$$\therefore \quad x = \frac{50s}{R_e} = \frac{50x}{28.28} = 1.77 \text{ s}$$

∴ To keep the body in equlibrium, a single force of 28.28 N must act at a perpendicular distance of 1.77 s from A, making an angle of 45° with the horizontal as shown in Fig. 3.20. **[Ans]**

Problem 9: Fig. 3.21 shows a body under the action of four coplanar forces. Determine the magnitude, direction and position of the resultant forces of the given force system.

Solution: Fig. 3.21. shows the various forces acting on the body.

Let 'R' resultant of the given force system.

Resolving all the forces horizontally and vertically,

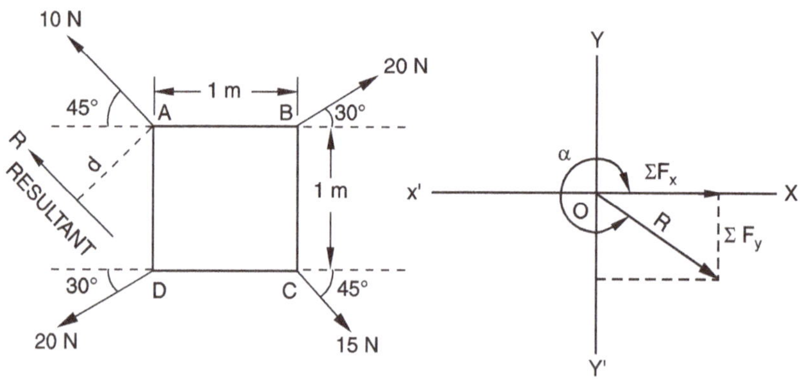

Fig. 3.21

$$\Sigma F_x = -10 \cos 45° + 20 \cos 30° + 15 \cos 45° - 20 \cos 30°$$
$$= -10 \times 0.707 + 20 \times 0.866 + 15 \times 0.707 - 20 \times 0.866$$
$$= 3.535 \text{ N}$$

$$\Sigma F_y = 10 \sin 45° + 20 \sin 30° - 15 \sin 45° - 20 \sin 30°$$

∴ $= -3.535$ N

The magnitude of the resultant 'R' can be dtermine as,

$$R = \sqrt{(\Sigma F_x)^2 + (\Sigma F_y)^2}$$

$$= \sqrt{(3.5535)^2 + (-3.535)^2} = \sqrt{25} = 5\text{N}$$ **[Ans]**

$$\tan\alpha = \frac{\Sigma F_y}{\Sigma F_x} = \frac{-3.535}{3.535} = -1$$

∴ $\alpha = \tan^{-1}(-1)$, = **315° with the axis OX as shown [Ans]**

Now, the position of the resultant 'R' about A = Algebric sum of moment of the relation,

Moment of resulant 'R' about A = Algebric sum of moment of the reactangular components of all the forces about A.

or $-R \times d = 10 \times 0 + 20 \cos 30° \times 0 + 20 \sin 30° \times 1 + 15 \cos 45° \times 1 - 15 \sin 45° \times 1 - 20 \cos 30° \times 1 + 20 \sin 30° \times 0$

$$-5 \times d = 0 + 20 \times 0.5 \times 1 + 15 \times 0.707 \times 1 - 15 \times 0.707 \times 1 - 20 \times 0.866 \times 1 + 0$$
$$= -7.32$$

$$\therefore \quad d = \frac{-7.32}{-5} = \textbf{1.464 m form A as shown [Ans]}$$

Problem 10: Fig. 3.22 shows a frame in the form of a right angled triangle ABC having sides AB = 4m, BC = 3 m and CA = 5 m. The forces 30 N, 50N and PN act on the above sides respectivley. Find the magnitude of the force P, if the resultant of these forces passes through a point D on AB produced so that BD = 8 m.

Solution: First of all draw DE perpendicular to AC produced.

Let $\angle BAC = \theta, \quad \sin\theta = 3/5$

But from rt. $\angle d\,\Delta AED$

$$\sin\theta = DE/AD$$
$$\text{or} \quad DE = AD \sin\theta$$
$$\therefore \quad DE = AD \times 3/5 = 12 \times 3/5 = 7.2 \text{ m}$$

Now, taking moments about D

$$50 \times 8 = P \times DE = 7.2\,P$$
$$\therefore \quad P = \frac{50 \times 8}{7.2} = \textbf{55.6 N [Ans]}$$

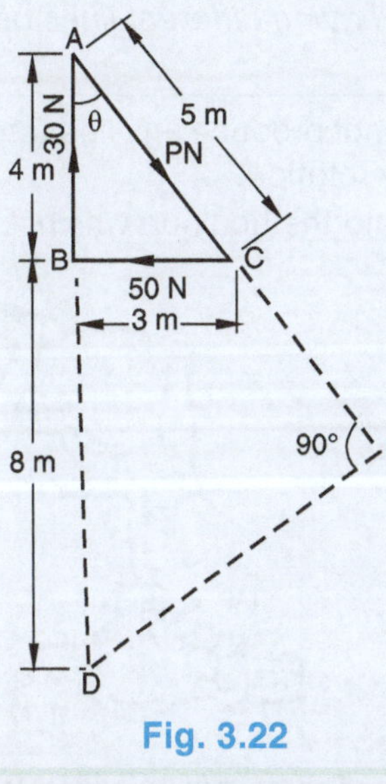

Fig. 3.22

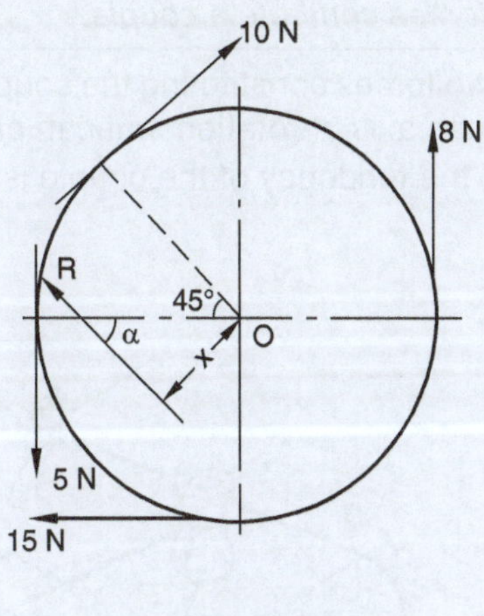

Fig. 3.23

Problem 11: Determine the resultant of four forces tangent to the circle of radius 0.3 m as shown in Fig. 3.23. What will be its location with respect to centre of the circle.

Solution: Resolving the force horinontally.

$$\Sigma H = -15 + 10 \cos 45° = -7.93 \text{ N}$$

Similarly, resolving the forces vertically, we get

$$\Sigma V = -5 + 10 \sin 45° + 8 = 10.07 \text{ N}$$

$$\therefore \quad \text{Resultant, } R = \sqrt{\Sigma H^2 + \Sigma V^2}$$

$$= \sqrt{(-7.93)^2 + (10.07)^2}$$

$$= \sqrt{164.31} = 12.82\text{N}$$

Let the resultant act an angle α with the horizontal.

$$\tan\alpha = \frac{\Sigma V}{\Sigma H} = \frac{10.07}{7.93} = 1.270$$

$$\therefore \quad \alpha = \tan^{-1} 1.270 = 51° \, 46'$$

Let R = resultant which acts at distance x from 0.

Taking moments about 0 = ΣM

$$-12.82\,x = -10 \times 0.3 + 8 \times 0.3 - 15 \times 0.3 + 5 \times 0.3$$

$$\therefore \quad x = 0.28 \text{ m}$$

$\therefore$ R = 12.85 N, which act clockwise, from the centre 0 and at 51° 46′ with the horizontal at a distance of **0.28 m [Ans]**

3.10. COUPLE

It has already been pointed out that two equal and unlike parallel force having different lines of action can not be replaced by a single force. In other words, the resultant force can not be determined. In fact, such a pair of forces consitute what is known as a couple.

> *When two equal and unlike parallel force having different lines of action act on a body, the forces contitute a couple.*

The two forces constituting the couple cannot produce any linear motion in a body fixed about an axis of rotation, but can produce rotation.

Thus, the tendency of the couple is to rotate the body on which it acts.

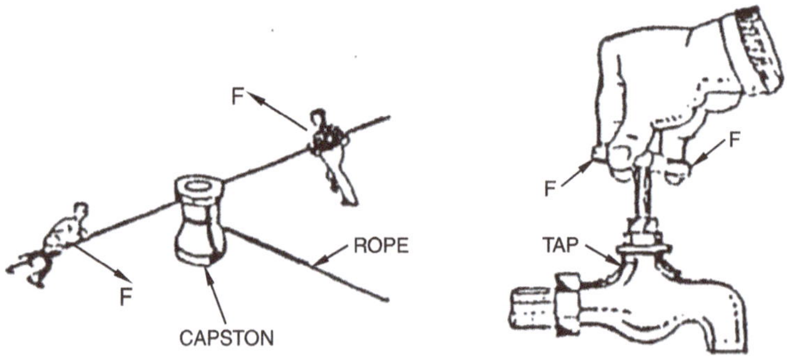

Capstan being rotated by two men *Tap handle being rotated by a person*

Fig. 3.24.

Examples : (i) Consider two men applying equal and unlike parallel forces to the arms of a capstan as shown in Fig. 3.24. This constitutes a couple and rotates the capstan.

(ii) When a person opens or closes a water tap, he applies two equal and unlike parallel forces, thus substituting a couple (see Fig. 3.24). This couple rotates the tap handle.

Fig. 3.25. *A couple is a pair of forces applied on the key of a lock*

(iii) Other examples of couple are:

(a) Key of a lock (see fig. 3.25)

(b) Winding of a watch

(c) Opening or closing the cap of a pen

(d) Rotation of a screw cap.

(e) Moving the stearing of a car etc.

3.11. MOMENT OF A COUPLE

The product of either force and the perpendicular distance between the lines of action of the two forces is called moments of a couple.

A couple has the tendency to rotate and hence it has moment.

i.e. Moment of couple = Either force × Perpendicular distance between the lines of action of the forces.

Consider two equal and unlike parallel forces each equal of F and acting at points A and B of a rigid body as shown in Fig. 3.26.

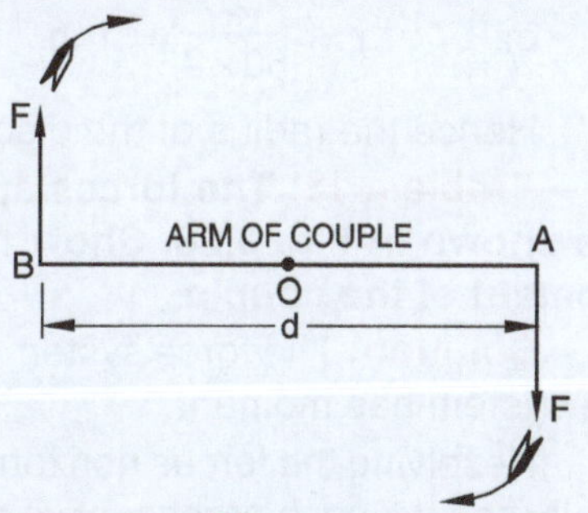

Fig. 3.26

Let d be the perpendicular distance between the lines of the forces, then

Moment of couple = F × d

Proof: The fact that of a couple is F × d can be easily established.Consider a point O on the line AB.

Moments of F at B about O = F × OB (anti-clockwise)

Moments of F at A about O = F × OA (anti-clockwise)

Resultant moments of two forces about O = F × OB + F × OA

= F × (OB + OA)

= F × AB = F × d

But the resultant moment of the two forces is the moment of couple.

∴ Moment of couple = F × d (anti-clockwise)

It should be noted that moment of couple is the same whatever be the postion of axis of rotation w.r.t. the point of application of the forces.

Arm of the copule: *The perpendicular distance between the lines of action of the two forces forming couple is called arm of the couple (see Fig. 3.26).*

3.12. PROPERTIES OF A COUPLE

A couple has the following important properties:

(i) Two equal and unlike parallel forces having different lines of action constitute a couple.

(ii) A couple has no resultant force but it has resultant moment.

In other words, a couple does not produce linear motion but it produces rotation.

(iii) The moment of a couple is equal to either force multiplied by arm of the couple. The anti-clockwise moment is positive and clockwise moment is negative.

(iv) Any number of coplanar couples can be represented by a single couple, the moment of which is equal to the algebraic sum of the moments of all the couples.

(v) A couple can be balanced only by a couple of equal and opposite moment.

(vi) A couple cannot be balanced by a single force. However, a single force can be replaced by an an equal and like parallel force through any other point and a couple.

Problem 12: A couple of moment 120 Nm acts on a disc as shown in fig. 3-27. Find the radius of the disc If P = 60 N.

Solution: Let r m be the radius of the disc

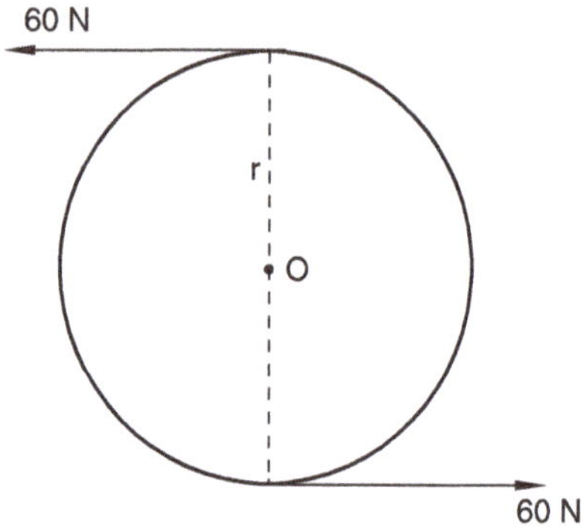

Fig. 3.27

Now arm of given couple

= 2 r m

Moment of the couple

= P × Arm of the couple

or $120 = 60 \times 2r$

or $r = \dfrac{120}{60 \times 2} = 1$ m

Hence the radius of the disc = **1 [Ans]**

Problem 13: The forces applied to the corners of a square plate 0.2 m side are shown in Fig. 3.28. Show that these forces reduce to a couple and find the moment of the couple.

Solution: The force system will constitute a couple if the resultant is zero and still the system has moment.

Resolving the forces horizontally and vertically and taking their respective algebric sum,

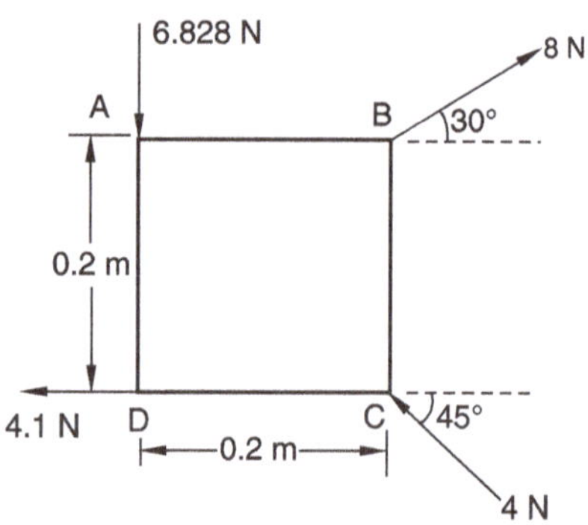

Fig. 3.28

$\Sigma F_x = 8 \cos 30° - 4 \cos 45° - 4.1$

$= 8 \times 0.866 - 4 \times 0.707 - 4.1 = 0$

$\Sigma F_y = 8 \sin 30° + 4 \sin 45° - 6.828$

$= 8 \times 0.866 \times 4 \times 0.707 - 6.828 = 0$

$\therefore R = \sqrt{\Sigma F_x^2 + \Sigma F_y^2} = 0$

Hence the system constitutes a couple.

Taking moments about point B,

Moment of couple = Algebric sum of moments of all forces.

$= 6.828 \times 0.2 - 4 \times BE - 4.1 \times 0.2$

[∵ Moment of 8 N in zero]

$= 1.3656 - 4 \times 0.2/\sqrt{2} - 0.82$

$= 1.3656 - 0.4 \times \sqrt{2} - 0.82$ = **– 0.02 Nm [Ans]**

Problem 14: ABCD is a square whose side is 2 m. Along AB, BC, CD and DA act forces equal to 1, 2, 8 and 5 N along AC and DB act forces equal to $5\sqrt{2}$ and $2\sqrt{2}$ N. Show that they are equivalent to a couple whose moments is equal to 16 Nm.

Solution: The force system as shown in Fig. 3.29. will constitute a couple if the resultant is zero and the system has moment.

Resolving the foces horizontally and vertically and taking their repective algebric sum,

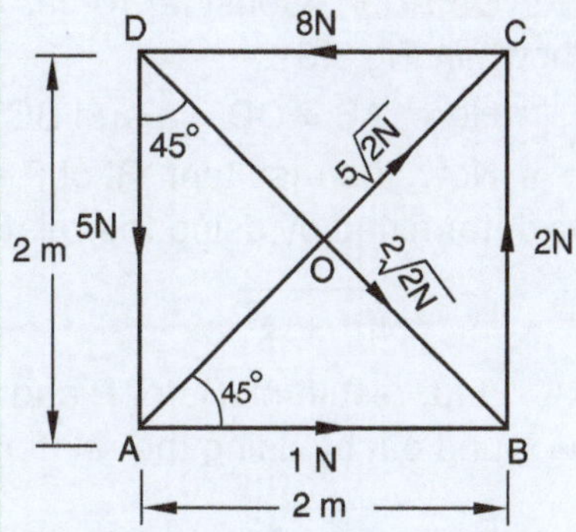

Fig. 3.29

$$\Sigma F_x = 1-8+5\sqrt{2}\cos 45^\circ + 2\sqrt{2}\cos 45^\circ$$

$$= 1-8+5\sqrt{2}\times 10\sqrt{2}+2\sqrt{2}\times 1/\sqrt{2} = 0$$

$$\Sigma F_y = 2-5+5\sqrt{2}\sin 45^\circ - 2\sqrt{2}\sin 45^\circ$$

$$= 2-5+5\sqrt{2}\times 1/\sqrt{2}-2\times\sqrt{2}\times 1\sqrt{2} = 0$$

Hence the system consitutes a couple.

Taking moments about point C

Moment of couple = Algebric sum of moments about C

$$= 1\times BC + 5\times CD + 2\sqrt{2}\times OC$$

$$= 1\times 2 + 5\times 2 + 2\sqrt{2}\times 2\sqrt{2}/2$$

$$= 2 + 10 + 4 \qquad [\because AC - \sqrt{2^2+2^2} = 2+2\sqrt{2}]$$

$$= \mathbf{16\ Nm\ [Ans]} \qquad [\because OC = AC/2 = 2\sqrt{2}/2]$$

Problem 15: The forces applied across the sides of triangle ABC (AB = 20 mm, BC = 50 mm and CA = 40 mm) are shown in Fig. 3.30. Show that these forces reduces to a couple and find the moment of the couple.

Solution: In ΔABC

$BC^2 = AB^2 + AC^2$

or $50^2 = 30^2 + 40^2$

$\therefore \angle BAC = 90^\circ$

Resolving the forces horizontally and vertically

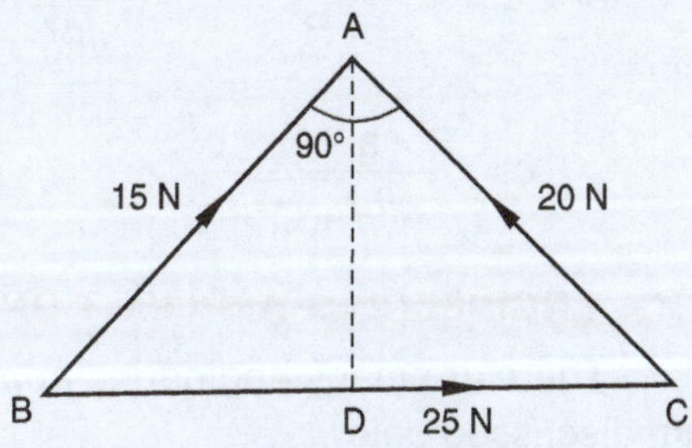

Fig. 3.30

ΣF_x = Algebric sum of horizontal components of 15 N, 25 N and 20 N

$= -DB + BC - CD$

$= BC - (DB + CD)$

$= BC - BC = 0$

$\Sigma F_y = -AD + 0 + DA = 0$

$$\therefore \quad R = \sqrt{\Sigma F_x^2 + \Sigma F_y^2} = 0$$

Resultant is zero and hence the force system consitutes a couple.

Moment of couple about B = 20 × 30 = **600 mm** **[Ans]**

Problem 16: ABCD is a rectangle such that AB = CD = a and BC = DA = b. Forces P act along AD and CB and forces Q act along AB and CD respectively. Prove that the perpendicular distance between the resultant of P and Q at A and resultant of P and Q at C is $-\dfrac{\mathbf{P\times a - Q\times b}}{\sqrt{\mathbf{P^2+Q^2}}}$.

Solution: Consider forces P and Q act along the sides of a rectangle ABCD as shown in Fig 3.31.

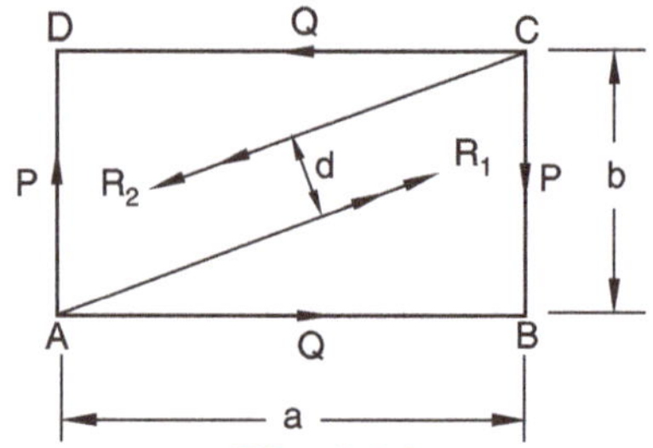

Fig. 3.31

Here, AB = CD = a and BC = DA = b

Now, the resultant 'R' of P and Q at A can be determined by using the relation,

$$R_1 = \sqrt{P^2 + Q^2} \quad ...(i)$$

and, resultant 'R_2' of P and Q about C can be found out by using the relation.

$$R_2 = \sqrt{P^2 + Q^2} \quad ...(ii)$$

Let 'd' = perpendicular distance between the lines of action of resulant R_1 and R_2.

Since R_1 and R_2 are equal, parallel and act In opposite direction, they consitute a couple, having moment = Rd ...(iii)

Forces P and P act in oposite driction along opposite side CB and AD of the rectangle ABCD, they constitute a couple having moment, $M_1 = P \times a$ (– ve)

[∵ It is clockwise]

Similarly, forces Q and Q acting along the sides AB and CD of rectnagle also constitute a couple having moment, $M_2 = Q \times b$ (– ve)

[∵ It is anti-clockwise]

$\therefore$ Net moment of the two couples = $Q \times b - P \times a$ ...(iv)

$R \times d = Qb - P \times a$

$$\text{or} \quad d = \frac{Q \times b - P \times a}{R} = \frac{Q \times b - P \times a}{\sqrt{P^2 + Q^2}}$$

$$\therefore \quad d = \frac{P \times a - Q \times b}{\sqrt{P^2 + Q^2}}$$

3.13. ENGINEERING APPLICATIONS OF MOMENTS

There are many enginering applications of moments and a few such applications are discussed below:

1. Lever
2. Steel Yard
3. Lever safety value
4. The balance

1. LEVER

A lever is a rigid rod or bar straight or bent having fixed point about which it can freely trun.

In Fig. 3.32, a man is moving a stone with a simple lever.

Fulcrum: The fixed point about which the level can turn is called fulcrum (see Fig. 3.32).

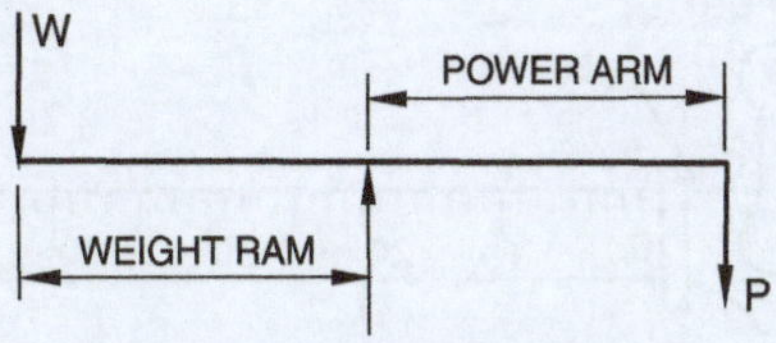

Fig. 3.32. *Simple lever*

Power arm: The perpendicular distance between the fulcrum and the point of application of force is called power arm of the lever (see Fig. 3.32).

Weight arm: The perpendicular distance between fulcrum and the point of applicaton of load/weight is called weight arm of the lever.

Types of levers: Levers are mainly divided into two types:

1. Simple levers
2. Compound levers

1. Simple levers: *A lever which has only one fulcrum is called a simple lever.*

By changing the positions of effort P, load W and fulcrum F. the following three types of levers can be obtained as shown in Fig. 3.33.

1. Firtst type of lever
2. Second type of lever
3. Third type o lever.

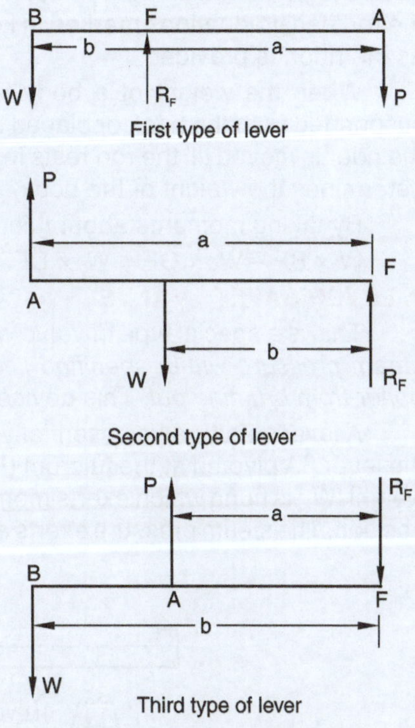

Fig. 3.33. *Types of levers*

In the first type of lever, fulcrum F is between W and P. In the second type of lever W is between fulcrum F and effort P and in the third type of lever P is between W and fulcrum F.

The following points must be noted in each type of lever:

(a) There is a reaction R at the fulcrum

In the first type of lever, R = W + P

In the second type of lever, R = W – P

In the third type of lever, R = P – W

(b) The sum of clockwise moments about the fulcrum is equal to the sum of anti-clockwise moments (principle of moments) i.e.

$$P \times AF = W \times BF$$

(c) By increasing the length of lever arm AF and by decreasing the lever arm BF. we can lift a load* W by means of a small effort P.

2. Compound levers: *Levers which have more than one fulcrum are called compound levers.* in fact, a compound lever is a combination of simple levers.

2. STEEL YARD

It is a machine for weighing goods. It is used on the working principle of 1st type of simple lever and is thus a useful application of the Law of Moments".

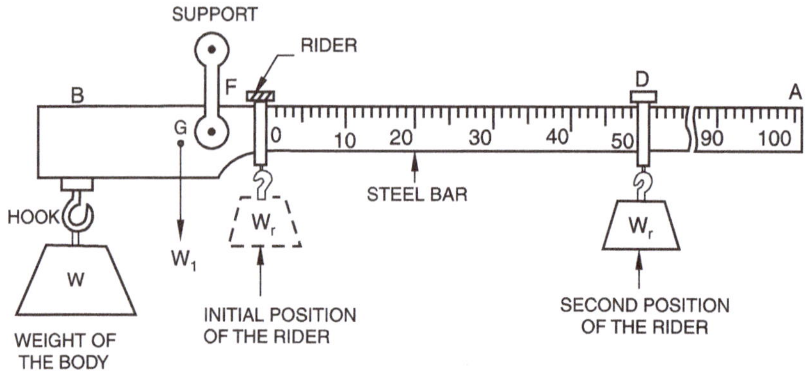

Fig. 3.34. *Steel yard*

A common type of steel yard consists of a straight steel rod BA movable about a fixed fulcrum (F) which is always kept near one of the end of the rod as shown in Fig. 3.34. The steel rod is made heavier on the shorter side BF, so that the centre of gravity 'G' of the whole rod along with hook attached to it near the end B. The longer side FA of the rod has graduations marked in kg or N on which a small movable weight 'Wr' known as the rider, is provided.

When the weight of a body is to be determined by this machine, the body is suspended from the hook or placed in the pan, provided near the end B of the rod. Then the rider is moved till the rod rests in a horizontal position. Thus, the position of the rider determines the weight of the body as

By taking moments about fulcurm, F,

$W \times BF + W_1 \times GF = W_r \times DF$

3. LEVER SAFETY VALVE

This is a special type of valve which is mounted on the boiler. *It is used to keep the steam pressure within specified limits in the boiler automatically and thus saves the boiler from brusting out.* This device is also based on the "Law of moments".

A lever safety valve essentially consists of a valve 'V' which is rigidly connected to the lever FA pivoted at the fulcrum (F) as shown in Fig. 3.35. At the end A of the lever, a weight 'W_1' is hung which exerts moment on to the valve 'V' to keep it on its seat in closed position. The steam pressure exerts a counter moments on the valve 'V' about fulcrum (F).

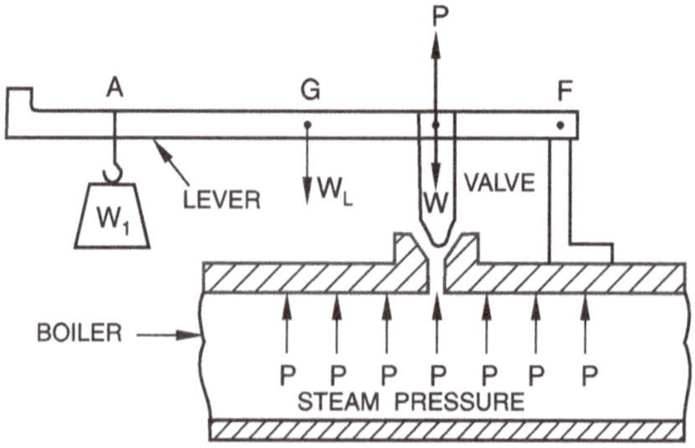

Fig. 3.35. *Lever safety valve*

When the pressure of steam inside the boiler increases beyond the safe value, the valve gets lifted up from its seat to release the excessive steam pressure into the amtosphere. When the steam pressure inside the boiler falls down to safe value, the valve automatically comes down to occupy its seat. At this time, the lever becomes horizontal and the valve 'V' stops further escape of the steam. In this way, steam pressure is maintained at a constant value automatically.

The least weight which will just keep the valve on its seat against the steam pessure can be found out as follows:–

Let P = Safe intensity of steam pressure inside the boiler = $p \times a$

a = Area of the valve

W_v = Weight of the valve

W_L = Weight of the lever acting at G

W_1 = Weight attached at the end A

Now taking moments about the function (F), we get

$$W_1 \times FA + W_L \times FG + W_V \times FV = p \times FV$$

or
$$W_1 = \frac{p \times a \times FV - W_L \times FG - W_V \times FV}{FA} \qquad [\because P = p \times a]$$

Since the values of all other quantities except W_1 are known, therefore, the value of W_1 can be found out.

Further, the reaction at fulcrum, $F = pa - W_L - W_V - W_1$

This will act downwards when it is lesser than the downwards forces which include the self weight of the lever, weight of the value and load attached at A.

4. THE BALANCE

It is a device used for weighing goods. The common balance is used on the working principle of 1st type of simple lever and is thus a useful application of the "Law of Moments".

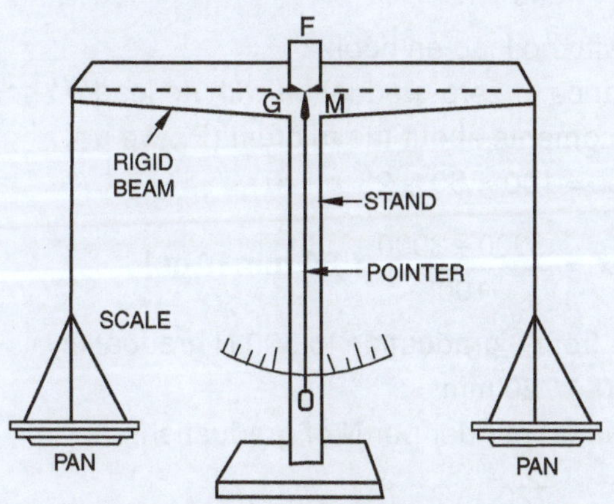

Fig. 3.36. *Balance*

The common balance consists of a rigid beam having two scale pans suspended from each end. The beam can turn freely about the fulcrum (F) which coincides with the middle point of the beam. The fulcrum is provided on a stand to which a scale with the zero reading in the middle is attached. A pointer is attached to the beam at its middle point, which moves over the scale as shown in Fig. 3.36.

Problem 17: The lever of common steel yard without hook connections weighs 100N, acting at 50 mm to the left of the fulcrum. The hook which carries the load, weighs 25 N and is also attached on the left hand side of the fulcrum at a distance of 120 mm. If the graduation mark 800 mm from the fulcrum reads 600 N, find :–

(a) Weight of the rider;

(b) How far is the zero graduationn from the fulcrum?

(c) What is the moment of rider per N of graduation?

Solution: (a) Here, weight of the steel rod W_1 = 600 N and weight of the hook alongwith the suspended weight,

$$W = 600 + 25 = 625 \text{ N}$$

Let weight of the rider = 'Wr'

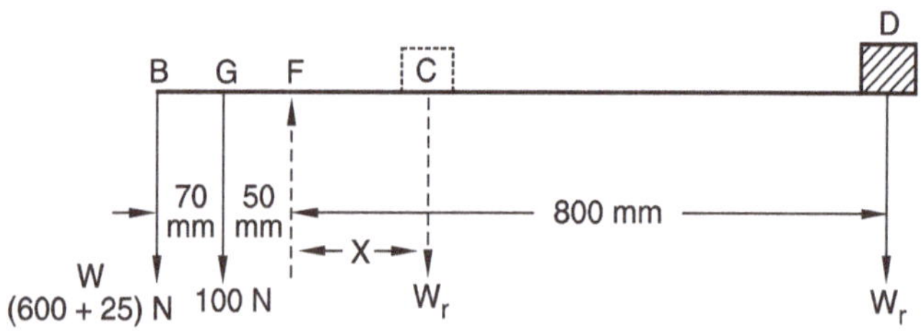

Fig. 3.37.

Taking moments about the fulcrum (F), we get

$$Wr \times 800 = 100 \times 50 = (600 + 25) \times 120$$

$$\therefore \quad Wr = \frac{5000 + 75000}{800} = \textbf{100 N[Ans]}$$

(b) Graduation with no load on hook

Let x be distance of zero graduation with no load on hook

Now taking moments about the fulcrum (F), we get

$$100\,x = 100 \times 50 + 25 \times 120$$

$$\therefore \quad x = \frac{5000 + 3000}{100} = \textbf{80 mm [Ans]}$$

(c) The distance from 0 graduation to 600 N graduation

$$800 - 80 = 720 \text{ mm}$$

∴ The movement of rider per N of graduation

$$= \frac{720}{600} = \textbf{1.2 mm [Ans]}$$

Problem 18: A lever safety valve 100 mm diameter is just on the point of blowing off steam. The weight of the lever is 300 N. The weight of the valve is 50 N and its centre lies at a distance of 200 mm from the fulcrum. If weight suspended from the lever is at a distance of 1000 mm from the fucrum is 320 N, determine the pressure of steam in the boiler.

Solution: Diameter of the safety value, d = 100 mm

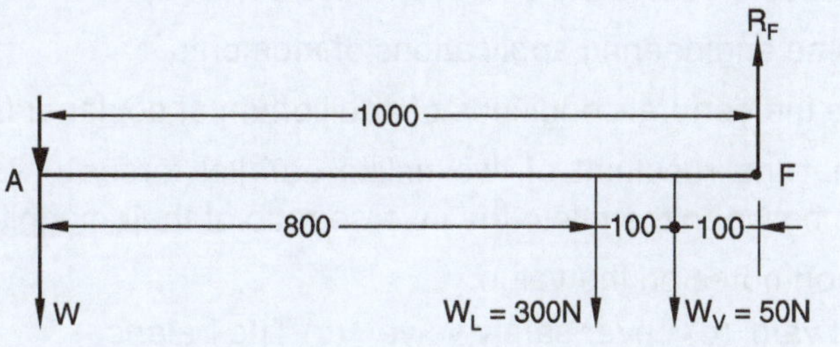

Fig. 3.38

$$\text{Weight of the lever, } W_L = 300 \text{ N}$$
$$\text{Weight of valve, } W_V = 50 \text{ N}$$
$$\text{Weight on the lever at A, } W = 320 \text{ N}$$

Also FG = 200 mm, FV= 100 mm and FA = 100 mm

Now, area of the safety valve,

$$a = \frac{\pi}{4} \times 100^2 = 7854 = 7854 \text{ mm}^2$$

Now, 'P' be the intensity of steam pressure acting on the valve,

$\therefore$ Upward thrust, $P = p \times a = p \times 7854 = 7854$ pN

Now, taking moments about the fulcrum (F), we get

$$7854 \text{ p} \times 100 = 320 \times 1000 + 300 \times 200 + 50 \times 100$$

or
$$p = \frac{3200 + 6000 + 500}{7854} = \frac{38500}{7854} = \textbf{4.902 N/mm}^2 \textbf{ [Ans]}$$

Problem 19: The arms of a blance are 200 mm and 220 mm respectively. The goods weighted being suspended from the long arm. Find the real weight of the goods whose apparent weight is 120 mm.

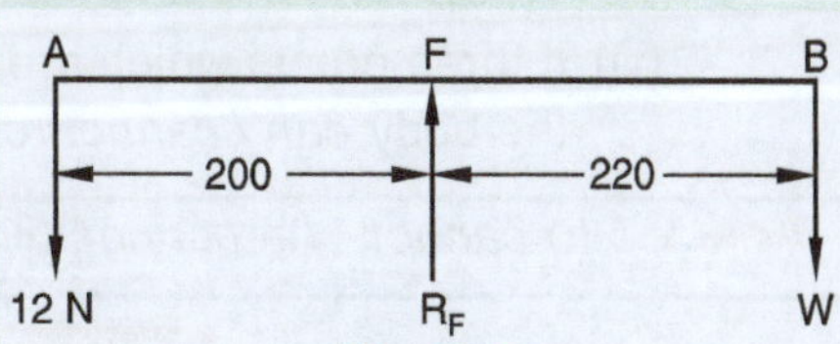

Fig. 3.39

Solution: Let 'W' = Real weight of the goods on the longer arm which is balanced by 120 mm weight on the shorter arm.

Now taking moments about fulcrum (F), we get

$$W \times 220 = 120 \times 200$$

$$W = \frac{120 \times 200}{220} = \textbf{109.10 mm [Ans]}$$

MODEL TEST QUESTIONS

1. Define moment of a force. What is its effect?
2. State and explain the principle of moments.
3. State and prove Varignon's theorem.

4. What do you mean by parallel forces? Discuss their classification.
5. What is couple? State the characteristics of a couple.
6. State some engineering applications of moments.
7. Describe the general conditions of equlilbrium of coplaner foces.
8. Show that the resultant of two unlike parallel forces divides externally the distance between them into the inverse ratio of their magnitude.
9. Write short notes on the value:
   (a) Steel yard (b) Lever safely valve (c) The balance

QUESTIONS FOR SELF EXAMINATION

1. **Fill in the blanks:–**
   (i) The turning effect of a force is called ....
   (ii) The moment of a force about a point through which its line of action passes is ....
   (iii) When body is in rotational equilibrium, sum of .... is equal to sum of .... about any point.
   (iv) Two equal and unlike parallel forces consitute a ....
   (v) For a body to be in equilibrium, it should have no .... and ....motion.

Ans. 1. (i) moment; (ii) zero; (iii) clockwise moments, anticlock-wise moments; (iv) couple; (v) rotational, linear.

2. **Pick up the correct words: —**
   (i) Concurrent forces *can/cannot* produce rotation.
   (ii) A couple *can / can not* be replaced by a single force.
   (iii) The resultant of like parallel forces P and Q = $P + Q/P - Q$.
   (iv) With the help of a lever, *we can / cannot* get output more than input.
   (v) Three concurrent forces *will / will not* be in equilibrium.
   (vi) If three points which are not in the same straight line in the body be fixed, the body *can / cannot rotate* under the action of forces.

Ans. 2. (i) *cannot*; (ii) *cannot*; (iii) $P + Q$; (iv) *cannot* ; (v) *will*; (vi) *cannot*;

MODEL TEST QUESTIONS

1. Force of 30N 50N 30N and 50 N act along the sides AB, CB, DC and DA respectively of a square ABCD of side 40 mm. Find the algebric sum of their moments about corners B and C.

   [**Ans.** – 08 N mm, – 32 N mm]

2. A rectangular lamina ABCD of size 2m × 3m shown in Fig. 3.40 is acted upon by six forces as shown. Find the resultant moment about B.

   [**Ans.** 17 Nm]

3. Find the equilibrant of the system of forces in magnitude and direction as shown in Fig. 3.41.

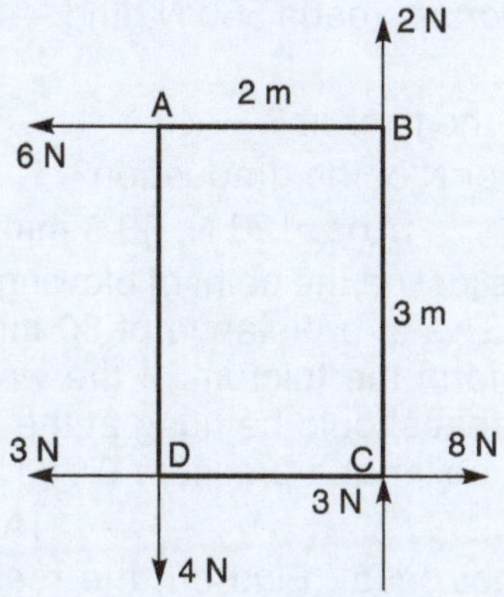

Fig. 3.40

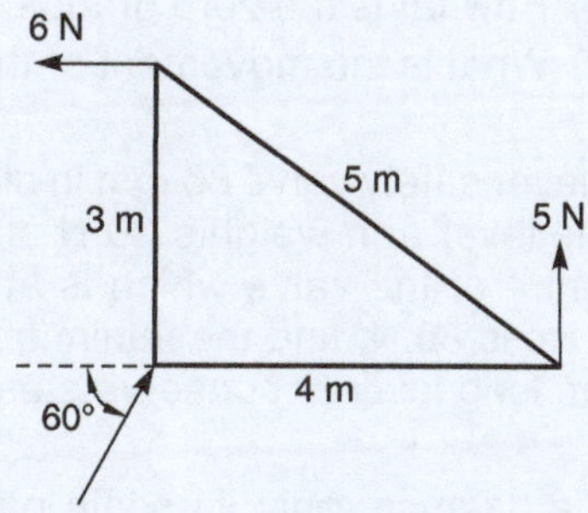

Fig. 3.41

4. Like parallel force 8N and 12 N act at point A and B, which are 300 mm apart. Find the magnitude of their resultant and the point at which it cuts AB. Also find the magnitude and position of the resultant when the forces are unlike.

[**Ans.** 20 downwards at 180 mm from A, 4N upwards]

5. Two men, one stronger than the other, have to cary a weight of 160 N by means of a light rod 4 m long. The weaker man can carry a load of 55 N. From where the weight must be suspended from the rod so as to allow man to take his full share.

[**Ans.** 1.375 m form stonger man side]

6. Forces acting along the sides AB, CB, DE, EF and FA of a regular hexagon ABCDEF are 1, 2, 3, 4, 5, and 6 N respectively. Find the resultant moment about the corner A. Take each side of hexagon equal to 1.2m.

[**Ans.** 5.11 Nm]

7. A square ABCD of 60 mm side subjected to forces of 10N, 20N, 30 N, 40N and $16\sqrt{2}$ N along AB, AD, CB, CD and BD. Find the magnitude, direction and position of the resultant force with respect to A.

[**Ans.** 46.6 N : 172° 6′ and 33.3 mm]

8. Prove that the three forces represented in magnitude, direction and line of action be the sides of a triangle taken in order are equivalent to a couple.

9. ABCD is a rectangle such that AB = CD = a and BC = DA = b. Forces P act along AD and CB and froces Q act along AB and CD. Prove that the perpendicular distance between the resultant of the forces P, Q at A and the resultant of the foces P, Q at C is

$$\frac{Pa - Qb}{\sqrt{P^2 + Q^2}}$$

10. ABCD is a square whose side is 2m. Along AB, BC, CD and DA act forces equal to 1, 2, 8 and 5N along AC and DB forces equal to $5\sqrt{2}$ and $2\sqrt{2}$ N. Show that they are equivalent to a couple whose moment is equal to 16 Nm.

11. The lever of a common steel yard without hook connection weights 75 N and acts at 60 mm to the left of the fulcrum. The hook carring the load, weights 40 N is also attached on the left hand side at a distance of 150 m form the fulcrum. If graduation mark, 700 mm from the fulcrum, reads 560 N, find:—
   (i) Weight of he rider;
   (ii) How far is the zero graduation from the fulcrum?
   (iii) What is the movement of the rider per N of the graduation?

[**Ans.** 126 N, 83.3 mm, 11.9 mm]

12. A lever safety valve 60 mm in diameter is just on the point of blowing off steam. The lever arm weights 30 N and its C.G. is at a distance of 80 mm from the centre of the valve which is at 70 mm form the fulcrum. If the weight of the valve is 80 N, find the minimum weight that should be hung at the end of 750 mm long lever to conserve steam in the boiler at a pressure 0.75 N/mm^2.

[**Ans.** 173 N]

13. In a balance, only its scale pans are not equal. Find out the real weight of goods which appears to weight 12 N, when placed in one scale pan and 10 N when placed in the other. Also find the difference between the weight of the scale pans. [**Ans.** 11N, 1N]

Chapter 4

Friction

INTRODUCTION

In practice the surfaces of bodies are never perfectly smooth. If we see a smooth surface under a microscope, it will be found that the surface has roughness and irregularities, which are not felt by ordinary touch or to the naked eyes. Therefore, when a body moves or tends to move over another body, the irregularities of the surface of one body get interlocked with that of the surface of another body (see fig. 4.1). This interlocking action of the irregularities produces a force which opposes the motion of one body over the other. Such an opposing force which comes into play along the surfaces of contact of the two bodies is called force of friction or simply friction.

4.1 FRICTION

The opposing force which comes into play in between the surfaces of contact of two bodies, when one body moves over the surface of another body is called the force of friction or simply friction.

The force of friction is tangential to the contacting surfaces and always acts opposite to the direction of motion of the body (see fig. 4.1).

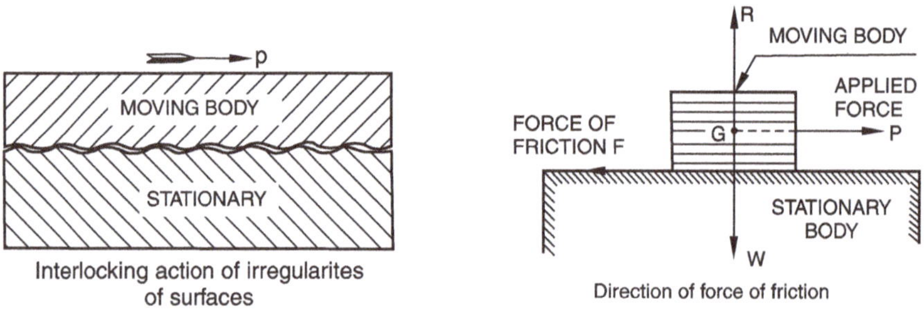

Fig. 4.1. *Concept of Friction*

The magnitude of the force of friction between the two surfaces in contact depends upon:

(a) The nature of the material of which the bodies are made.

(b) The pressure between the bodies at the point of contact.

Notes: *(i) Smoother the surfaces, lesser will be the friction (force of friction) and on the other hand rougher the surfaces, the greater the frictional force.*

(ii) The friction can be reduced by proper lubrication. When the libricating fluid is not used, the friction existing is called dry friction.

4.2 IMPORTANCE OF FRICTION IN ENGINEERING APPLICATION

In engineering application, friction is desirable as well as undesirable. There are number of devices such as belts, ropes, brakes, friction cluches, jib and cotter joints, nuts and bolts in which friction is desirable and efforts are made to maximise it. On the contrary, the friction is undesirable in moving parts of a machine because it causes the loss of power. Thus, a great force is required to cause motion of a body and efficiency of such a machine is reduced. In order to improve the efficiency of the machine the friction force is to be reduced to minimum possible by a suitable method of lubrication.

But for friction, there would have been a great mess and confusion in this universe. Persons would not have been able to walk, skate and drive automobiles without friction. Hence, *friction is said to be a 'necessary evil' in engineering application.*

4.3 USEFUL AND HARMFUL EFFECTS OF FRICTION

Friction plays most advantageous role in the solution of many mechenical problems in daily life. But on the other side, it produces harmful effects also. Some of the useful and harmful effects of friction are given below:

(a) Useful effects: 1. It helps in walking and skating by the persons safely on the ground.

2. It helps in smooth and safe movement of vehicles.

3. It helps in efficient working of appliances and devices such as belts and ropes, nuts and bolts, brakes, jib and cotter joints etc.

4. It helps in holding the nails into the wood in position.

5. It helps in holding the writing medium such as chalk, pencil, pen etc.

(b) Harmful effects: 1. It causes loss of power in machines.

2. It causes wear and tear of moving parts of machines, cutting tools, road surfaces, rails, etc.

3. It causes more tractive resistance to the movement of vehicles thus causing more consumption of the fuel.

4.4 METHODS OF REDUCING FRICTION

The following are the different methods of reducing friction:

1. By Polishing: In this method, a fine layer of the material is deposited on the surface of the body. This material fills up the space between the projection on the surface and makes it smooth.

2. By Lubricant: A lubricant (oil or grease) is a substance which forms thick layer between the surfaces of two bodies in contact. It fills up the depressions in the surfaces of contact and thus reduces the friction.

3. By use of ball or roller bearings: By this method, ball or roller bearings are used to convert sliding friction into rolling friction. Rolling friction is always less than the sliding friction.

4. By use of alloys: Alloys (anti-friction materials) have low coefficient of friction. Thus friction is reduced by lining the moving parts by alloys.

5. By avoiding the moisture: Friction is always more in the presence of moisture, so by avoiding the moisture between the two contacting surfaces, friction can be reduced.

4.5 TYPES OF FRICTION

In general, the friction is of the following two types:-

1. Static friction
2. Dynamic friction.

1. Static friction: *The friction offered by the surfaces subjected to external force until there is no motion between them is called static friction.*

For example, consider a block lying on a table. If starting from zero, the block is pulled with increasing force, it will be noted that the block does not move at first. Since the force is trying to move the block over the surface of table, therefore, friction comes into play. This force of friction between the surfaces of block and table without any motion of the block is the static friction.

It should be noted that if the force on the block is slowly increased, the force of friction also increases and a state reaches when the block is just ready to move.

2. Dynamic friction: *The friction experienced by a body when it is in motion over another body is called dynamic friction.*

The dynamic friction is also known as *kinetic friction.* It is always less than static friction.

In general, dynamic friction comes into play when the applied force increses to just over the firctional force. This phenomenon can be clearly understood from the graph drawn between applied force and the friction (i.e. force of friction) as shown in Fig. 4.2.

From O to A, the body is at rest i.e. the frictional force is just equal to the applied force. At point A, the applied force just overcomes the frictional force which drops (A to B) and develops motion in the body. At point B, the friction is refered as dynamic friction.

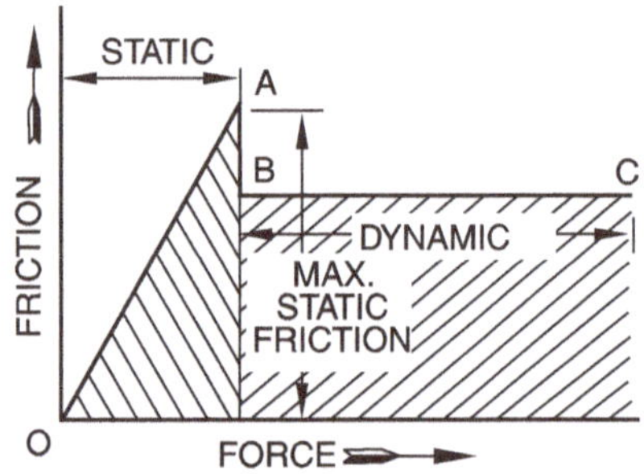

Fig. 4.2

Dynamic friction is further divided into the following two types:

(i) Sliding friction.

(ii) Rolling friction.

(i) **Sliding friction:** *The friction experienced by a body when it slides over another body is called sliding friction.*

For example, when a piston moves in the cylinder of the engine.

(ii) **Rolling friction:** *The friction experienced by a body, when it rolls over another body is called rolling friction.*

For example, when wheels of train move on the railway track.

4.6 LAWS OF FRICTION

The various laws of friction, given by Prof. Coulomb have been grouped into the following heads:

1. Laws of static friction
2. Law of dynamic or kinetic friction.

1. Laws of static friction: The following are the laws of static friction:-

1. The force of friction always acts in a direction opposite to that in which the body tends to move.

2. The magnitude of the force of friction is exactly equal to the applied force which tends to move the body.

3. The magnitude of the limiting friction bears a constant ratio to the normal reaction between the two surface in constant i.e. $F/R = \mu$.

4. The force of friction is independent of the area of contact between the two surfaces.

5. The force of friction depends upon the roughness of the surfaces.

This rock climber uses the static friction between her hands and feet with rock walls

2. Laws of kinetic or dynamic friction: The following are the laws of kinetic or dynamic friction:

1. The force of friction always acts in a direction opposite to that in which the body is moving.

2. The force of friction is directly proportional to the normal reaction between the two contact surfaces. But the ratio is slightly less than in case of limiting friction.

3. The force of friction remains constant for moderate speeds. But it decreases slightly with the increase of speed.

4.7 LIMITING FRICTION

The maxmimum value of frictional force which comes into play, when a body is just on the point of motion over the surface of another body is called limiting friction.

In actual practice, limiting friction is the maximum static friction.

Friction is self adjusting force up to a certain limit i.e. the force of friction goes on increasing as the force on the body is increased till it just starts moving over the surface of another body.

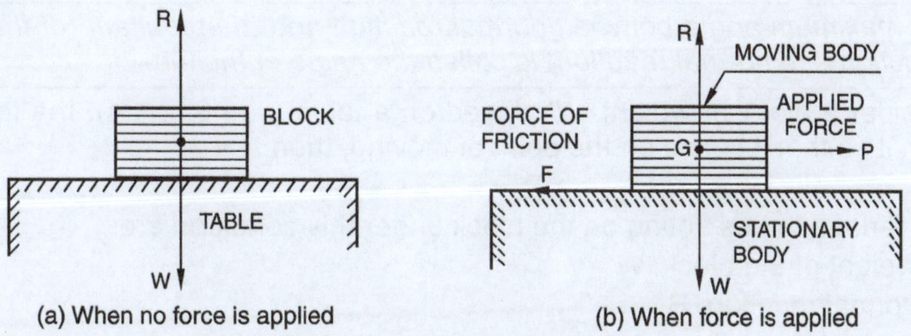

Fig. 4.3. *Concept of limiting friction*

To understand, consider a block placed over the surface of a table [see Fig. 4.3 (a)]. When no force is applied on the block, there is no force of friction between the surfaces in contact and the block is under the action of two forces:

1. Weight of the block 'W' acting vertically downwards.
2. Normal reaction 'R' acting vertically upwards.

The above two forces balance each other. However when some force is applied to pull the block, friction comes into play and the block does not move because an equal and opposite friction force is developed at the surfaces of contact. If the pulling force is gradually increased, the force of friction also increase. When the block is just on the point of moving [see fig. 4.3 (b)], the various acting on the block are:

1. Weight of the block 'W'
2. Normal reaction 'R'
3. Pulling force 'P'
4. Force of friction 'F'

Since the block is in equilibrium under the above four force, we have,

$W = R$ i.e. Weight of the body = Normal reaction

and $P = F$ i.e. Pulling force = Force of friction

Now, when the pulling force is slightly increased than that of limiting friction, the body just starts moving. Then, the dynamic friction comes into play. It may be noted that when the pulling force is less than the limiting friction, the block remains at rest and the friction comes into play is called static friction.

4.8 CO-EFFICIENT OF FRICTION (μ)

> *The ratio of limiting friction and the normal reaction is called co-efficient of friction.*

According to laws of friction, force of limiting friction is directly proportional to the normal reaction i.e.

$$F \propto R$$

or $$F = \mu R$$

Where μ is a constant of proportionality and is called co-efficient of friction.

$$\therefore \quad \mu = \frac{F}{R}$$

$$\text{Co-efficient of friction} = \frac{\text{Limiting friction}}{\text{Normal reaction}}$$

4.9 ANGLE OF FRICTION (ϕ)

> *The maximum angle between normal reaction and the resultant (of the force of limiting friction and normal reaction) is called the angle of friciton.*

Consider a block of weight 'W' placed on a table (see Fig. 4.4). If a force 'P' is applied on it so that it is just on the point of moving, then

$$P = F$$

The various forces acting on the block under this condition are:

1. Weight of the block, W
2. Normal reaction, R
3. Applied force, P
4. Force of friction, F

If we combine 'F' and 'R', we get the resultant 'S' inclined at angle ϕ. With R. Since the block is just on the point of moving, therefore the value of angle is also maximum.

This maximum value of angle ϕ is called limiting angle of friction or simple angle of friction.

From Fig. 4.4 we have

$$\tan\phi = \frac{F}{R}$$

$$\left[\text{But } \frac{F}{R} = \mu\right]$$

or $$\tan\phi = \mu$$

$$\therefore \text{ Angle of friction, } \phi = \tan^{-1}\mu$$

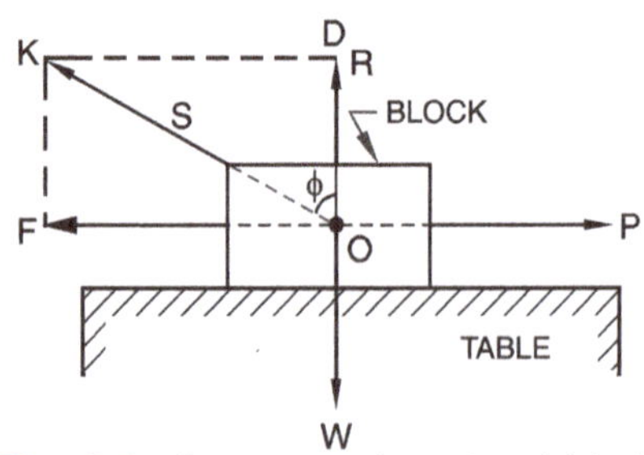

Fig. 4.4. *Concept of angle of friction*

Thus, the tangent of the angle of friction is always equal to the co-efficient of friction.

The value of the co-efficient of friction μ and angle of friction ϕ for various material are given in the table.

Co-efficient of Friction (μ) and Angle of Friction of Material (φ)

Sr. No.	Material	Co-efficient of friction (μ)	Angle of friction (φ) in degrees
1.	Wood on wood	0.2 to 0.5	11 to 26
2.	Wood on leather	0.2 to 0.5	11 to 26
3.	Wood on metal	0.2 to 0.6	11 to 31
4.	Cast iron on cast iron	0.3 to 0.4	17 to 22
5.	Metal on leather	0.3 to 0.6	17 to 31
6.	Steel on Steel	0.4 to 0.8	22 to 39
7.	Rubber on concrete	0.6 to 0.8	31 to 39

4.10. ANGLE OF REPOSE (α)

The angle which an inclined plane makes with the horizontal, when a body placed on it is just on the point of moving down is called angle of repose.

Consider a block placed on an inclined plane. If inclination is gradually increased, it will be observed that at certain inclination the body is just to move downwards.

Let α be the angle of inclination when the body is just to move downwards as shown in Fig. 4.5. The various forces acting on the block under such condition are:-

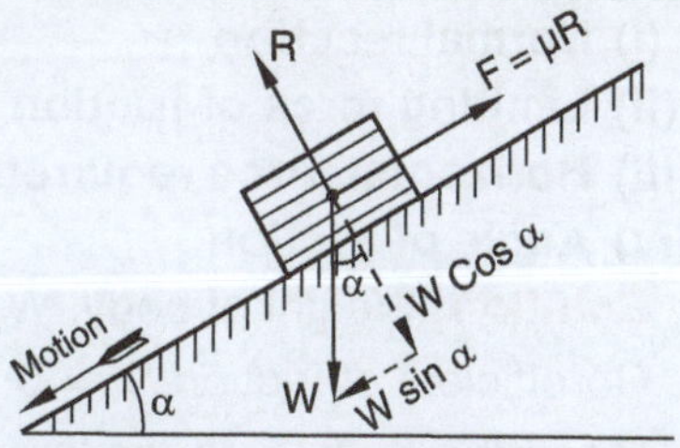

Fig. 4.5. *Concept of angle of repose*

1. Weight of block 'W' actng vertically downwards.
2. Normal reaction 'R' perpendicular to the inclined plane.
3. The force of limiting friction 'F' acting up the plane.

Now the weight W can be resolved into two components as:-

(i) Component W $\cos\alpha$ acting perpendicular to the inclined plane. This component blances the normal reaction 'R' i.e.,

$$R = W\cos\alpha \qquad \text{... (i)}$$

(ii) Component W $\sin\alpha$ acting downwards along the plane. This component balances the limiting friction 'F' i.e.,

$$F = W\sin\alpha \qquad \text{... (ii)}$$

By dividing equation (ii) by (i), we get

$$\frac{F}{R} = \frac{W\sin\alpha}{W\cos\alpha} = \tan\alpha$$

But $\frac{F}{R} = \mu$

$\therefore$ $\mu = \tan\alpha$

or $\tan\alpha = \tan\phi$ $[\because \mu = \tan\phi]$

$\therefore$ $\alpha = \phi$

i.e., Angle of repose = Angle of friciton

Hence, the angle of respose is equal to angle of friction.

4.11 CONE OF FRICTION

The cone with point of contact as the vertex and the normal at the point of contact as the axis, having semi-vertex angle ϕ (angle of friction) is known as cone of friction.

If we describe a cone with point of contact as the vertex and the common normal R at the point of contact as the axis where semi vertex angle is ϕ, then for equilibrium, the resultant must lies within or upon this cone. This is called cone of friction (see Fig. 4.6).

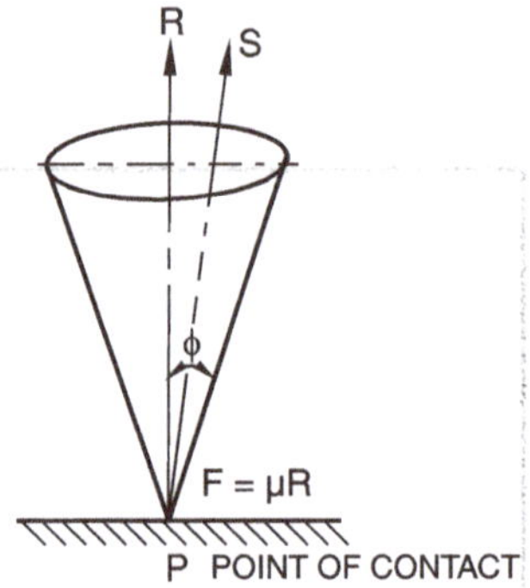

Fig. 4.6. *Concept of cone of friction*

SELECTED PROBLEMS

Problem 1: A body of weight 100 N lies on a horizontal plane for which $\mu = 0.75$. Determine:

(i) Normal reaction

(ii) Limiting force of friction

(iii) Horizontal force required to move it

(iv) Angle of friction.

Solution: Weight of body, W = 100 N

Co-efficient of friction, $\mu = 0.75$

Since the body is in limiting equilibrium under the forces as shown in fig. 4.7, therefore,

(i) Normal reaction, R = W = **100 N [Ans]**

(ii) Limiting force of friction F = μ R = 0.75 × 10 = **75 N [Ans]**

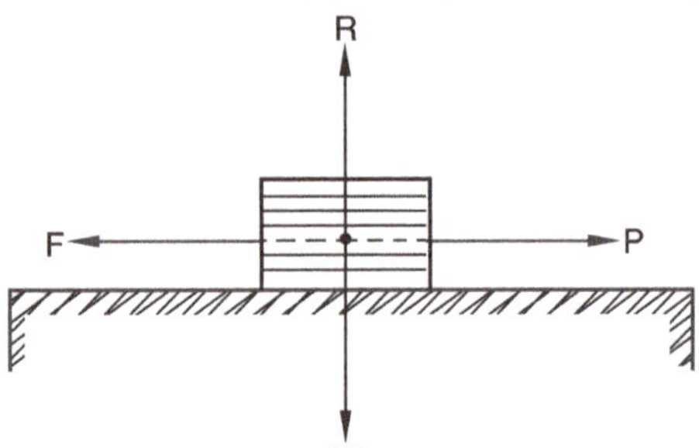

Fig. 4.7

(iii) Horizontal force required to move the body will be equal to the limting force of friction i.e.,

$$P = F$$

$$\therefore \quad P = \mathbf{75\ N\ [Ans]}$$

(iv) Angle of friction can be determined by using the relation,

$$\tan\phi = \mu$$

or $\phi = \tan^{-1}\mu = \tan^{-1} 0.75$

$\therefore$ ϕ = **36° – 52′ [Ans]**

Problem 2: A block of wood weighs 25 N. It is just drawn along a horizontal table by a horizontal force of 12 N. Find (a) Coeff. of friction, μ.

(b) If the block is then loaded with another 10 N load, what least force would be able to move the block?

Solution: (a) When only the block of wood is drawn [Refer Fig. 4.7]:

Weight of block, W = 25 N
Horizontal force, P = 12 N

Let 'μ' be the co-efficient of friction.

Since the block is in limiting equilibrium.

$\therefore$ Normal reaction, R = W = 25 N

$\therefore$ Limiting force of friction, F = P = 12 N

Now the value of co-efficient of friction can be determined by using the relation,

μ = F/R = 12/25 = **0.48 [Ans]**

(b) When the block is loaded with another 10 N load :

Weight of the block and the load added, W = 25 + 10 = 35N

Co-efficient of friction μ = 0.48

We know, normal reaction, R = W = 35 N [∵ μ is same]

Now the least force required to move the block can be found out by using the reaction,

$\therefore$ F = μ R = 0.48 × 35

= **16.8 N [Ans]**

Problem 3: A body is resting on rough horizontal plane. It requires a pull of 18 newtons inclined at 30° to the plane just to move it, while a push of 22 newtons inclined at 30° to the plane just to move it. Determine weight of the body and the coefficient of friction.

Soluiton: case (i): When the body is pulled [see Fig. 4.8 (a)]

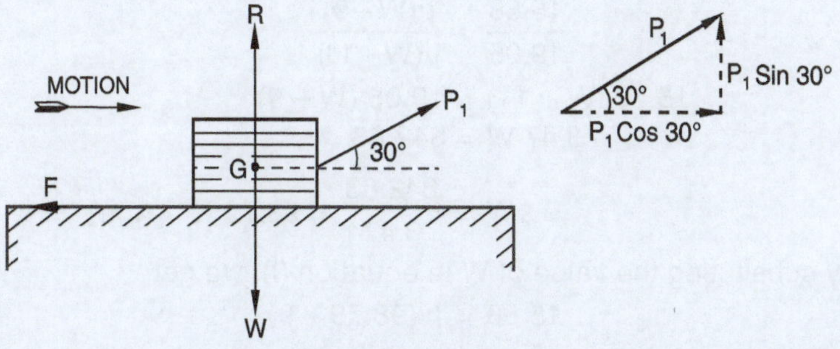

Fig. 4.8. (*a*) *When the body is pulled*

Pull, P_1 = 18 N

Inclination, θ = 30°

The body is in limiting equilibrium, therefore, resolving all the force acting on the body horizontally and vertically, we get

$$F = P_1 \cos 30° = 18 \times 0.866 = 15.58 \text{ N}$$

[Horizontal component along the plane]

$$R = W - P_1 \sin 30° = W - 18 \times 0.5 = (W - 9) \text{ N}$$

[Vertical components along the plane]

Using the friction $F = \mu R$

$$\therefore \; 15.58 = \mu (W - 9) \quad \text{... (i)}$$

Case (ii): When the body is pushed [see Fig. 4.8 (b)]

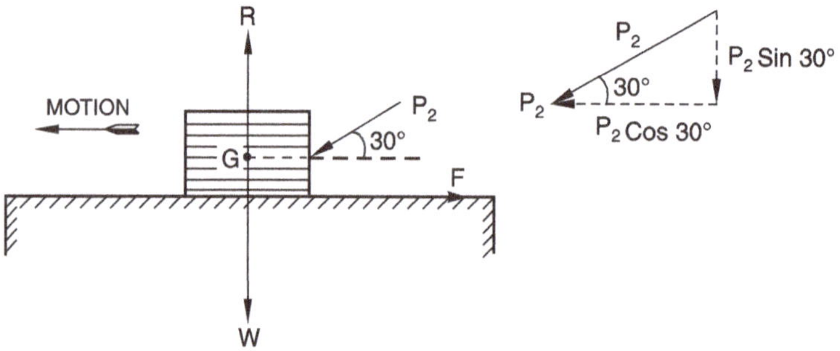

Fig. 4.8. *(b) When the body is pushed*

Push, P_2 = 22 N and Inclination, θ = 30°

Resolving the forces horizontally and vertically, we get

$$F = P_2 \cos 30° = 22 \times 0.866 = 19.05 \text{ N}$$

[Horizontal components along the plane]

$$R = W + P_2 \sin 30° = W + 22 \times 0.5 = W + 1]$$

[Vertical components along the plane]

Again we may use the relation, $F = \mu R$

$$\text{or} \quad 19.05 = \mu (W + 11) \quad \text{... (ii)}$$

Now, dividing equation (i) by (ii), we get

$$\frac{15.58}{19.05} = \frac{\mu(W-9)}{\mu(W+11)}$$

or $$15.58\,(W + 11) = 19.05\,(W - 9)$$

or $$3.47\,W = 342.83$$

$$W = \frac{342.83}{3.47} = \mathbf{98.79\ N} \quad \textbf{[Ans]}$$

Then by subsituting the value of W in equation (i), we get

$$15.58 = \mu (98.79 - 9)$$

$$\therefore \; \mu = \frac{15.58}{89.79} = \mathbf{0.1735} \; \textbf{[Ans]}$$

Problem 4: If the static co-efficient of friction of rubber tyres of a car on the concrete road is 0.6. What is the limiting slope down which the car may run at uniform speed.

Solution: Co-efficient of friction, $\mu = 0.6$

The condition for the car to move down the slope with uniform speed is that the maximum angle α (slope of the road) should be such that its $\tan\alpha$ is equal to the co-efficient of friction i.e. $\tan\alpha = \mu$.

But, under limiting equilibrium $\alpha = \phi$

or $\tan\alpha = \tan\phi$

or $\tan\alpha = \tan\phi = \mu$ $[\because \mu = \tan\phi]$

or $\tan\alpha = 0.6$

$\therefore$ $\alpha = \tan^{-1} 0.6$ = **30.96° [Ans]**

4.12. EQUILIBRIUM OF A BODY ON ROUGH INCLINED PLANE

In practice, one comes across such situations where a body is kept in equilibrium on a rough inclined plane. If the inclination (θ) of the inclined plane is less than the angle of friction (ϕ), the body will remain in equilibrium automatically [see Fig. 4.9 (i)]. However, if inclination (θ) is greater than angle of friction ϕ, then force (P) will have to be applied to keep the body in equilibrium [see Fig. 4.9 (ii)]. It means that force (P) is always needed up the plane to keep the body in equilibrium. The applied force may be

(i) along the plane;
(ii) horizontal;
(iii) inclined at an angle (β) to the inclined plane.

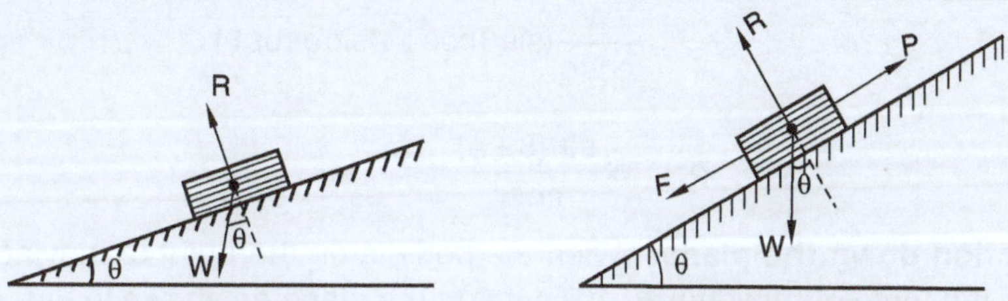

(i) Angle θ < angle ϕ (ii) Angle θ > angle ϕ

Fig. 4.9

The procedure for solving such problems is to resolve all the forces acting on the body along the plane and perpendicular to the plane and finally by using the relation, $F = \mu R$

4.13 EQUILIBRIUM OF A BODY ON AN INCLINED PLANE, WHEN FORCE IS ACTING PARALLEL TO PLANE

(a) Motion up the plane: Consider a body of weight W placed on a rough inclined plane, making an angle θ with the horizontal. Let a force P acting along the plane is required to keep the body in equilibrium.

The various forces acting on the body are shown in Fig. 4.10 (a).

Now, forces acting parallel to the the plane

$$P = W \sin\theta + F \quad \text{... (i)}$$

Forces perpendicular to the plane

$$R = W \cos\theta \quad \text{... (ii)}$$

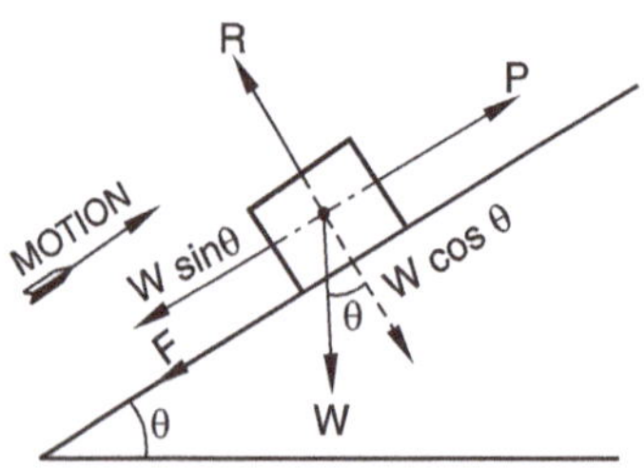

Fig. 4.10 (a)

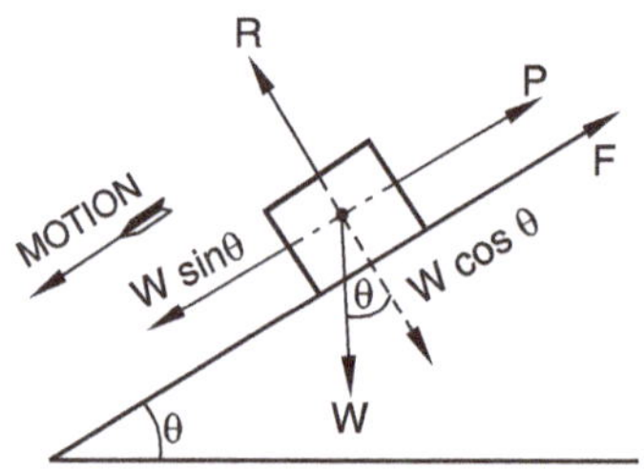

Fig. 4.10 (b)

Then, from equation (i), we get

$$P = W \sin\theta + \mu R \qquad [\because F = \mu R]$$

$$= W \sin\theta + \mu(W\cos\theta) \qquad [\because R = W\cos\theta]$$

$$= W(\sin\theta + \mu\cos\theta)$$

$$= W(\sin\theta + \tan\phi\cos\theta) \qquad [\because \mu = \tan\phi]$$

$$= W(\sin\theta + \frac{\sin\phi}{\cos\phi} \times \cos\theta)$$

$$= \frac{W}{\cos\phi}(\sin\theta\cos\phi + \sin\phi\cos\theta)$$

$$\therefore \quad \mathbf{P = W\frac{\sin(\theta+\phi)}{\cos\phi}}$$

(ii) **Motion down the plane:** When the body is on the point of moving down the plane, then the force of friction $F = \mu R$ acts up the plane as shown in Fig. 4.10 (b).

Now for limiting equilibrium of the body, we get

$$P + F = W\sin\theta \quad \text{... (i)}$$

$$\text{and,} \quad R = W\cos\theta$$

Then from equation (i), we get

$$P = W\sin\theta - F$$

$$= W\sin\theta - \mu R \qquad [\because F = \mu R]$$

$$= W\sin\theta - \mu(W\cos\theta) \qquad [\because R = W\cos\theta]$$

$$= W(\sin\theta - \mu\cos\theta)$$

$$= W(\sin\theta - \tan\phi\cos\theta) \qquad [\because \mu = \tan\phi]$$

$$= W(\sin\theta - \frac{\sin\phi}{\cos\phi}\cos\theta)$$

$$= \frac{W}{\cos\phi}(\sin\theta\cos\phi - \sin\phi\cos\theta)$$

$$\therefore \mathbf{P = W\frac{\sin(\theta - \phi)}{\cos\phi}}$$

Hence, when the force acts along the plane, the following condition should be fulfilled to keep the body in equilibrium:-

$$\frac{W\sin(\theta-\phi)}{\cos\phi} < P = \frac{W\sin(\theta+\phi)}{\cos\phi}$$

4.14. EQUILIBRIUM OF A BODY ON AN INCLINED PLANE WHEN FORCE IS ACTING HORIZONTALLY

Let us consider a body of weight 'W' placed on a rough inclined plane, the angle of inclination of which is θ, where θ is greater than ϕ (angle of friction). Now, we may suppose that a force 'P' acting horizontally, is required to keep the body in equilibrium. Then there can be the following two situation:-

(a) Motion of the body up the plane

(b) Motion of the body down the plane.

(a) **Motion of th body up the plane:** Let us suppose that the body is on the point of moving up the plane as shown in Fig. 4.11 (a). Then the force of friction $F = \mu R$ acts down the plane.

The various forces acting are:-

(i) Weight 'W' acting vertically downwards.

(ii) Normal reaction 'R' acting perpendicular to the plane.

(iii) Force of friction $F = \mu R$ acting down the plane.

(iv) Force applied 'P' acting horizontally.

The weight 'W' and force P can be resolved into the following components [see Fig. 4.11 (a)]:-

W $\cos\theta$ and P $\sin\theta$ perpendicular to the plane, and

W $\sin\theta$ and P $\cos\theta$ down the plane

Since the body is in equilibrium therefore, forces acting along the plane and perpendicular to the plane, must balance each other.

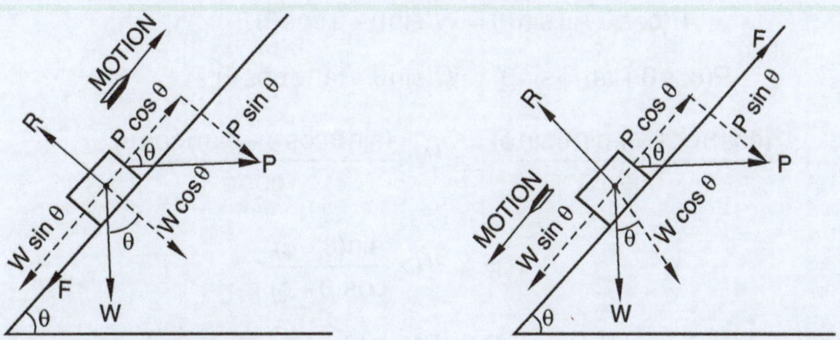

Fig. 4.11. (a) Fig. 4.11. (b)

$$\text{Thus,} \quad P\cos\theta = F + W\sin\theta \quad \text{... (i)}$$

$$\text{and} \quad R = P\sin\theta + W\cos\theta \quad \text{... (ii)}$$

Now from equation (i), we get

$$P\cos\theta = \mu R + W\sin\theta \quad [\because R = P\sin\theta + W\cos\theta]$$

$$P\cos\theta = \mu(P\sin\theta + W\cos\theta) + W\sin\theta$$

$$P\cos\theta - \mu P\sin\theta = \mu W\cos\theta + W\sin\theta$$

$$P(\cos\theta - \mu\sin\theta) = W(\mu\cos\theta + \sin\theta)$$

$$P(\cos\theta - \tan\phi\sin\theta) = W(\tan\phi\cos\theta + \sin\theta) \quad [\because \mu = \tan\phi]$$

$$P \times \frac{(\cos\theta\cos\phi - \sin\theta\sin\phi)}{\cos\phi} = W \times \frac{(\sin\phi\cos\theta + \cos\phi\sin\theta)}{\cos\phi}$$

$$\text{or} \quad P\cos(\theta + \phi) = W\sin(\theta + \phi) \quad [\because \tan\phi = \frac{\sin\phi}{\cos\phi}]$$

$$\text{or} \quad P = W\frac{\sin(\theta + \phi)}{\cos(\theta + \phi)}$$

$$\mathbf{P = W\tan(\theta + \phi)}$$

(b) **Motion of body down the plane:** When the body is on the point of moving down the plane, then the force of friction $F = \mu R$ acts up the plane as shown in Fig. 4.11 (b).

Now for equilibrium of the body, we have

$$P\cos\theta + F = W\sin\theta \quad \text{... (i)}$$

$$\text{and} \quad R = P\sin\theta + W\cos\theta \quad \text{... (ii)}$$

Then from equation (i), we get

$$P\cos\theta = W\sin\theta - F$$

$$= W\sin\theta - \mu R \quad [\because F = \mu R]$$

$$= W\sin\theta - \mu(P\sin\theta + W\cos\theta)$$

$$\text{or} \quad P\cos\theta + \mu P\sin\theta = W\sin\theta - \mu W\cos\theta$$

$$P(\cos\theta + \mu\sin\theta) = W(\sin\theta - \mu\cos\theta)$$

$$P(\cos\theta + \tan\phi\sin\theta) = W(\sin\theta - \tan\phi\cos\theta) \quad [\because \mu = \tan\phi]$$

$$P \times \frac{(\cos\theta\cos\phi + \sin\theta\sin\phi)}{\cos\phi} = W \times \frac{(\sin\theta\cos\phi - \sin\phi\cos\theta)}{\cos\phi}$$

$$P = W \times \frac{\sin(\theta - \phi)}{\cos(\theta + \phi)}$$

$$\therefore \quad \mathbf{P = W\tan(\theta - \phi)}$$

4.15. EQUILIBRIUM OF A BODY ON AN INCLINED PLANE, WHEN FORCE IS ACTING INCLINED TO PLANE

Let us consider a body of weight 'W' placed on a rough inclined plane, the angle of inclination of which is θ, where θ greater than ϕ (angle of friction). Now, we may suppose that a force 'P' acting at an angle β with the inclined plane, is required to keep the body in equilibrium. Then, there can be following two situation:-

(a) Motion of the body up the plane,
(b) Motion of the body down the plane.

(a) Motion of the body up the plane: Let us suppose that the body is on the point of moving up the plane as shown in Fig. 4.12 (a). Then the force of friction, F = μ R acts down the plane. In this situation, the various forces acting on the body are :

(i) Weight 'W' acting vertically downwards.
(ii) Normal reaction 'R' acting perpendicular to the plane.
(iii) Force of friction 'F' = μ R acting down the plane.
(iv) Force applied 'P' at angle β with the plane.

The weight 'W' can be resolved into two component:

$W \cos\theta$ perpendicular to the plane, and

$W \sin\theta$ down the plane as shown in Fig. 4.12 (a).

Also, the force 'P' can be resolve into two component:

$P \cos\beta$ up the plane, and

$P \sin\beta$ perpendicular to the plane as shown.

Now, for limiting equilibrium of the body, we get

$$P\cos\beta = F + W\sin\theta \qquad \text{... (i)}$$

And,

$$R = W\cos\theta - P\sin\beta \qquad \text{... (ii)}$$

Then from equation (i), we get

$$P\cos\beta = \mu R + W\sin\theta \qquad [\because F = \mu R]$$

$$= \mu(W\cos\theta - P\sin\beta) + W\sin\theta$$

$$[\because R = W\cos\theta - P\sin\beta]$$

$$= \mu\, W\cos\theta - \mu\, P\sin\beta + W\sin\theta$$

$$P\cos\beta + \mu P\sin\beta = \mu W\cos\theta + W\sin\theta$$

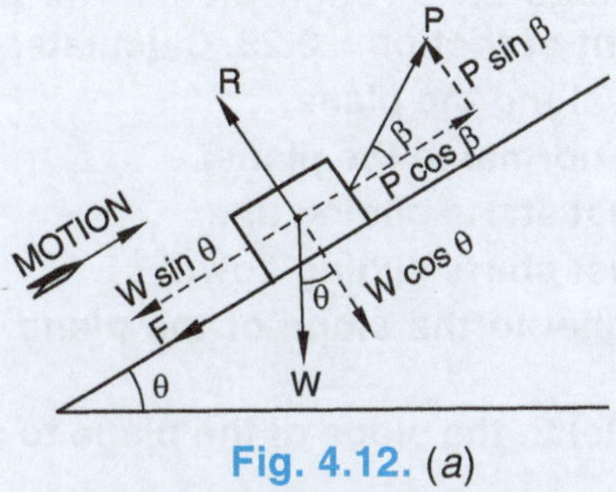

Fig. 4.12. (a)

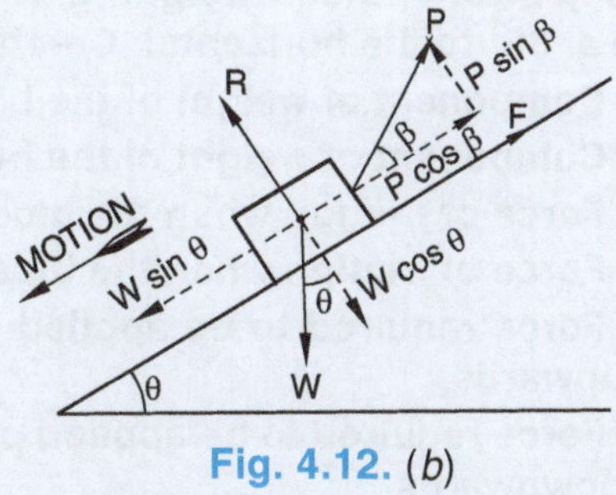

Fig. 4.12. (b)

$$P(\cos\beta + \mu\sin P) = W(\mu\cos\theta + \sin\theta)$$

$$P(\cos\beta + \frac{\sin\phi}{\cos\phi}\sin\beta) = W(\frac{\sin\phi}{\cos\phi}\cos\theta + \sin\theta) \qquad [\because \mu = \tan\phi = \frac{\sin\phi}{\cos\phi}]$$

or $$P \times \frac{(\cos\beta\cos\phi + \sin\beta\sin\phi)}{\cos\phi} = W \times \frac{(\sin\phi\cos\theta + \sin\theta\cos\phi)}{\cos\phi}$$

or $$P\cos(\beta - \phi) = W\sin(\theta + \phi)$$

$\therefore$ $$\mathbf{P = W\frac{\sin(\theta + \phi)}{\cos(\beta - \phi)}}$$

(b) Motion down the plane: When the body is on the point of moving down the plane, then the force of friction F = μ R act up the plane as shown in Fig. 4.12 (b)

Now for limiting equilibrium of the body, we get

$$P\cos\beta = W\sin\theta - F \qquad \text{... (i)}$$

$$R = W\cos\theta - P\sin\beta \qquad \text{... (ii)}$$

Then from equation (i) we get

$$P\cos\beta = W\sin\theta - \mu R \qquad [\because F = \mu R]$$

$$= W\sin\theta - \mu(W\cos\theta - P\sin\beta)$$

$$= W\sin\theta - \mu W\cos\theta + \mu P\sin\beta$$

or $$P\cos\beta - \mu P\sin\beta = W\sin\theta - \mu W\cos\theta$$

$$P(\cos\beta - \mu\sin\beta) = W(\sin\theta - \mu\cos\theta)$$

$$P(\cos\beta - \frac{\sin\phi}{\cos\phi}\sin\beta) = W(\sin - \frac{\sin\phi}{\cos\phi}\cos\theta) \qquad [\because \mu = \tan\phi = \frac{\sin\phi}{\cos\phi}]$$

$$P \times \frac{(\cos\beta\cos\phi - \sin\beta\sin\phi)}{\cos\phi} = W \times \frac{(\sin\theta\cos\phi - \cos\theta\sin\phi)}{\cos\phi}$$

$$P\cos(\beta + \phi) = W\sin(\theta - \phi)$$

$$\therefore \quad \mathbf{P = W\frac{\sin(\theta - \phi)}{\cos(\beta + \phi)}}$$

SELECTED PROBLEMS

Problem 5: A block weighing 10 N rests on a rough plane. The plane is inclined at 30° to the horizontal. Co-efficient of friction = 0.25. Calculate:-

(a) Component of weight of the block along the plane,

(b) Component of weight of the block normal to the plane,

(c) Force of friction when the block just starts sliding up,

(d) Force of friction when the block just starts sliding down,

(e) Force required to be applied parallel to the slope of the plane to start sliding upwards,

(f) Force required to be applied parallel to the slope of the plane to prevent sliding downwards.

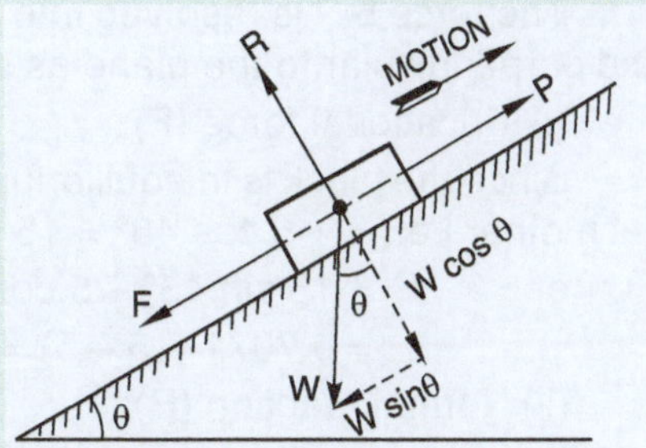

Fig. 4.13

Solution: W = 10 N, μ = 0.25 and θ = 30°

(a) Component of weight of the block along the slope of the plane

$= W\sin\theta = 10 \times 30°$

$= 10 \times 0.5 =$ **5 N [Ans]**

(b) Component of weight of block normal to plane

$= W\cos\theta = 10 \times \cos 30°$

$= 10 \times 0.866 =$ **8.66 N [Ans]**

(c) Force of friction when block just starts sliding up the plane

i.e., $F = \mu R = \mu \times W\cos\theta$

$= 0.25 \times 10 \times \cos 30° = 0.25 \times 10 \times 0.866 =$ **2.165 N [Ans]**

(d) Force of friction when the block just starts sliding down the plane will also be **2.165 N acting upwards [Ans.]**

(e) Force required to be parallel to the slope of the plane to start sliding upwards,

$P_1 = F + W\sin\theta = 2.165 + 10 \times \sin 30°$

$= 2.165 + 10 \times 0.5 =$ **7.165 [Ans]**

(f) Force required to be applied parallel to the slope of the plane to prevent sliding downwards,

$P_2 = W\sin\theta - F = W\sin 30° - F$

$= 10 \times 0.5 - 2.165 =$ **2.835 N [Ans]**

Problem 6: A block of weight 15 N is a rectangular prism resting on a rough inclined plane as shown in fig 4-14 (i). The block is tied up by a string which has a tension of 5 N. Find,

(i) Frictional force on the block,

(ii) Normal reaction of the inclined plane, and

(iii) Co-efficient of friction between the surfaces of contact.

Solution: Weight of block, W = 15 N

Force of string, P = 5 N

Inclination of plane, θ = 45°

Resolve all the inclined foces parallel to the plane and perpendicular to the plane.

The weight 15 N is resolved into two components (i) 15 sin 45° and (ii) 15 cos 45°, parallel and perpendicular to the plane as shown in Fig. 4.14 (ii)

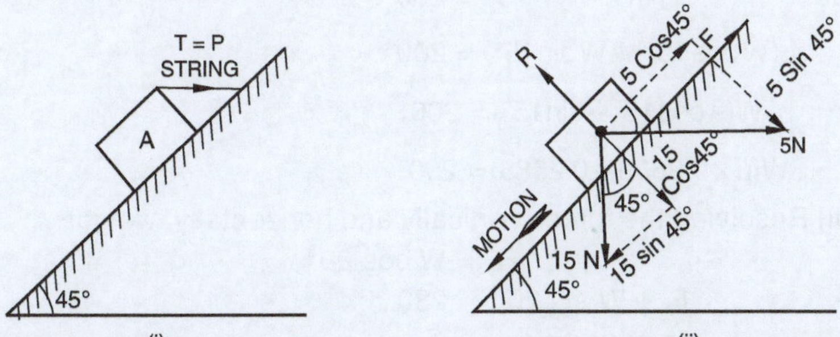

Fig. 4.14

The force 5 N is resolved into two components (i) 5 cos 45° and 5 sin 45°, parallel and perpendicular to the plane as shown in Fig. 4.14 (ii).

(i) Fricational force (F):

Since the block is in equilibrium, therefore, the force along the plane must balance each other i.e., F + 5 cos 45° = 15 sin 45°

or $F = 15 \sin 45° - 5 \cos 45° = 15 \times 0.7071 - 5 \times 0.7071$

$= 0.7071\ (15 - 5) =$ **7.071 N [Ans]**

(ii) Normal reaction (R):

Force perpendicular to the plane must balance each other, i.e.,

$R = 5 \sin 45° + 15 \cos 45° = 5 \times 0.7071 + 15 \times 0.7071$

$= 0.7071\ (5 + 15) = 0.7071 \times 20 =$ **14.142 N [Ans]**

(iii) Coefficient of friction (μ) :

$$\mu = \frac{F}{R} = \frac{7.071}{14.142} = \mathbf{0.5}\ \ \textbf{[Ans]}$$

Problem 7: An effort of 200 N is required just to move a certain body up an inclined plane of 15°, the force acting parallel to the plane. If the angle of inclination of the plane is made 20° and effort required parallel to the plane is found to be 230 N. Find the weight of the body and coefficient of friction.

Solution: Effort P_1= 200 N; Angle of inclination, $\theta_1 = 15°$

Effort P_2= 230 N; Angle of inclination, $\theta_2 = 20°$

Fig. 4.15

Coefficient of friction (μ) :

Case (i) Resolving the forces vertically and horizontally, we get;

$$R_1 = W \cos 15° \quad \text{... (i)}$$

Also, $F_1 + W \sin 15° = 200$

$$\mu\ R_1 + W\ 15° = 200 \quad [\because F_1 = \mu R_1]$$

$$\mu(W\cos 15°) + W\sin 15° = 200 \quad [\because R_1 = W\cos 15°]$$

$$W(\mu\cos 15° + \sin 15°) = 200$$

$$W(\mu \times 0.9659 + 0.2588) = 200 \quad \text{... (ii)}$$

Case (ii) Resolving the forces vertically and horizontally, we get

$$R_2 = W \cos 20° \quad \text{... (iii)}$$

Also $F_2 + W \sin 20° = 230$

$$\mu R_2 + W\sin 20° = 230 \quad [\because F_2 = \mu R_2]$$

$$\mu W \cos 20° + W \sin 20° = 230 \qquad [\because R_2 = W \cos 20°]$$

$$W(\mu \cos 20° + \sin 20°) = 230$$

or $$W(\mu \times 0.9397 + 0.3420) = 230 \qquad \text{... (iv)}$$

Dividing equation (iv) by equation (ii), we get

$$\frac{W(\mu \times 0.9397 + 0.3420)}{W(\mu \times 0.9659 + 0.2588)} = \frac{230}{200}$$

or $$200(\mu \times 0.9397 + 0.3420) = 230\ (\mu \times 0.9659 + 0.2588)$$

$$187.94\mu + 68.4 = 222.157\mu + 59.524$$

$$34.217\mu = 8.876$$

$$\therefore \quad \mu = \mathbf{0.2594} \quad \textbf{[Ans]}$$

Subsituting the value of μ in equation (ii), we get;

$$W(0.2594 \times 0.9659 + 0.2588) = 200$$

$$\therefore \quad W = \mathbf{392.65\ N} \quad \textbf{[Ans]}$$

Problem 8: Find the least force required to move a body of 30 N up a rough inclined plane, the force being applied parallel to the plane. The inclination of the plane is such that when the same body is kept on a perfectly smooth plane inclined at that angle, a force of 6 N applied at an inclination of 30° to plane would keep the same load in equilibrium. The coefficient of friction is 0.3.

Solution: Weight of body, W = 30 N, $\mu = 0.3$

Case (i) When the body is resting on smooth surface [see Fig. 4.16 (i)]

Fig. 4.16

Inclination of force, $\beta = 30°$ and Force, P = 6 N

Force of friction on smooth surface, F = 0

Let, θ be the angle of inclination of plane.

Since, the load is in equilibrium, therefore, the forces acting along the plane balance each other i.e., $6 \cos 30° = W \sin\theta$

$$\text{or } 6 \times 0.866 = 30 \sin\theta$$

$$\text{or} \quad \sin\theta = \frac{6 \times 0.866}{30}$$

$$\therefore \quad \theta = 9°\text{-}58'$$

Case (ii) When load is resting on a rough surface [see Fig. 4.16 (ii)]

Angle of inclination, $\theta = 9°58'$

Let, P = least foce required

F = Force of friction and R = Normal reaction

Since, the load is in equilibrium, therefore, the forces acting along the plane and perpendicular to the plane balance each other i.e.,

$$P = F + W\sin\theta \qquad [\because F = \mu R]$$

or $$P = 0.3\,R + 30 \sin 9°\text{-}58' \qquad \ldots \text{(i)}$$

And $$R = 30\cos\theta \qquad \ldots \text{(ii)}$$

Subsituting the value of R in equation (i), we get;

$P = 0.3 \times 30 \cos 9°\ 58' + 30 \sin 9°\ 58'$

$= 0.3 \times 30 \times 0.9849 + 30 \times 0.1782 =$ **14.21 N [Ans]**

Problem 9: A mass of 500 kg is lying on a wooden plank with an incline of 1 : 2 (1 in 2). The co-efficient of friction is 0.3. State if the load moves down, what force parallel to the plank is necessary to hold it back.

Solution: Mass, $m = 500$ kg

$\therefore$ Weight, $W = m \times g$

$= 500 \times 9.81 = 4905$ N

Co-efficient of friction, $\mu = 0.3$

or $\tan\phi = \mu = 0.3$

$\therefore \quad \phi = \tan^{-1} 0.3 = 16°\text{-}41'$

Also $\tan\theta = 1/2 = 0.5$

$\therefore \quad \theta = \tan^{-1} 0.5 = 26°\text{-}33'$

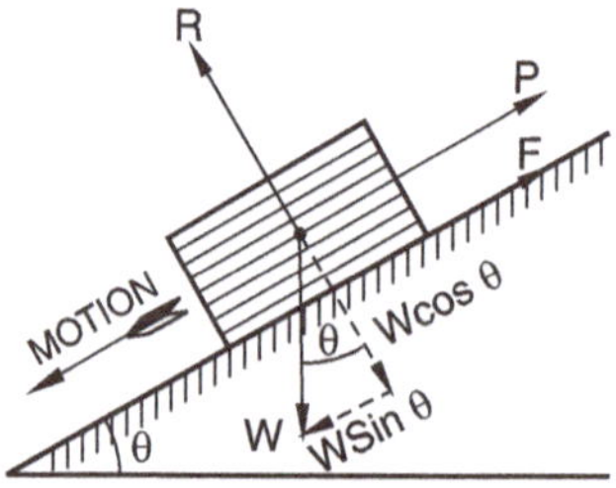

Fig. 4.17

As the angle of inclination of wooden plank **(θ)** is more the angle of friction **(ϕ)**, the load will move down the plane. **[Ans]**

Let P = Force applied to hold back the load. Now, resolving the forces parallel to plane.

$$P + \mu R = 4905 \sin 26°\text{-}33'$$

$$P + 0.3 \times R = 4905 \times 0.447 = 2192.53 \qquad \ldots \text{(i)}$$

Resolving the forces perpendicular to the plane

$$R = 4905 \cos 26°\text{-}33' = 4905 \times 0.894$$

$$\therefore \quad R = 4385.07 \text{ N}$$

Subsituting the value of R in equation (i), we get

$P + 0.3 \times 4385.07 = 2192.53$

$\therefore \quad P = 2192.53 - 1315.52 =$ **877.01 N [Ans]**

Problem 10: A block weighing 360 N is resting on a rough inclined plane having inclination of 30°. A force of 120 N, inclined at 10° to the plane, is applied to the block up the plane and the block is just on the point of moving down the plane. Find the value of co-efficient of friction.

Solution: Weight of block,

Inclination of plane, $\theta = 30°$

Force, $P = 120$ N

Inclination of force, $\beta = 10°$

Let 'μ' be the co-efficient of friction.

As the block is on the point of moving downwards, therefore, the force of friction 'F' acts up the plane.

The weight W = 360 N can be resolved into two components.

360 cos 30° perpendicular to the plane and 360 sin 30° along the plane.

Also, the force = 120 N can be resolved into two components, i.e, 120 cos 10° along the plane and 120 sin 10° perpendicular to the plane.

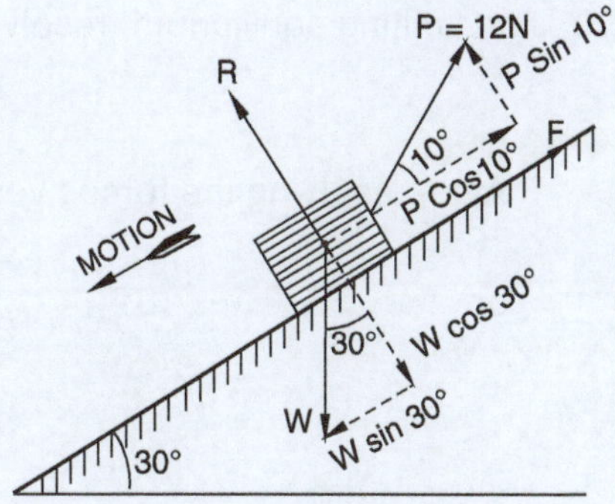

Fig. 4.18

Since the block is in limiting equilibrium therefore, we have

$$120 \cos 10° + F = 360 \sin 30° \qquad \text{... (i)}$$

$$\text{and,} \quad R + 120 \sin 10° = 360 \cos 30°$$

Now from equation (i), we get $120 \cos 10° + \mu R = 360 \sin 30°$

$$120 \cos 10° + \mu (360 \cos 30° - 120 \sin 10°) = 360 \sin 30°$$

$$\text{or } 120 \times 0.9848 + \mu (360 \times 0.866 - 120 \times 0.1736) = 360 \times 0.5$$

$$\text{or } \mu (360 \times 0.866 - 120 \times 0.1736) = 360 \times 0.5 - 120 \times 0.9848$$

$$\therefore \quad \text{Co-efficient of friction} \quad \mu = \frac{6.1824}{29.0928}$$

$$= \mathbf{0.2125\ [Ans]}$$

4.16. LADDER FRICTION

The friction produced by the ends of ladder on the floor and wall surfaces is called ladder friction.

Let us consider a ladder AB resting on the floor and leaning against a wall as shown in Fig. 4.19.

Let 'W' be the weight of the ladder and θ be its inclination to the horizontal.

When the lower end of the ladder tends to slip away from the wall, the direction of force of friction, between the ladder and the floor i.e., F_A will be towards the wall. Similarly, when the upper end of the ladder tend to slip downwards, the direction of the force of friction between the ladder and the wall i.e. F_B will be upwards as shown in Fig. 4.19.

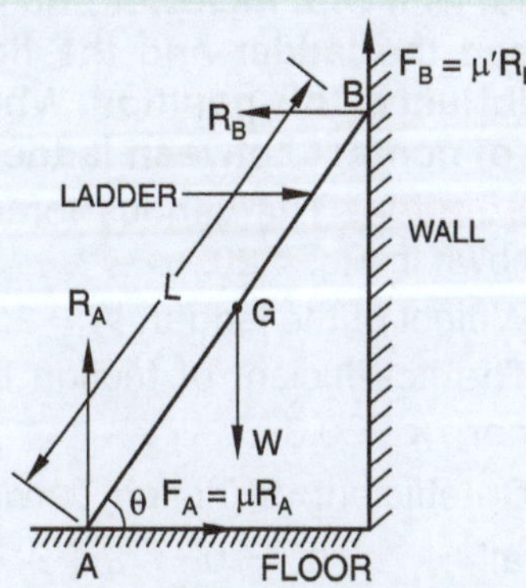

Fig. 4.19

Now the various forces acting on the ladder will be as follows:-

(i) Weight of ladder, W acting vertically downwards, through G. of the ladder.

(ii) The normal reaction R_A at A,

(iii) The force of friction $F_A = \mu R_A$ acting horizontal at A,

(iv) The normal reaction R_B at B,

(v) The force of friction $F_B = \mu' R_B$ acting horizontal at B.

In limiting equilibrium, resolving the forces horizontally, we get

$$\Sigma H = 0$$

or $$R_B = R_A = \mu R_A \quad \text{... (i)}$$

Also, resolving the forces vertically, we get

$$\Sigma V = 0$$

or $$W = R_A + F_B$$
$$= R_A + \mu' R_B$$
$$= R_A + \mu(\mu' R_B) \quad [\because R_B = \mu' R_A]$$

$\therefore$ $$W = R_A(1 + \mu\mu') \quad \text{... (ii)}$$

Now taking moments about A, We get

$$W \times \frac{L}{2}\cos\theta - R_B \times L\sin\theta - \mu' R_B \times L\cos\theta = 0$$

or $$R_A(1+\mu\mu') \times \frac{L}{2}\cos\theta - \mu R_A \times L\sin\theta - \mu'\mu R_A \times L\cos\theta = 0 \quad [\because W = R_A(1+\mu\mu')]$$

or $$(1+\mu\mu')\cos\theta - 2\mu\sin\theta - 2\mu\mu'\cos\theta = 0 \quad [\text{Eliminating } R_A \text{ and } 1/2]$$

or $$\cos\theta[1-\mu\mu'] = 2\mu\sin\theta$$

$\therefore$ $$\mathbf{\tan\theta = \frac{1-\mu\mu'}{2\mu}}$$

Hence, the inclination of the ladder to the horizontal, when it is about the point of slipping can be calculated as follows:

$$\theta = \tan^{-1} \times \frac{1-\mu\mu'}{2\mu}$$

Problem 11: A 10 m ladder, weighing 200 N, is placed against a smooth vertical wall with its lower end 4 metres from the wall. The co-efficient of friction between the ladder and the floor is 0.3. Show that the ladder shall remain in equilibrium in this position. What is the frictional force acting on the ladder at the point of contact between ladder and floor.

Solution: The various forces acting on ladder are shown in Fig. 4.20.

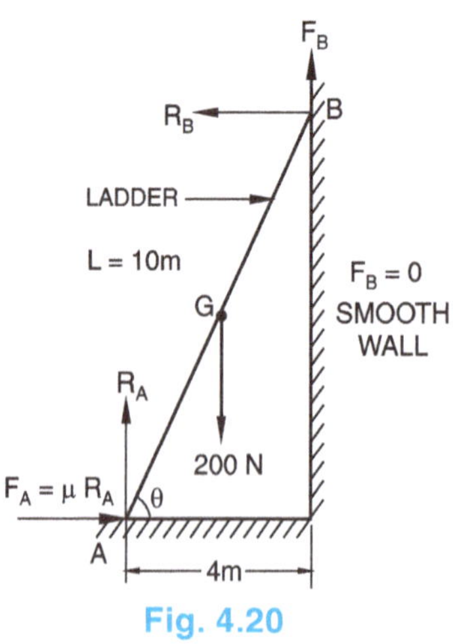

Fig. 4.20

Weight of the ladder, W = 200 N

The co-efficient of friction between ladder and the floor, $\mu = 0.3$

Co-efficient of friction between the ladder and the wall, $\mu' = 0$ [$\because$ The wall is smooth]

The inclination of the ladder to the horizontal.

$\theta = \cos^{-1} 4/10 = \cos^{-1} 0.4 = 66°\text{-}25'$

Resolving the forces horizontally, we get

$$F_A = R_B \quad \text{... (i)}$$

Now, resolving the forces vertically, we get

$$R_A = 200 \text{ N} \quad \text{... (ii)}$$

Taking moments of all forces about A,

$$200 \times 5 \cos\theta - R_B \times 10 \sin\theta = 0$$

$$\therefore \quad R_B = \frac{200 \times 5 \times 0.4}{9.5} = 43.71\,N$$

$\therefore$ From equation (i), we get $F_A = R_B = 43.71$ N

Maximum force of friction available act at the point of contact between ladder and floor = $\mu\ R_A = 0.3 \times 200 = 60$ N.

As the force of friction 60 N available at the point of contact is more than the force of friction 43.71 i.e. **ladder shall remain in equilibrium [Ans]**

$\therefore$ Frictional force acting at A i.e. point of contact between ladder and floor = **43.71 N [Ans]**

Problem 12: A uniform ladder of 7 metre length rests against a vertical wall with which it makes an angle of 45°, the co-efficient of friction between the ladder and the wall is 1/3 and that between ladder and floor is 1/2. If a man whose weight is one-half of that of the ladder, ascends it, how will he be when the ladder slips?

Solution: Length of the ladder, L = 7 m

Angle of inclination of ladder, $\theta = 45°$

Co-efficient of friction between the ladder and the wall, $\mu' = 1/3$ and

Co-efficient of friction between the ladder and floor, $\mu = 1/2$

Let 'W' weight of the ladder AB acting at G, W/2 is the weight of man who has ascended the length x metre of the ladder from A is shown in Fig. 4.21.

(i) In limiting equilibrium, resolving the forces horizontally, we get

$$R_B + F_A = 0$$

$$R_B - \mu\ R_A = 0$$

$$\therefore \quad R_B = \frac{1}{2} R_A$$

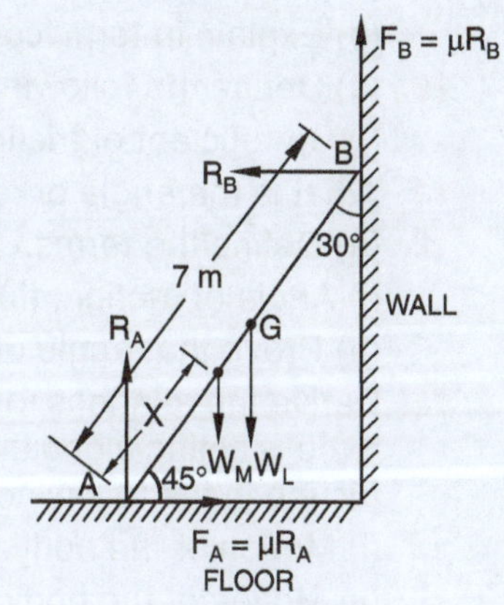

Fig. 4.21

Also resolving the forces vertically, we get

$$R_A + F_B - W - \frac{W}{2} = 0$$

$$R_A + \mu' R_B - W - \frac{W}{2} = 0$$

$$R_A + 1/3 \times 1/2\ R_A - 3/2\ W = 0 \quad [\because R_B = 1/2 R_A]$$

$$R_A + 1/6\ R_A = 3/2\ W$$

or $$R_A(1 + 1/6) = 3/2\ W$$

or $$R_A = \left(\frac{3}{2} \times \frac{6}{7}\right)W$$

$$\therefore R_A = \frac{9}{7}W \quad \text{and} \quad R_B = \frac{9}{14}W \qquad [\because R_B = 1/2\ R_A]$$

Now taking moments about A, we get

$$W \times \frac{7}{2}\cos 45° + \frac{W}{2} \times \cos 45° - R_B \times 7\sin 45° \ - \mu' R_B \times 7\cos 45° = 0$$

$$W \times 3.5\cos 45° + \frac{W}{2} \times \cos 45° \ - \frac{9}{14} W \times 7\sin 45° \ - \frac{1}{3} \times \frac{9}{14} W \times 7\cos 45° = 0$$

Dividing both the sides by W sin 45° which is also equal to W cos 45°.

$$3.5 + X/2 = 9/2 + 3/2$$

or $\quad X = (6 - 3.5)) \times 2$

$\therefore \quad X =$ **5 m [Ans]**

Hence, the ladder will slip away when the man ascends 5 m of the ladder.

MODEL TEST QUESTIONS

1. What do you understand by the term friction? Explain clearly why it comes into play.
2. Friction is a necessary evil, discuss.
3. Give the useful and harmful effects of friction.
4. Friction may be objectionable or useful in a machine. Mention one case of each.
5. State various type of friction.
6. State the difference between static and dynamic friction, which of the two is more?
7. (a) State clearly the laws of friction between solid surfaces.
   (b) Explain limiting friction with example.
8. (a) State the laws of limiting friction.
   (b) Explain $\mu = \tan\phi$ = F/R (with usual notation).
9. (a) State the laws of static friction.
   (b) Explain in term 'co-efficient of friction' and 'angle of friction'.
10. (a) Define the following terms:
    (i) Co-efficient of friction; (ii) Limiting force of friction.
11. What is the angle of repose and what is its relation with the angle of friction?
12. (a) Define the terms:-
    (i) Angle of friction; (ii) Angle of repose
    (b) Prove that angle of friction is equal to angle of respose.
13. Derive the relations for eqilibrium of a body on a rough inclined plane when the force is acting along the plane and the angle of inclination is more than angle of friction in the following cases:-
    (i) Motion of the body is down the plane.
    (ii) Motion of the body is up the plane.
14. Write short notes on the following:-
    (a) Friction is a necessary evil;
    (b) Useful and harmful effects of friction;
    (c) Laws of static friction;
    (d) Laws of dynamic friction.

QUESTIONS FOR SELF EXAMINATION

1. **Fill in the blanks:**
   (i) The opposite force to motion is called ....
   (ii) Friction acts in the direction .... to the motion.

(iii) Sliding friction is always .... than rolling friction.
(iv) Static friction is .... than dynamic friction.
(v) Co-efficient of friction is more in ....
(vi) A piston moving in the cylinder of a car engine is an example of .... friction
(vii) Lubricant is used to .... friction.
(viii) In the absence of lubricant, friction is referred as ....
(ix) .... is a self adjusting.
(x) Angle of repose is .... angle of friction.

Ans. 1. (i) force of friction; (ii) opposte; (iii) more; (iv) greater; (v) wood and wood; (vi) sliding; (vii) decrease; (viii) dry friction (ix) friction; (x) equal.

2. State whether the following statementsare true and false:
(i) Resistance to motion is called the force of friction.
(ii) Force of friction is not equal to the force which tends to move the body.
(iii) Limiting friction is maximum when the applied force is less than the limiting friction.
(iv) Force of friction depends upon speed of motion.
(v) Frictional force depends upon the material of the surface in contact.
(vi) Friction in steel is more than that in cast iron.
(vii) Sliding friction is always greater than rolling friction.
(viii) Co-efficient of friction is the ratio of frictional resistance and angle of repose.
(ix) More force is required for pulling load upon an inclined plane than on horizontal plane.

Ans. 2. (i) True; (ii) False; (iii) False; (iv) False; (v) True; (vi) False; (vii) True; (viii) False; (ix) True.

PROBLEMS FOR PRACTICE

1. A body weighing 150 N is resting on a horizontal plane and can be just moved by a force of 50 N applied horizontally. Find the co-efficient of friction, and the magnitude of the resultant reaction of the plane.
[**Ans**. 0.33, 158 N]

2. A block weights 250 N is just drawn along a horizontal table by a horizontal force of 120 N. If the block is then loaded with another 100 N load, what least force would be able to move the block?
[**Ans**. 168 N]

3. A block resting on a horizontal plane is pulled by two horizontal ropes at an angle 90° to each other. If the tensions in the ropes are 150 N and 200 N find:
(i) Resultant force acting on the body.
(ii) Weight of the block, if $\mu = 0.25$. [**Ans**. 250, 1000 N]

4. A box weighing 100 N is resting on a horizontal plane, the co-efficient of friction being 0.4. Find the least force acting (i) horizontally; (ii) at an angle of 30° with the horizontal which would move the box.
[**Ans**. (i) 40 N, (ii) 46.20 N]

5. A body resting on a rough horizontal plane required a pull of 20 N inclined at 25° to the plane to move it. It was also found that a push of 25 N inclined at 25° to the plane just moved the body. Find the weight of the body and co-efficient of friction. [**Ans**. 84.55 N, 0.138]

6. An effort of 200 N is required just to move a certain body up an inclined plane 15°, the force acting parallel to the plane. If the angle of inclination of the plane is 20°, the effort required, again applied parallel to the plane is found to be 230 N. Find the weight of the body and co-efficient of friction.[**Ans**. 392.2 N. 0.26]

7. A weight of 250 N is pulled up a rough inclined plane, whose inclination to the horizontal is 30°, by a force of 180 N acting parallel to the plane. Find the co-efficient of friction. [**Ans.** 0.254]

8. A block resting on a rough inclined plane whose inclination is 20° with the horizontal. The maximum force which will just move the block up is 100 N. If the co-efficient of friction between the block and the plane is 0.23, find:

   (i) Weight of the block;

   (ii) Minimum force required to move the block down the plane.

   [**Ans.** 179.2 N, 22.5 N]

9. A body weighing 150 N is resting on a rough inclined plane which rises 250 mm. in every metre. Find the least horizontal effort required just to move the body up the plane. Take co-efficient of friction between the plane and the body as 0.2. [**Ans.** 72.4 N]

10. A wooden block weighing 360 N is resting on a rough inclined plane having inclination of 30° with the horizontal. A force of 120 newtons is applied at an angle of 10° up the plane and the block is just on the point of moving down the plane. Find the co-efficient of friction. [**Ans.** 0.212]

11. A uniform ladder rests with one end on a horizontal floor and the other end against a vertical wall. The co-efficient of friciton for the floor and wall are 4/9 and 1/4 respectively. Find the inclination of the ladder, when it is about to slip.

    [**Ans.** 45°]

12. A ladder 4 metres long is supported by a horizontal floor and vertical wall. The co-efficient of friction at the wall is 0.25 and the floor is 0.50. The weight of the ladder is 20 newtons. The ladder also support a vertical load of 100 newtons at a distance of 1 m from the wall and side. Determine the least inclination of ladder without slipping towards left. [**Ans.** 53°-21′]

13. A ladder 6 metre long rests on a horizontal ground and leans against a smooth vertical wall at an angle of 20° with the vertical. Its weight is 900 N acting at its middle. The ladder is on the point of slipping when a man weighing 750 N stands on a rang 2 metres from the foot of ladder. Calculate the co-efficient of friction. [**Ans.** $\mu = 0.1542$]

Chapter 5

Centre of Gravity

INTRODUCTION

Everybody is attracted towards the centre of earth due to gravity. The force with which the body is attracted towards earth is called weight of the body. The whole weight of the body is supposed to act at a point in the body and this point is known as the *centre of gravity*. It is of great importance to know the position of centre of gravity in a number of engineering applications. Thus for the design of dam, the stability of dam can only be ascertained if we know the position of centre of gravity. This necessitates a full knowlege of this chapter.

5.1. CENTRE OF GRAVITY

The point at which the whole weight of the body is supposed to act is called the centre of gravity.

Everybody is made up of infinite number of small particles. These particles are attracted towards the centre of earth. The weight of each particle is due to the force of attraction exerted by the earth and acts vertically downwards. Since the distance of the centre of earth is very large from these particles, it can be safely assumed that the lines of action of these particles are parallel to each other as shown in Fig. 5-1.

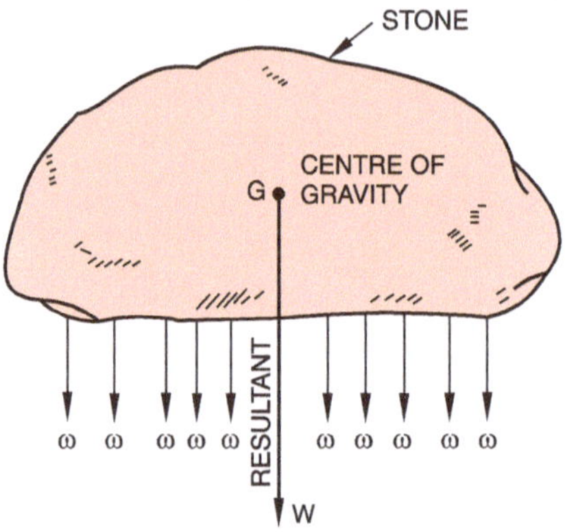

Fig. 5.1. *Concept of centre of gravity*

Thus the weights of all particles of a body form a system of parallel forces. The point through which the resultant (Total weight of the body, W) of these parallel forces acts is known as centre of gravity.

5.2. CENTROID

The point at which the whole area of lamina is supposed to act is called centroid.

The term centre of gravity is used for solid bodies. If we imagine that the thickness of this body is uniformly reduced, the position of the centre of gravity will clearly remain unchanged, although the total weight is reduced. If now the body is assumed to be infinitely thin, it becomes a figure having an area (lamina) but no weight. Under such conditions, the whole figure may be supposed to be made up of a large number of small areas. The point through which the resultant of all such areas acts is called centroid or centre of area of that body.

Notes: *(i) The term centre of gravity applies to soilds and centroid to areas.*

(ii) The term centre of gravity is also used for centroid sometimes e.g., centroid or centre of gravity of plane figure but the term centroid is not used for soilds.

5.3. A BODY HAS ONLY ONE CENTRE OF GRAVITY

A body has one and only one centre of gravity. Consider a rigid body as shown in fig. 5-2. If possible, let us assume that there are two centres of gravity G_1 and G_2. Then in all possible positions of the body, the line of action of its weight will pass through both the points G_1 and G_2. This is only possible if the line G_1 and G_2 is vertical. But if the body is rotated so that G_1G_2 is horizontal, the weight of the body should act along G_1G_2 which becomes a horizontal force.

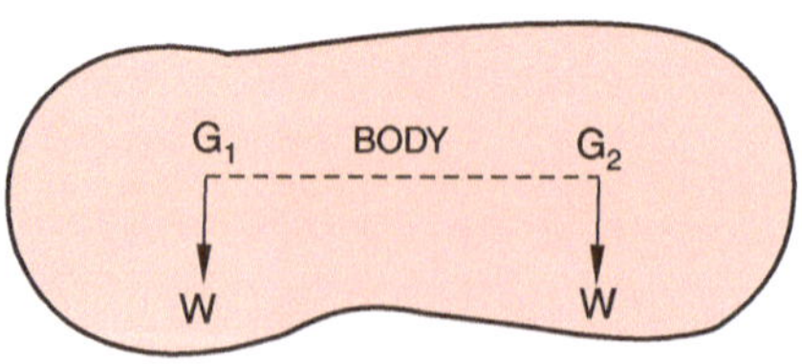

Fig. 5.2

But this is imposible., because the weight of the body always acts vertically downwards. Hence a body has one and only one C.G.

5.4. METHODS OF DETERMINING CENTRE OF GRAVITY

The centre of gravity of a solid body or lamina can be determined by the following methods:—

1. By geometrical method
2. By the method of moments
3. By graphical mehtod
4. By intergraion method.

1. By Geometrical method: The positon of centre of gravity of many bodies (e.g., solid bodies of regular shape of common geometrical section such as rectangular, circular, triangular etc). can be found by inspection. In general, If a body is symmetrical about a point, then the centre of gravity lies at that point.

The position of centriod for various sections and position of centre of gravity of smiple solid bodies are given in the following tables.

Position of centroid (*G*) of plane sections

No.	Section	Shape	Position of G	Area
1.	Straight line	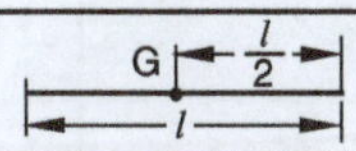	At mid point	—
2.	Rectangle	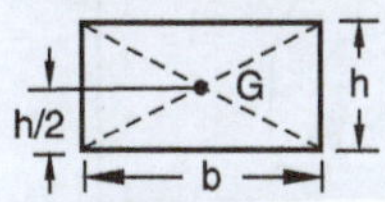	At intersection of diagonals	b × h
3.	Square	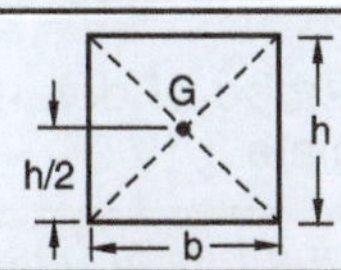	At intersection of diagonals	b × h
4.	Parallelogram	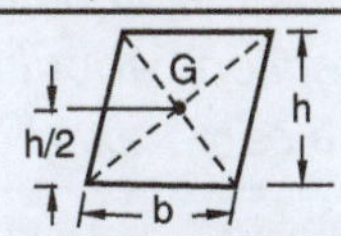	At intersection of diagonals	b × h
5.	Cricle	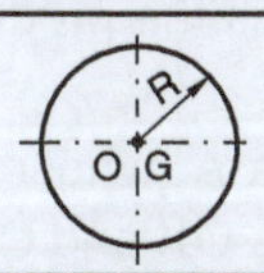	At intersection of two diameters i.e. centre	πR^2
6.	Semicircle	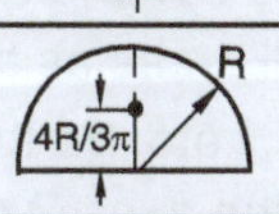	$\frac{4R}{3\pi}$ on a line perpendicular to diameter	$\frac{\pi R^2}{2}$
7.	Traingle	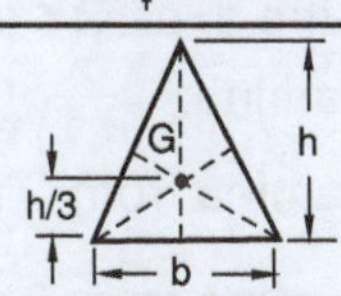	At intersection of median (*h*/3 from base along axis	$\frac{1}{2} \times b \times h$
8.	Sector	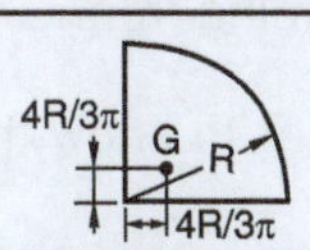	$\frac{4R}{3\pi}$ as shown	$\frac{\pi R^2}{4}$
9.	Trapezium	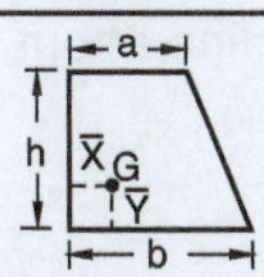	$\overline{X} = \frac{a^2 + b^2 + ab}{3(a+b)}$ $\overline{Y} = \frac{(2a + b)h}{3(a+b)}$	$(a + b)\frac{h}{2}$

Position of centre of gravity (C. G.) of simple solids

No.	Solid	Shape	Position of C.G.	Volume
1.	Cylinder	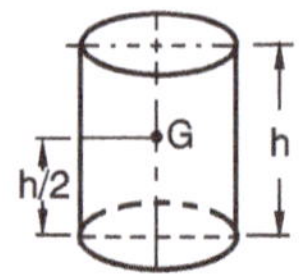	$\frac{h}{2}$ from base along axis	$\pi R^2 \times h$
2.	Cone	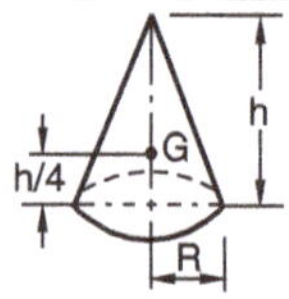	$\frac{h}{4}$ from base along axis	$\pi R^2 \times \frac{h}{3}$
3.	Sphere	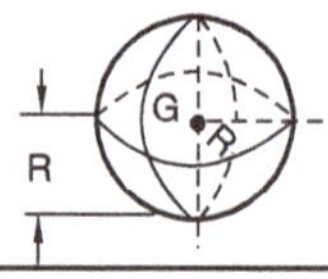	At its centre	$\frac{4}{3}\pi R^3$
4.	Hemisphere	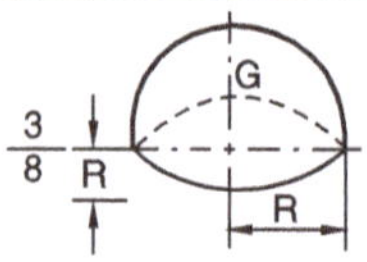	$\frac{3}{8}$ R from base along axis	$\frac{2}{3}\pi R^3$

By the method of moments: In this method of determining the C.G. the principle of moments is used. According to this principle,

> *"If a lamina is a combination of a number of basic section, the algebraic sum of the moments of the individual areas of sections about any axis is equal to the moment of the whole area of the lamina about the same axis."*

The moment of area about any axis is the product of area and the perpendicular distance of its centriod from that axis.

Consider an irregular lamina of area A. whose centroid is to be determine with respect to two mutually perpendicular axis OX and OY as shown in fig. 5-3. Divide the whole lamina into a large number of small areas as a_1, a_2, a_3 ...etc.

Let x_1, x_2, x_3 ... etc. and y_1, y_2, y_3 ... etc. be the perpendicular distances of the centroids of areas a_1, a_2, a_3 ... etc. from the axes OY and OX respectively.

Let G be the centroid of the whole lamina.

$\overline{X}$ = Distance of centroid G of the lamina from OY axis

$\overline{Y}$ = Distance of centroid G of the lamina from OX axis.

It is desired to find $\overline{X}$ and $\overline{Y}$ in order to know the position of centroid G of the lamina.

Moments of all small areas of the lamina about OY axis

$$= a_1x_1 + a_2x_2 + a_3x_3 + \ldots\ldots \qquad \ldots \text{(i)}$$

Moment of total area A of the lamina about OY axis = $A\overline{X}$ ...(*ii*)

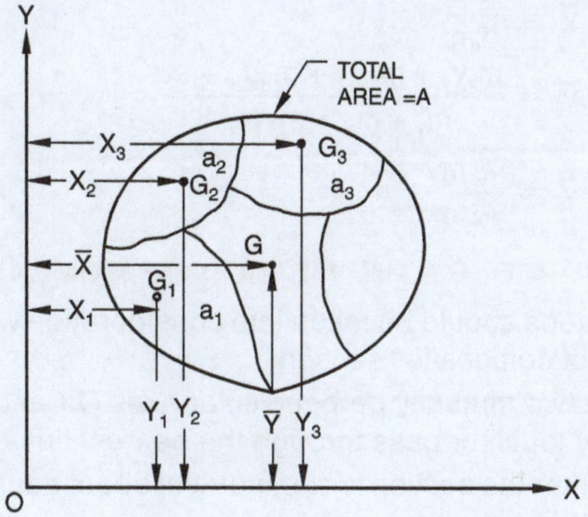

Fig. 5.3.

According to principle of moments,

Equation (ii) = Equation (i)

$$\therefore \quad A\overline{X} = a_1x_1 + a_2x_2 + a_3x_3 + ... \qquad ...(iii)$$

$$\text{or} \quad \overline{X} = \frac{a_1x_1 + a_2x_2 + a_3x_3 + ...}{A}$$

$$\overline{X} = \frac{a_1x_1 + a_2x_2 + a_3x_3 + ...}{a_1 + a_2 + a_3 + ...} \qquad [\because A = a_1 + a_2 + a_3 + ...]$$

$$\text{or} \quad \overline{X} = \frac{\Sigma ax}{\Sigma a}$$

Similarly, moments of all small area of the limina about OX axis

$$= a_1y_1 + a_2y_2 + a_3y_3 + ...$$

Moment of total area of the lamina about ox OX axis = $A\overline{Y}$ ...(*iv*)

According to principle of moments,

Equation (iv) = Equation (iii)

$$\therefore \quad A\overline{Y} = a_1y_1 + a_2y_2 + a_3y_3 + ...$$

$$\text{or} \quad \overline{Y} = \frac{a_1y_1 + a_2y_2 + a_3y_3 + ...}{A} \qquad [\because A = a_1 + a_2 + a_3 + ...]$$

$$\therefore \quad \overline{Y} = \frac{a_1y_1 + a_2y_2 + a_3y_3 + ...}{a_1 + a_2 + a_3 + ...}$$

$$\text{or} \quad \overline{Y} = \frac{\Sigma ay}{\Sigma y}$$

Centre of gravity of soild bodies : In case of a solid body having mass m, the whole mass can be supposed to consist of small masses m_1, m_2, m_3, ... etc. The position of centre of gravity can be found out in the similar manner as discussed above.

$$\therefore \qquad \bar{X} = \frac{m_1x_1 + m_2x_2 + m_3x_3 + \ldots}{m_1 + m_2 + m_3 + \ldots}$$

$$\text{or} \qquad \bar{X} = \frac{\Sigma mx}{\Sigma m}$$

$$\text{and} \qquad \bar{Y} = \frac{m_1y_1 + m_2y_2 + m_3y_3 + \ldots}{m_1 + m_2 + m_3 + \ldots}$$

$$\text{or} \qquad \bar{Y} = \frac{\Sigma my}{\Sigma m}$$

5.5. STEPS TO FIND C.G. BY THE METHOD OF MOMENTS

The following steps should be taken into consideration while solving the problems on centre of gravity of composite* section:

1. Choose the two mutually perpendicular axes OX and OY. The axes should be so selected that they touch or pass through the nearest line of the figure.

2. Check whether the section is symmetrical about any X-X axis or Y-Y axis, then C.G. will lie on that axis.

3. The composite figure should be divided into small figures such as rectangles, triangles, circles, etc.

4. Then find the position off C.G. by calculating the $\bar{X}$ and $\bar{Y}$.

5. If any area is removed from the figure such as a hole, it should be subtracted from the original area and the moment should be treated as negative.

SELECTED PROBLEMS

Problem 1: Find the position of centroid of a 150 mm × 200 mm × 50 mm T-section.

Solution: Choose the refrence axes OX and OY as shown in fig. 5-4.

Divide the composite section into two rectangles ① and ②

As the composite section is symmetrical about vertically axis y-y.

$$\therefore \quad \bar{X} = \frac{150}{2} = \mathbf{75\ mm\ [Ans]}$$

To find the value of $\bar{Y}$ proceed as under:

Areas: $a_1 = 150 \times 50 = 7500\ mm^2$

$a_2 = 50 \times 150 = 75000\ mm^2$

Distance of centroids of two rectangles ① and ② from OX axis,

$y_1 = 150 + 50/2 = 175$ mm

$y_2 = 150/2 = 75$ mm

$$\therefore \quad \bar{Y} = \frac{a_1y_1 + a_2y_2}{a_1 + a_2}$$

$$= \mathbf{125\ mm\ [Ans]}$$

$$= \frac{7500 \times 175 + 7500 \times 75}{7500 + 7500}$$

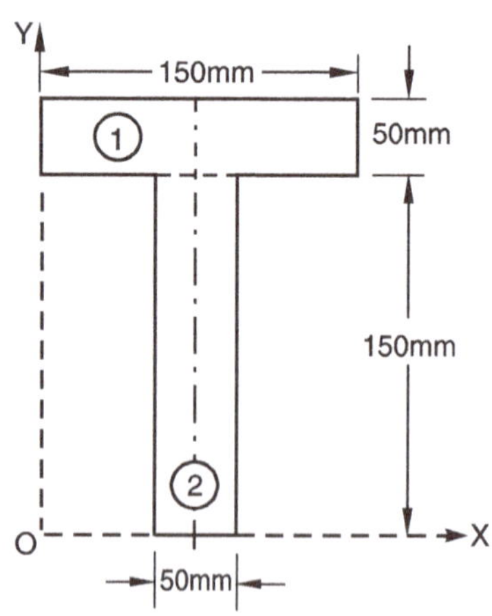

Fig. 5.4

∴ The centroid lies at 75 from OY axis 125 mm form OX axis.

* A section consisting of more than one regular shapes of common geometrical sections such as rectangle, etc. is called composite section.

Problem 2: Find the position of the centroid of 80 mm × 160 mm × 20 mm channel section.

Solution: The channel section is shown in fig 5-5.

Choose the refrenece axes OX and OY.

Divide the composite section into three rectangles ①, ② and ③. about horizontal axis.

As the section is symmetrical about horizontal axis

$$\therefore \quad \overline{Y} = 160/2 = \mathbf{80\ mm\ [Ans]}$$

To find the value of $\overline{X}$ proceed as under:

Areas : $a_1 = 80 \times 20 = 1600\ mm^2$

$a_2 = 120 \times 20 = 2400\ mm^2$

$a_3 = 80 \times 20 = 1600\ mm^2$

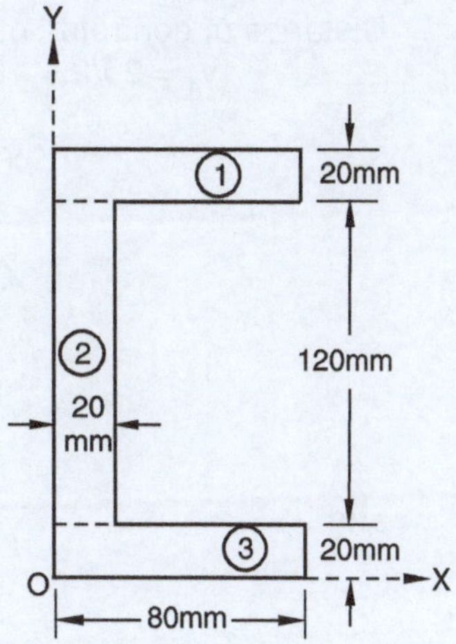

Fig. 5.5

Distance of centroids of three rectangles ①, ② and ③ from OY axis.

$x_1 = 80/2 = 40$ mm

$x_2 = 20/2 = 10$ mm

$x_3 = 80/2 = 40$ mm

Using the relation to find $\overline{X}$,

$$\therefore \quad \overline{X} = \frac{a_1x_1 + a_2x_2 + a_3x_3}{a_1 + a_2 + a_3}$$

$$\therefore \quad \overline{X} = \frac{1600 \times 40 + 2400 \times 10 + 1600 \times 40}{1600 + 2400 + 1600}$$

$$= \frac{152000}{5600} = \mathbf{27.14\ mm\quad [Ans]}$$

∴ The centroid lies at 80 mm from OX axis and 27.14 mm from OY axis.

Problem 3: Two rectangles 100 × 20 mm and 120 × 20 mm are cut from a plane sheet and are connected at right angles to form a right angle. 100 mm piece is horizontal and lies on the top of 120 mm piece which is vertical. Find the postion of the centroid of assembly from the top surface.

Solution: Choose the reference axes OX and OY as shown in fig. 5-6.

Divide the composite section into two rectangles ① and ②.

As the section is not symmetrical about any axis, therefore, to find the value of $\overline{X}$ and $\overline{Y}$ proceed as under: —

Area of rectangle, a_1

$= 100 \times 20 = 2000\ mm^2$

Area of rectangle, a_2

$= 120 \times 20 = 2400\ mm^2$

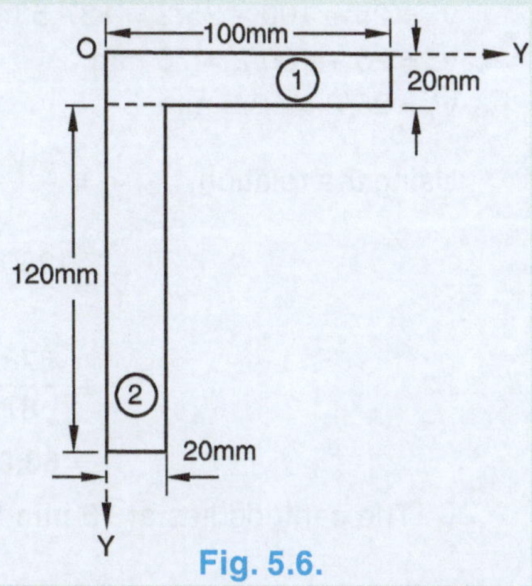

Fig. 5.6.

Distance of centroids of two rectangles ① and ② from OY axis,

$x_1 = 100/2 = 50$ mm
$x_2 = 20/2 = 10$ mm

Distance of centroids of two rectangles ① and ② from OX axis,

$y_1 = 20/2 = 10$ mm; $y_2 = 20 + 120/2 = 80$ mm

$$\bar{X} = \frac{a_1x_1 + a_2x_2}{a_1 + a_2}$$

$$\therefore \quad \bar{X} = \frac{2000 \times 50 + 2400 \times 10}{2000 + 2400}$$

$$= \frac{124000}{4400}$$

$= \mathbf{28.18\ mm}$ **[Ans]**

and
$$\bar{Y} = \frac{a_1y_1 + a_2y_2}{a_1 + a_2}$$

$$= \frac{2000 \times 10 + 2400 \times 80}{2000 + 2400} = \frac{212000}{4400}$$

$= \mathbf{48.18\ mm}$ **[Ans]**

∴ Centroid lies at 28.18 mm from OY axis and 48.18 mm from OX axis.

Problem 4: Find the cetroid of an I-section whose dimensions are; top flange 100 m × 25 mm, web 100 mm × 25 mm and bottom flange 150 × 25 mm.

Solution: The I-section is shown in fig 5.7

Divide the composite I-section into three rectangles ①, ② and ③.

As the section is symmetrical about vertical axis

$\therefore \quad \bar{X} = 150/2 = \mathbf{75\ mm}$ **[Ans]**

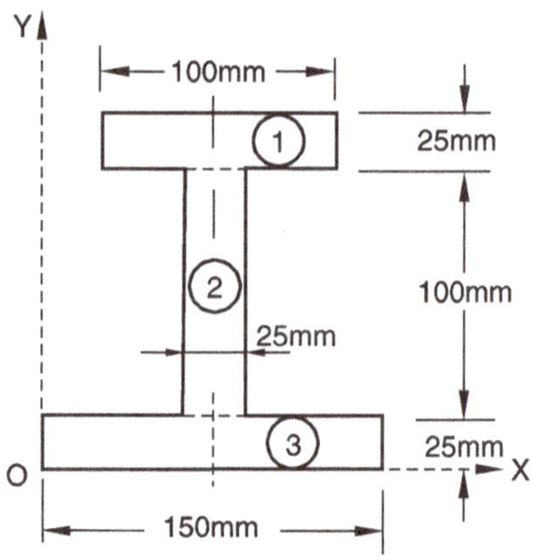

Fig. 5.7.

To find the value of $\bar{Y}$ proceed as under:

$a_1 = 100 \times 25 = 2500$ mm^2
$a_2 = 100 \times 25 = 2500$ mm^2
$a_3 = 150 \times 25 = 3750$ mm^2

Distance of centroids from OX axis

$y_1 = 25 + 100 + 25/2 = 137.5$ mm ..(for area a_1)
$y_2 = 25 + 100/2 = 75$ mm ..(for area a_2)
$y_3 = 25/2 = 12.5$ mm ..(for area a_3)

Using the relation,
$$\bar{Y} = \frac{a_1y_1 + a_2y_2 + a_3y_3}{a_1 + a_2 + a_3}$$

$$\therefore \quad \bar{Y} = \frac{2500 \times 137.5 + 2500 \times 75 + 3750 \times 12.5}{2500 + 2500 + 3750}$$

$$= \frac{578125}{8750}$$

$= \mathbf{66.07\ mm}$ **[Ans]**

∴ The centriod lies at 75 mm from OY axis and 66.07 mm from OX axis.

Problem 5: ***A*** **base plate 15 mm thick is welded symmetrically to the base of 120 mm × 200 mm × 15 mm I-section. Find the length of the welded plate so that the centroid of the composite section lies at 79.74 mm from the bottom edge of the base plate.**

Solution: The I-section with base plate is shown in Fig. 5.8. Choose the two refrence axes OX and OY.

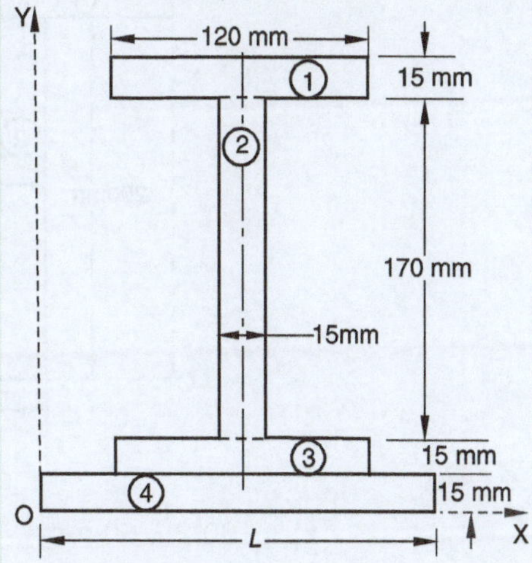

Fig. 5.8

Divide the whole section into four rectangles ①, ②, ③ and ④.

Let L be the length of the welded base plate.

As the section is symmotrical about vertical axis.

$$\therefore \quad \overline{X} = \frac{L}{2}$$

To find $\overline{Y}$ proceed as under:

$a_1 = 120 \times 15 = 1800 \text{ mm}^2$
$a_2 = 170 \times 15 = 2550 \text{ mm}^2$
$a_3 = 120 \times 15 = 1800 \text{ mm}^2$
$a_4 = L \times 15 = 15 \text{ L mm}^2$

Distance of centrolds from OX axis.

$y_1 = 15 + 15 + 170 + 15/2 = 207.5 \text{ mm}$ ...(for area a_1)
$y_2 = 15 + 15 + 170/2 = 115 \text{ mm}$ ...(for area a_2)
$y_3 = 15 + 15 + 15/2 = 22.5 \text{ mm}$ ...(for area a_3)
$y_4 = 15/2 = 7.5 \text{ mm}$ ...(for area a_4)

Using the relation with usual notations,

$$\overline{Y} = \frac{a_1y_1 + a_2y_2 + a_3y_3 + a_4y_4}{a_1 + a_2 + a_3 + a_4}$$

We know $\overline{Y} = 79.74 \text{ mm}$

$$\therefore \quad 79.74 = \frac{1800 \times 207.5 + 2550 \times 115 + 1800 \times 22.5 + 15L \times 7.5}{1800 + 2550 + 1800 + 15L}$$

$(6150 + 15 \text{ L}) \times 79.74 = 373500 + 29320 + 40500 + 112.5 \text{ L}$

$490401 + 1196.1 \times \text{L} = 707250 + 112.5 \text{ l}$

$1083.6 \text{ L} = 216849$

$\therefore$ L = **200 mm approx. [Ans]**

Problem 6: Find the centrold of Z-section as shown in fig. 5-9.

Solution : Choose the two refernece axes OX and OY as shown in fig. 5-9.

Divide the whole section into three rectangles ①, ② and ③.

As the section is not symmetrical about any axis so to find $\overline{X}$ and $\overline{Y}$ proceed as under:

$a_1 = 100 \times 20 = 2000 \text{ mm}^2$
$a_2 = (200 - 40) \times 20 = 3200 \text{ mm}^2$
$a_3 = 200 \times 40 = 8000 \text{ mm}^2$

Distance of centrolds from OY axis.

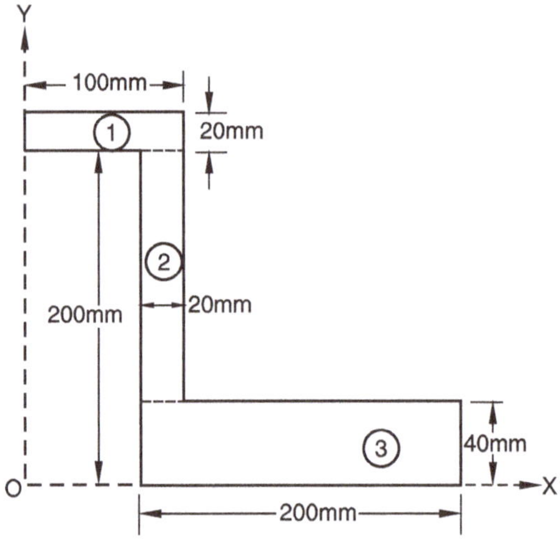

Fig. 5.9

$x_1 = 100/2 = 50$ mm

$x_2 = (100 - 20) + 20/2 = 90$ mm

$x_3 = (100 - 20) + 200/2 = 180$ mm

Distance of centroids from OX axis,

$y_1 = 200 + 20/2 = 210$ mm

$y_2 = 40 + (200 - 40)/2 = 120$ mm

Using the relation,

$$\left[\bar{X} = \frac{a_1 x_1 + a_1 x_2 + a_3 x_3}{a_1 + a_2 + a_3}\right]$$

$$\therefore \quad \bar{X} = \frac{2000 \times 50 + 3200 \times 90 + 8000 \times 180}{2000 + 3200 + 8000}$$

$$= \frac{1828000}{13200} = \mathbf{138.48\ mm} \quad \textbf{[Ans]}$$

$$\text{Also} \quad \bar{Y} = \frac{a_1 y_1 + a_2 y_2 + a_3 x_3}{a_1 + a_2 + a_3}$$

$$\therefore \quad \bar{Y} = \frac{2000 \times 210 + 3200 \times 120 + 8000 \times 20}{200 + 3200 + 8000}$$

$$= \mathbf{73.03\ mm} \quad \textbf{[Ans]}$$

$\therefore$ The centroid of section lies at 138.48 mm from OY axis and 79.03 mm from OX axis

Problem 7: A compound laminar body is formed of a square 60 mm side and an equilateral triangle described outwardly on one side of square. Find the distance of the centrold from the apex of the traingle.

Solutin : The compound laminar body is shown in fig. 5-10. Take the reference axes OX and OY as shown.

Divide the section into two section I.e. (1) square (2) triangle.

As the section is symmetrical about vertical axis.

$\therefore \quad \overline{X} = 60/2 = \mathbf{30\ mm}$ **[Ans]**

To find $\overline{Y}$ proceed as under:

$$a_1 = 60 \times 60 = 3600 \text{ mm}$$

$$a_2 = \frac{1}{2} \times 60 \times 60 \times \frac{\sqrt{3}}{2} = 1558.8 \text{ mm}^2$$

Distance of the centrolds from OX axis,

$$y_1 = \frac{60}{2} = 30 \text{mm}$$

$$y_2 = 60 + \frac{1}{3} \times 60 \times \frac{\sqrt{3}}{2} = 77.32 \text{ mm}$$

Using the relation,

$$\overline{Y} = \frac{a_1 y_1 + a_2 y_2}{a_1 + a_2}$$

$$\therefore \quad \overline{Y} = \frac{3600 \times 30 + 1558.8 \times 77.32}{3600 + 1558.8}$$

$$= \frac{228566.42}{5158.8} = 44.29 \text{ mm}$$

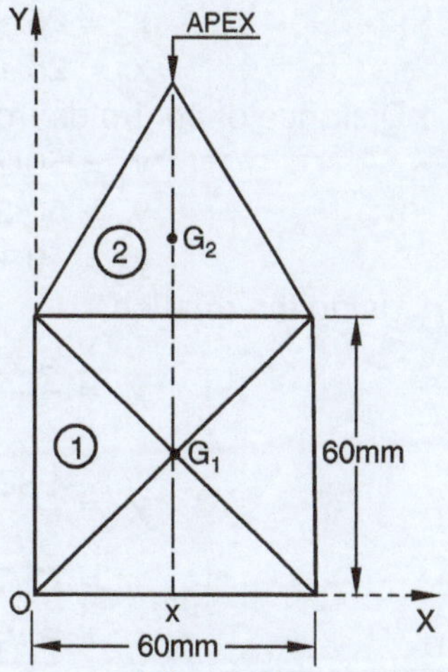

Fig. 5.10

Distance of centrold from apex along vertical axis = Total heightt − $\overline{Y}$

$$= 60 + \sqrt{3}/2 - 44.29 = \mathbf{67.67\ mm} \ \textbf{[Ans]}$$

∴ The centrold G of composite section lies 30 mm from OY axis and 67.67 mm from the apex.

Problem 8: Determine the centre of gravity of the plane uniform lamina as shown in Fig. 5-11.

Solution: Choose the two refrece axes OX and OY as shown in fig. 5-11.

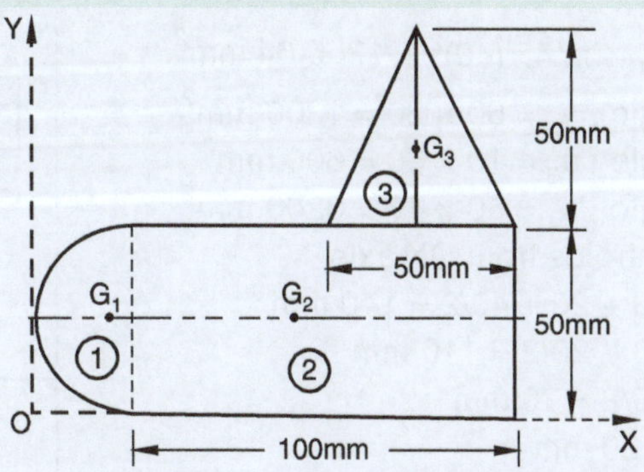

Fig. 5.11.

Divide the lamina into three section I.e. , semlclrcle ① rectangle ② and traingle ③. As the section is not symmetrical about any axis, therefore to find $\overline{X}$ and $\overline{Y}$ proceed as under:

Semicircle, $a_1 = \pi r^2 = \pi(25)^2 / 2 = 982 \text{ mm}^2$

Rectangle, $a_2 = 100 \times 50 = 5000 \text{ mm}^2$

Triangle, $a_3 = 1/2 \times 50 \times 50 = 1250\ mm^2$

Distance of centroids from OY axis.

$$x_1 = 25 - 4R/3\pi = 25 - 4 \times 25/3\pi = 14.4\ mm$$
$$x_2 = 25 + 100/2 = 75\ mm$$
$$x_3 = 25 + 50 + 25 = 100\ mm$$

Distance of centrolds from OX axis

$$y_1 = 50/2 = 25\ mm$$
$$y_2 = 50/2 = 25\ mm$$
$$y_3 = 50 + 50/3 = 66.7\ mm$$

Using the relation,

$$\overline{X} = \frac{a_1x_1 + a_1x_2 + a_3x_3}{a_1 + a_2 + a_3}$$

$$\therefore \quad \overline{X} = \frac{982 \times 14.4 + 5000 \times 75 + 1250 \times 100}{982 + 5000 + 1250}$$

= **71.1 mm [Ans]**

$$\overline{Y} = \frac{a_1y_1 + a_2y_2 + a_3y_3}{a_1 + a_2 + a_3}$$

$$\overline{Y} = \frac{982 \times 25 + 5000 \times 25 + 1250 \times 66.7}{982 + 5000 + 1250}$$

= **32 mm [Ans]**

Problem 9: Locate the position of centrold of the section shown in fig. 5-12.

Solution: Choose the refrence axes OX and OY as shown In fig. 5-12.

Divide the composite section into circle ① and three rectangles ②, ③ and ④.

As the section is symmetrical about vertical axis,

$$\therefore \quad \overline{X} = 80/2 = \mathbf{40\ m\ [Ans]}$$

To find the value of $\overline{Y}$ proceed as under:

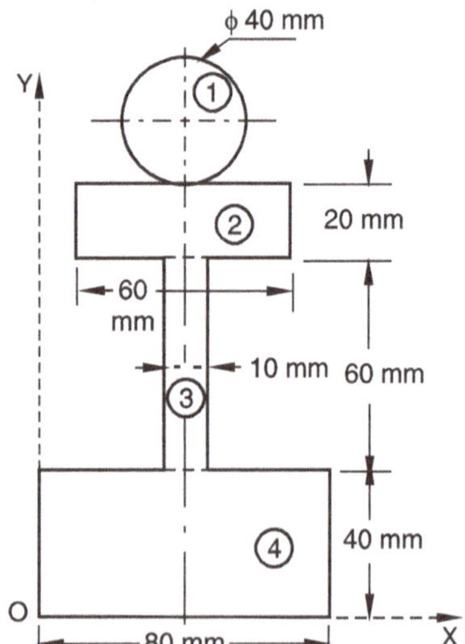

Fig. 5.12

Area of circle $a_1 = \pi/4 \times (40)^2 = 1256.64\ mm^2$.

Area of rectangle , $a_2 = 60 \times 20 = 1200\ mm^2$

Area of rectangle , $a_3 = 10 \times 60 = 600\ mm^2$

Area of rectangle , $a_4 = 80 \times 40 = 3200\ mm^2$

Distance of centroids from OX axis,

$$y_1 = 40 + 60 + 20 + 40/2 = 140\ mm$$
$$y_2 = 40 + 60 + 20/2 = 110\ mm$$
$$y_3 = 40 + 60/2 = 70\ mm$$
$$y_4 = 40/2 = 20\ mm$$

Using the relation,

$$\overline{Y} = \frac{a_1y_1 + a_2y_2 + a_3y_3 + a_4y_4}{a_1 + a_2 + a_3 + a_4}$$

$$\therefore \quad \overline{Y} = \frac{1256.64 \times 140 + 1200 \times 110 + 600 \times 70 + 3200 \times 20}{1256.64 + 1200 + 600 + 3200}$$

= **66.16 mm [Ans]**

∴ The centrold G of the section lies at 40 from OY axis and 66.16 mm from OX axis.

5.6. CENTRE OF GRAVITY (CENTROID) OF REMAINDER

When a portion of a body is removed, the remaining portion is known as remainder. In order to determine the C.G. of the remainder, the following steps should be followed:

(i) The removed portion (area or mass) is considered negative.

(ii) $\overline{X}$ and $\overline{Y}$ for the remainder are calculated as usual except taking the removed portions as negative in the expressions.

Illustration: Consider a circular disc of area a_1 from which area a_2 be removed as shown in fig. 5.13, Then

$$\therefore \quad \overline{X} = \frac{a_1x_1 - a_2x_2}{a_1 - a_2}$$

and

$$\overline{Y} = \frac{a_1y_1 - a_2y_2}{a_1 - a_2}$$

SELECTED PROBLEMS

Problem 10: A circular disc of 50 mm diameter is cut out from a circular disc of 100 mm diameter. Find the centroid of the remainder.

Solution: Take the reference axes OX and OY as shown in fig. 5-13.

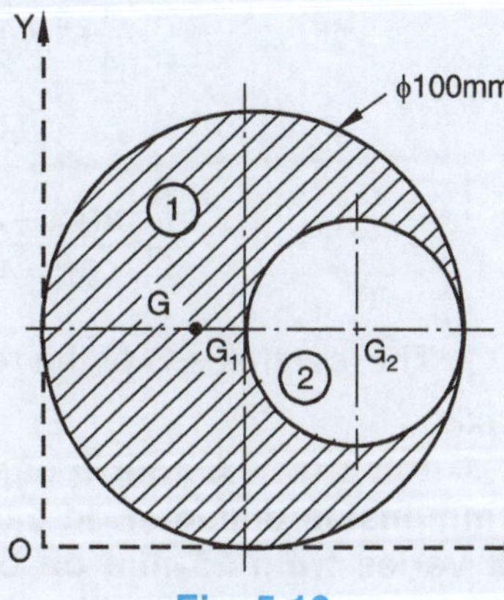

Fig. 5.13

As the section is symmetrical about horizontal axis,

$\therefore \quad \overline{Y} = 100/2 =$ **50 mm [Ans]**

To find the $\overline{X}$ proceed as under.

Area of bigger circle, a_1

$= \pi/4 \times (D)^2 = \pi/4 \times (100)^2 = 2500\,\pi\ mm^2$

Area of cut out circle, a_2

$$= \frac{\pi}{4} \times (d)^2 = \frac{\pi}{4} \times (50)^2 = \frac{2500}{4}\pi\ mm^2$$

Distance of centrolds from OY axis,

$$x_1 = 100/2 = 50\ mm$$

$$x_2 = 50 + 50/2 = 75\ mm$$

Using the relation,

$$\overline{X} = \frac{a_1x_1 - a_2x_2}{a_1 - a_2}$$

$$\therefore \quad \overline{X} = \frac{2500\pi \times 50 - 2500\pi/4 \times 75}{2500\pi - 2500\pi/4} = \mathbf{41.67\ mm} \quad \textbf{[Ans]}$$

$\therefore$ The centrold G lies 50 mm from OX axis and 41.66 mm from OY axis.

Problem 11: A square hole is punched out of a circular lamina, the diagonal of the square being the radius of the circle. Find the CG of the remainder, if 'a' is the diameter of the circle,

Solution: Choose the refernce axes OX and OY as shown in fig. 5-14.

As the section is symmetrical about horizontal axis,

$\therefore \quad \overline{Y} = \mathbf{a/2}$ **[Ans]**

To find $\overline{X}$ proceed as under,

Area of circular lamina,

$$a_1 = \pi/4 \times a^2 mm^2 \qquad [\because \text{ a is the diameter of circle}]$$

Let L be the length of each side of the square

Area of punched out square,

$$a_2 = L \times L = \frac{a}{2\sqrt{2}} \times \frac{a}{2\sqrt{2}} = \frac{a^2}{8} \qquad [\because \text{ length, } L = \frac{a}{2\sqrt{2}}]$$

Distance of the centroids from OX asis,

$$x_1 = \frac{a}{2}$$

$$x_2 = \frac{a}{2} + \frac{a}{4} = \frac{3a}{4}$$

Using the relation,

$$\overline{X} = \frac{a_1x_1 - a_2x_2}{a_1 - a_2}$$

$$\overline{X} = \frac{\frac{\pi}{4}a^2 \times \frac{a}{2} - \frac{a^2}{8} \times \frac{3a}{4}}{\frac{\pi}{4}a^2 - \frac{a^2}{8}}$$

$$= \mathbf{\frac{a(4\pi - 3)}{8\pi - 4}} \quad \textbf{[Ans]}$$

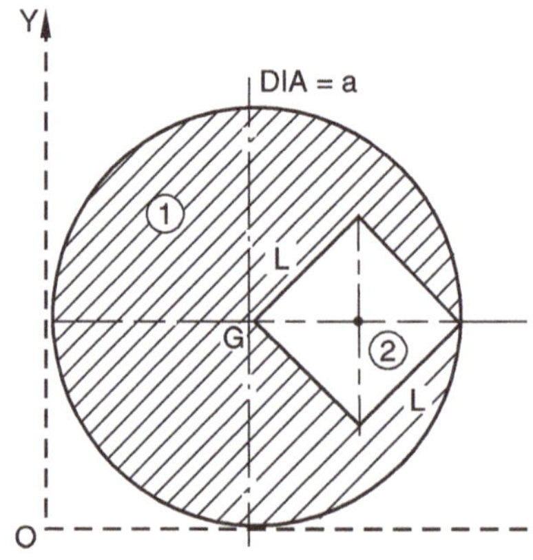

Fig. 5.14.

$\therefore$ The centroids G of the remainder lies at a/2 from OX axis and $\frac{a(4\pi - 3)}{8\pi - 4}$ from OY axis.

Problem 12: A hollow cast iron column has 300 mm external diameter and 225 mm inside diameter. In casting, the bore got eccentirc such that the thickness varies from 25 mm on one side and 50 mm on the other side. Find the position of centre of gravity of the section.

Solution : Choose two reference axes OX and OY (see fig.5-15).

As the section is symmetrical about horizontal axis.

$\therefore \quad \overline{Y} = 300/2 = \mathbf{150\ mm}$

To find $\overline{X}$ proceed as under:

Area of external circle,

$$a_1 = \pi D^2/4 = \pi \times (300)^2/4$$

$$= 22500\pi \text{ mm}^2$$

Area of internal circle,

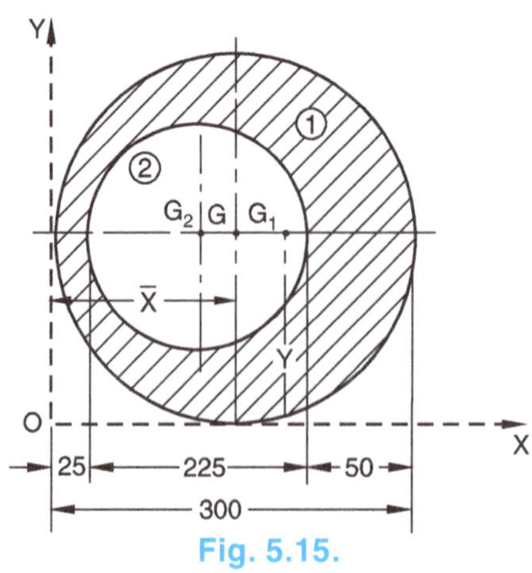

Fig. 5.15.

$a_2 = \pi d^2/4 = \pi \times (225)^2/4 = 12656.25\pi \text{ mm}^2$

$x_1 = 300/2 = 150$ mm

$x_2 = 25 + 225/2 = 137.5$ mm

Using the relation,

$$\overline{X} = \frac{a_1x_1 - a_2x_2}{a_1 - a_2}$$

$$\therefore \quad \overline{X} = \frac{22500\pi \times 150 - 12656.25\pi \times 137.5}{22500\pi - 12656.25\pi}$$

$= \mathbf{166.07\ mm}$ **[Ans]**

Hence, the centre of gravity of the column section will lie at a distance 166.07 mm from OY axis and at 150 mm from OX axis

Problem 13: From a trapezoid, a semi-circle is cut out as shown in fig. 5-16. Find the position of centroid of the remainder.

Solution: Choose the reference axes OX and OY (see fig 5.16)

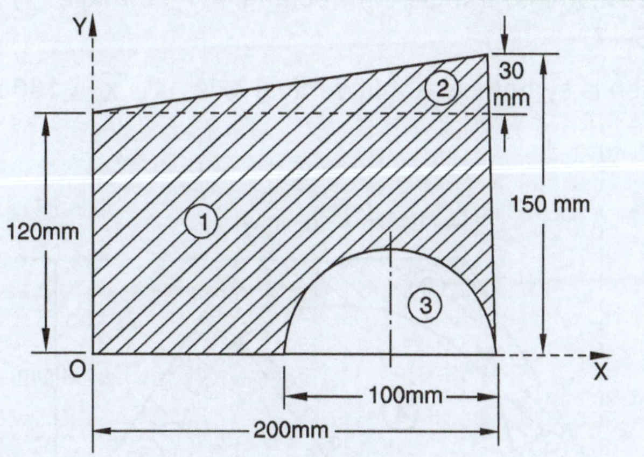

Fig. 5.16

Divide the section into rectangle ①, triangle ② and semi-circle ③

Area of rectangle ①, $a_1 = 200 \times 120 = 2400 \text{ mm}^2$

Area of triangle ②, $a_2 = 1/2 \times 200 \times 30 = 3000 \text{ mm}^2$

Area of semi-circle ③, $a_3 = \pi/8 \times (100)^2 = 3927 \text{ mm}^2$

Distance of centrods from OY axis,

$x_1 = 200/2 = 100$ mm ...(for area a_1)

$x_2 = 2/3 \times 200 = 400/3 = 133.3$ mm ...(for area a_2)

$x_3 = 100 + 50 = 150$ mm ...(for area a_3)

Distance of centroids from OX axis,

$y_1 = 120/2 = 60$ mm ...(for area a_1)

$y_2 = 120 + 30/3 = 130$ mm ...(for area a_2)

$y_3 = 4 \times 50/3\pi = 21.22$ mm ...(for area a_3)

Using the relation, $\overline{X} = \dfrac{a_1x_1 + a_2x_2 - a_3x_3}{a_1 + a_2 - a_3}$

$$\therefore \quad \overline{X} = \frac{24000 \times 100 + 3000 \times 133.3 - 3927 \times 150}{24000 + 3000 - 3927}$$

$= \mathbf{95.8\ mm\ [Ans]}$

$$\overline{Y} = \frac{a_1y_1 + a_2y_2 - a_3y_3}{a_1 + a_2 - a_3}$$

$$\therefore \quad \overline{Y} = \frac{24000 \times 60 + 3000 \times 130 - 3927 \times 21.22}{24000 + 3000 - 3927}$$

$= \mathbf{75.7\ mm\ [Ans]}$

∴ The centroid G of the section lies at 75.7 mm from OX axis and 96.8 mm from OY axis.

Problem 14: Determine the centre of gravity of the hatched area shown in fig. 5-17.

Solution: Choose the reference axes OX and OY.

Divide the section into trangle ①, rectangle ②, triangle ③ and semi-circle ④ as shown in fig. 5-17.

As the section is symetrical about vertical axis. ∴ $\overline{X}$ **= 160 mm [Ans]**

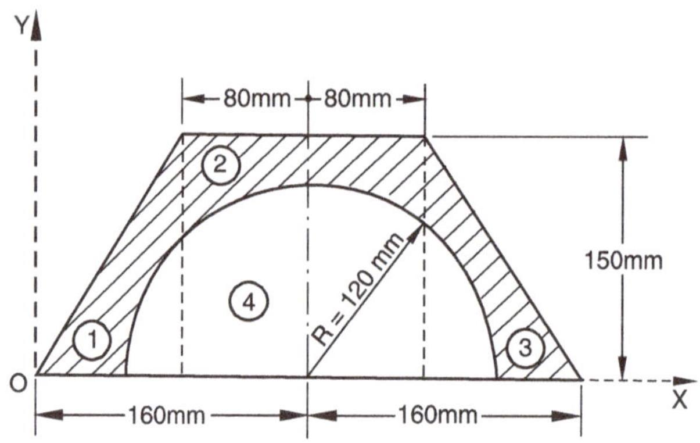

Fig. 5.17

To find $\overline{Y}$ proceed as under,

Area of traingle ①, $a_1 = 80 \times 150/2 = 6000\ mm^2$

Area of rectangle ②, $a_2 = 160 \times 150 = 24000\ mm^2$

Area of triangle ③, $a_3 = 80 \times 150/2 = 6000\ mm^2$

Area of semicircle ④, $a_4 = \frac{\pi}{8}(200)^2 = 22619.46\ mm^2$

Distance of centroids from OX axis,

$$y_1 = 150/3 = 50\ mm$$
$$y_2 = 150/2 = 75\ mm$$
$$y_3 = 150/3 = 50\ mm$$
$$y_4 = 4 \times 120/3\pi = 50.9\ mm$$

Using the relation, $\bar{Y} = \dfrac{a_1y_1 + a_2y_2 + a_3y_3 - a_4y_4}{a_1 + a_2 + a_3 - a_4}$

$= \dfrac{6000\times50 + 24000\times75 + 6000\times50 - 22619.46\times50.9}{6000 + 24000 + 6000 - 22619.46}$

$=$ **93.4 mm [Ans]**

∴ The C.G. of the section lies at 160 mm from OY axis and 93.4 mm from OX axis.

Problem 15: A right angle triangle base 400 mm and height 540 mm is cut in two pieces by a straight line parallel to base at mid height. Find The vertical distance of centre of gravity above the base of the bigger place.

Solution: Here, ABC is a right angled triangle. It is cut into two pieces ADE and DECB by straight line DE parallel to base at mid height. To find C. G. of the bigger piece DECB, proceed as under:

Let a_1 = area of the whole triangle ABC

$= 400 \times 540/2 = 108000\ mm^2$

a_2 = area of the removed triangle ADE

$= 200 \times 270/2 = 27000\ mm^2$

[∵ DE = 400/2 = 200 mm]

Distance of centroids from BC axis.

y_1 = distance of C. G. of removed traingle

$= 540/3 = 180$ mm

$y_2 = 270 + 270/3 = 360$ mm

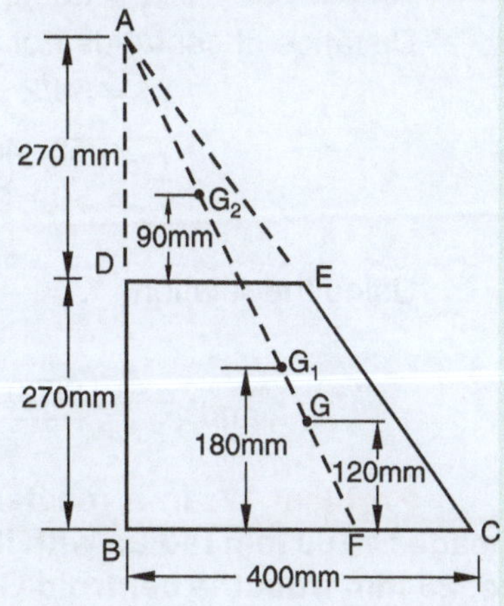

Fig. 5.18

Using the relation,

$\bar{Y} = \dfrac{a_1y_1 - a_2y_2}{a_1 - a_2}$

$= \dfrac{108000\times180 - 27000\times360}{108000 - 27000}$

$=$ **120 mm [Ans]**

∴ The C. G. lies at 120 mm from the base of the bigger place and on the axes of medians.

Problem 16: Determine the location of the centroid of the lamina as shown in fig. 5-19.

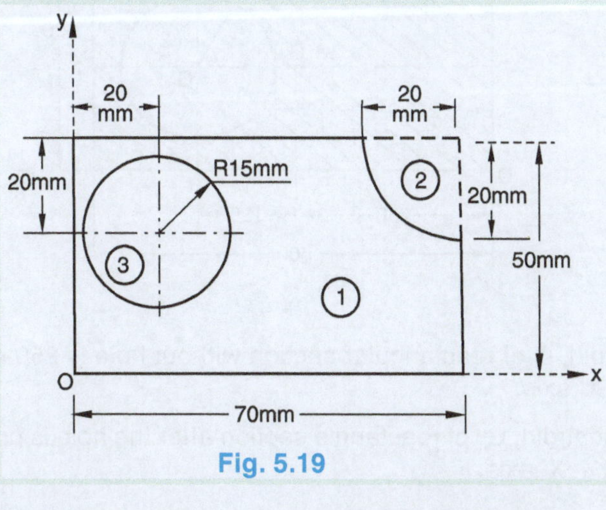

Fig. 5.19

Solution: Choose the two reference axes OX and OY

Divide the whole section into three sections, rectangle ①, quadrant ② and circle ③.

As the section is not symmetrical about any axis, therefore to find $\overline{X}$ and $\overline{Y}$ proceed as under:

$$a_1 = 70 \times 50 = 3500 \text{ mm}^2$$

$$a_2 = \pi r^2/4 = \pi(20)^2/4 = 314 \text{ mm}^2$$

$$a_3 = \pi r^2 = \pi(15)^2 = 706.8 \text{ mm}^2$$

Distance of centroids from OY axis,

$$x_1 = 70/2 = 35 \text{ mm}$$

$$x_2 = 70 - 4r/3\pi = 70 - 4 \times 20/3\pi = 61.5 \text{ mm}$$

$$x_3 = 20 \text{ mm}$$

Distance of centroids from OX axis,

$$y_1 = 50/2 = 25 \text{ mm}$$

$$y_2 = 50 - 4r/3\pi z = 41.5 \text{ mm}$$

$$y_3 = 50 - 20 = 30 \text{ mm}$$

Using the relation, $\overline{X} = \dfrac{3500 \times 35 - 314 \times 61.5 - 706.8 \times 30}{3500 - 314 - 706.8}$ = **35.9 mm [Ans]**

and, $\overline{Y} = \dfrac{3500 \times 25 - 314.5 \times 41.5 - 706.8 \times 30}{3500 - 314 - 706.8}$ = **21.5 mm [Ans]**

Problem 17: In a reactangle section 500 mm × 300 mm, a circular hole in made of 100 mm radius with Its centre on contriodal X′ – X′ axis and at a distance of 25 mm from the centrold G of the whole body. By how much distance the G of the remainder will shift on the controidal axis after the hole is punched.

Solution: Choose the two refrence axes OX and OY

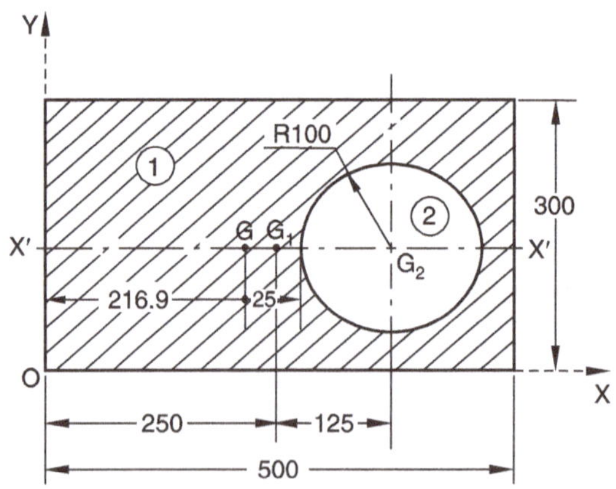

Fig. 5.20.

The centrold, G of reactangular section without hole is 250 mm from OY axis on centroidal X′ – X′ axis.

Let $\overline{X}$ = controld, G_1 of reactangle section after the hole is punched from OY axis on centroldal X′ – X′ axis.

Area of rectangle ① and area of removed circle ②

$a_1 = 500 \times 300 = 150000 \text{ mm}^2$ ...[area of rectangle ①]

$a_2 = \pi r^2 = \pi(100)^2 = 3141.6 \text{ mm}^2$ ...[area of circle ②]

Distance of centroid from OY axis

$x_1 = 250$ mm

$x_2 = 250 + 125 = 375$ mm

$$\therefore \quad \overline{X} = \frac{150000 \times 250 - 3141.6 \times 375}{150000 - 3141.6} = \mathbf{216.9 \text{ mm}} \quad \textbf{[Ans]}$$

∴ The distance of G of reactangular section after the hole is punched will shift on the controidal X′ – X′ axis,

= 250 – 216.9 = 33.1 mm towards OY axis **[Ans]**

Problem 18: The corner of the plane lamina is folded along the plane plate as shown in fig. 5-21. Find the centroid .

Solution: Choose the two refrence axes OX and OY.

Divide the lamina into three section i.e. square ①, trangle ② and triangle ③.

As the section is not symmetrical about any axis, therefore to find $\overline{X}$ and $\overline{Y}$ proceed as under:

a_1 (uncut square) = 100 × 100 = 10000 mm²

a_2 (corner cut ABC) = $\frac{1}{2} \times 30 \times 30 = 450 \text{ mm}^2$

a_3 (Folded corner ACD)

= 1/2 × 30 × 30 = 450 mm²

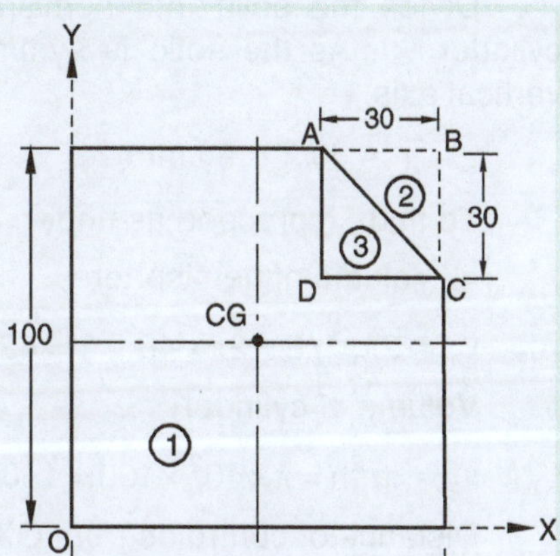

Fig. 5.21

Distance of centroids from OY axis

$x_1 = 100/2 = 50$ mm

$x_2 = 50 + 20 + 2/3 \times 30 = 90$ mm

$x_3 = 50 + 20 + 1/3 \times 30 = 80$ mm

Distance of centroids from OX axis

$y_1 = 100/2 = 50$ mm

$y_2 = 50 + 20 + 2/3 \times 30 = 90$ mm

$y_3 = 50 + 20 + 1/3 \times 30 = 80$ mm

$$\overline{X} = \frac{a_1x_1 - a_2x_2 + a_3x_3}{a_1 - a_2 + a_3}$$

$$= \frac{10000 \times 50 - 450 \times 90 + 450 \times 80}{10000 - 450 + 450} = \frac{495500}{10000} = \mathbf{49.55 \text{ mm}} \textbf{ [Ans]}$$

$$\overline{Y} = \frac{a_1y_1 - a_2y_2 + a_3y_3}{a_1 - a_2 + a_3} = \frac{10000 \times 50 - 450 \times 90 + 450 \times 80}{10000 - 450 + 450} = \frac{495500}{10000}$$

= **49.55 mm [Ans]**

Problem 19: Find the C.G. of lamina from point O, of 220 mm radius as shown in Fig. 5.22

Solution : As the lamina is symmetrical about y-y axis, therefore its C.G. lies on this axis. Let O be the reference point. From the geometry of the lamina,

$$\alpha = 30° = \frac{\pi}{6} \text{ rad.}$$

we know the distance between the reference point O and the C.G. of the lamina

$$\bar{Y} = \frac{2r}{3} \times \frac{\sin\alpha}{\alpha} = \frac{2 \times 220}{3} \times \frac{\sin 30°}{(\pi/6)} = \frac{440}{3} \times \frac{0.5}{(\pi/6)}$$

$= \mathbf{140\ mm\ [Ans]}$

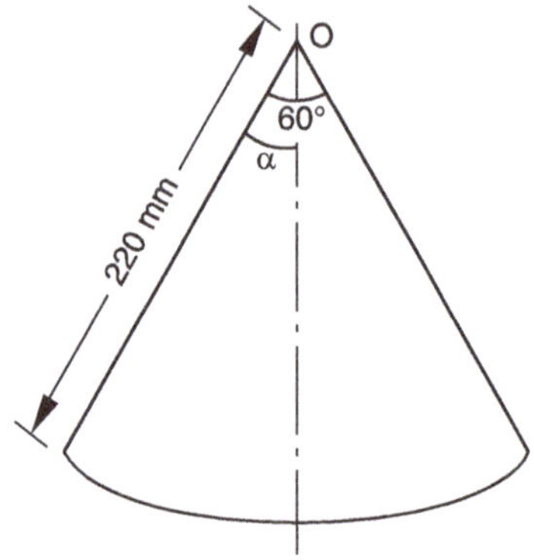

Fig. 5.22

CENTRE OF GRAVITY OF SOLIDS

Problem 20: A hemisphere of diameter 80 mm is placed on the top of a cylinder, whose diameter is also 80 mm. The height of the cylinder is 100 mm. Find the common C.G. of the composite body.

Solution: Choose the two reference axes OX and OY

Divide the solid into hemisphere ① and cylinder ②. As the solid is symmetrical about vertical axis.

$\therefore\ \bar{X} = 80/2 = 40$ mm

To find $\bar{Y}$ proceed as under

$\therefore$ Volume of hemisphere,

$$v_1 = 2\pi r^3/3 = 2\pi \times 40^3/3 = 42666.67\pi \text{ mm}^3$$

Volume of cylinder,

$$v_2 = \pi r^2 h = \pi \times 40^2 \times 100 = 160000\pi \text{ mm}^3$$

Distance of centroids from OX axis

$y_1 = 100 + 3\ r/8 = 100 + 3 \times 40/8 = 115$ mm

and $y_2 = 100/2 = 50$ mm

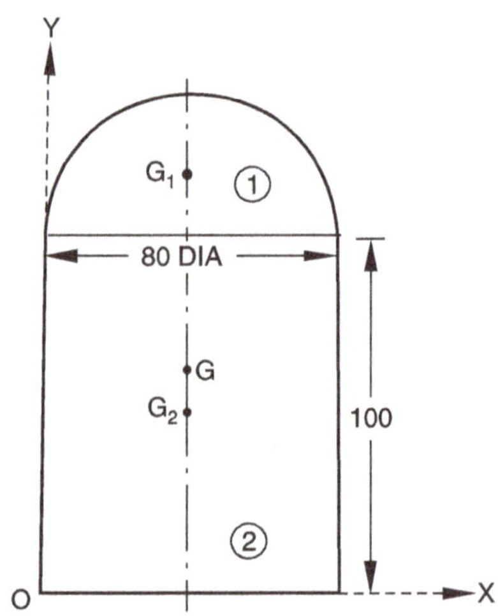

Fig. 5.23

Using the relation, $\bar{Y} = \dfrac{v_1 y_1 + v_2 y_2}{v_1 + v_2}$

$$= \frac{42666.67\pi \times 115 + 160000\pi \times 50}{42666.67\pi + 160000\pi} = \mathbf{63.08\ mm\ \ [Ans]}$$

Hence, centre of gravity, G of the composite body will be on its axis symmetrical and at a distance of 63.08 mm from OX axis.

Problem 20: A cone and a hemisphere made of same material are arranged so that cone sits on the hemisphere. The diameter of the base of the cone is 300 mm. What will be the height of the cone when the hemispherical surface rests on a horizontal plane in neutral equilibrium?

Solution: The equilibrium will be neutral when centre of gravity of the composite body lies at G, the center of the hemisphere i.e. $\overline{Y}$ = 150 mm.

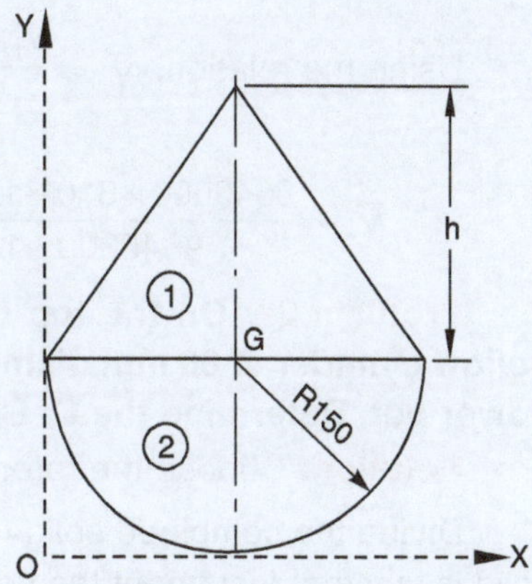

Fig. 5.24.

Let h be the height of the cone.

Divide the solid into cone ① and hemisphere ②

Voume of cone ①.

$$v_1 = \pi r^2 \times h/3$$

$$= \pi \times 150^2 \times h/3 = 7500\pi h \text{ mm}^3$$

$$v_2 = 2\pi r^3/3$$

$$= 2\pi \times 150^3 \times h/3$$

$$= 22{,}50{,}000\, \pi h \text{ mm}^3$$

Distance of C.G. from OX axis,

$$y_1 = 150 + \frac{h}{4} \text{ mm}$$

$$y_2 = 150 - 3/8 \times r = 150 - 3/8 \times 150 = 750/8 \text{ mm}$$

Using the relation, $\overline{Y} = \dfrac{v_1y_1 + v_2y_2}{v_1 + v_2}$

$$50 = \frac{7500\pi h \times (150 + \frac{h}{4}) + 225000\pi \times \frac{750}{8}}{7500\pi h + 2250000\pi}$$

∴ h = **259.8 mm [Ans]**

Problem 21: Fig. 5.25 shows a compound solid. Locate the C. G. of the solid.

Solution: Choose the refrence OX and OY as shown in Fig 5.25

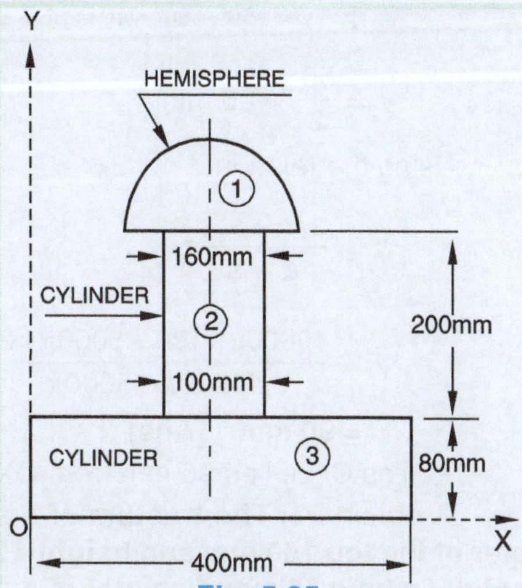

Fig. 5.25.

Divide the compound solid into hemisphere ① and two cylinder ② and ③.

∴ $\overline{X}$ = 400/2 = **200 mm [Ans]**

To find the $\overline{Y}$ proceed as under:

Volume of hemisphere ①,

$$v_1 = 2\pi r^3/3 = 2\pi(80)^3/3$$

$$= 9636000 \text{ mm}^3$$

Volume of cylinder ②,

$$v_2 = \pi r^2 h = \pi \times (50)^3 \times 200 = 1550000 \text{ mm}^3$$

Volume of cylinder (3),

$$v_3 = \pi r^2 h - \pi \times (200)^2 \times 80 - 10050000 \text{ mm}^3$$

Distance of centroids from OX axis,

$$y_1 = 80 + 200 + 3 \times 80/8 = 310 \text{ mm}$$

$$y_2 = 80 + 200/2 = 180 \text{ mm}$$

and $$y_3 = 80/2 = 40 \text{ mm}$$

Using the relation, $\overline{Y} = \dfrac{v_1y_1 + v_2y_2 + v_3y_3}{v_1 + v_2 + v_3}$

$$\overline{Y} = \frac{9646000 \times 310 + 157000 \times 180 + 1005000 \times 4}{9646000 + 157000 + 1005000 \times 4} = \mathbf{78\ mm}\quad \textbf{[Ans]}$$

Problem 22: On the top of a hollow hemisphere of 100 mm in diameter, a hollow cylinder of 60 mm diameter and 180 mm hight fixed axially so as to form a flower pot. Determine the C. G. of the combined solid.

Solution: Choose the reference axes OX and OY as shown in fig 5.26.

Divide the combined solid into cylinder ① and hemisphere ②. As the combined solid is symmetrical about the vertical axis,

$$\therefore \quad \overline{X} = 100/2 = 50 \text{ mm}$$

To find the value of $\overline{Y}$ proceed as under:

Since the cylinder and hemisphere are hollow, therefore their surface area are taken into consideration.

Surface areas:

$$a_1 = 2\pi rh = 2\pi \times (30) \times 180 = 10800\pi \text{ mm}^2$$

$$a_2 = 2\pi r^2 = 2\pi \times (50)^2 = 5000\pi \text{ mm}^2$$

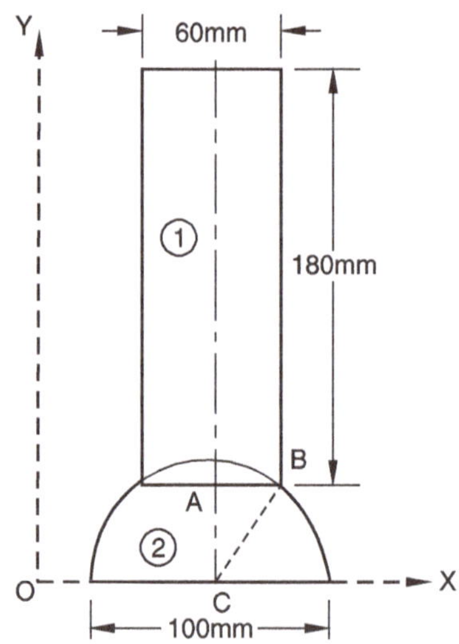

Fig. 5.26.

Distance of centroids from OX axis.

$$y_1 = AC + 180/2$$

$$y_1 = 40 + 180/2 = 130 \text{ mm}$$

$$[\because AC = \sqrt{50^2 - 30^2} = 10 \text{ mm}]$$

$$y_2 = \frac{50}{2} = 25 \text{ mm}$$

Using the relation,

$$\overline{Y} = \frac{a_1y_1 + a_2y_2}{a_1 + a_2}$$

$$\overline{Y} = \frac{10800\pi \times 130 + 5000\pi \times 25}{10800\pi + 5000\pi}$$

$$= \mathbf{90\ mm}\quad \textbf{[Ans]}$$

$\therefore$ The C. G. lies 90 mm from OX axis and on the axis of symmetry.

Problem 23: The frustum of a solid cone has base diameter 200 mm, diameter at the top 100 mm and height 150 mm. Determine the centre of gravity of the solid if a hole 50 mm diameter is cut axially from the frustum of solid cone.

Solution: The frustum of a solid cone is shown in Fig. 5.27 (i). Extend the frustum BCED in order to obtain the full cone ADE as shown in Fig. 5.27 (ii).

Choose the reference axes DX and DY.

Let v_1 = Volume of the full Cone ADE

v_2 = Volume of upper Cone ABC and

v_3 = Volume of hole.

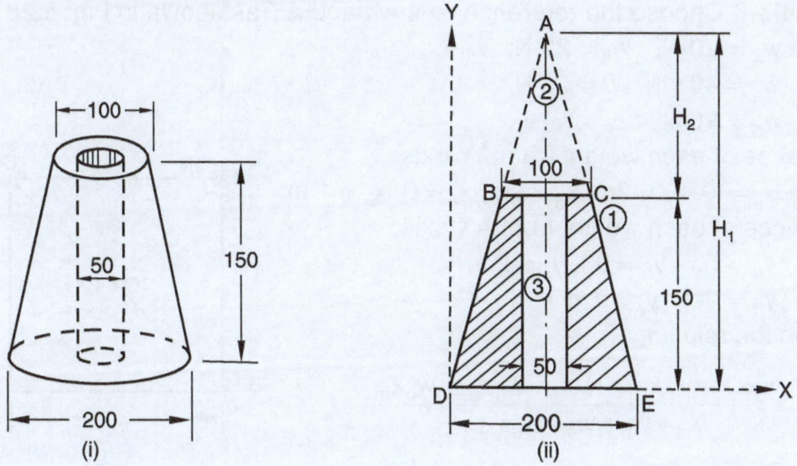

Fig. 5.27.

From the similar triangle ADE and ABC,

$$\frac{DE}{BC} = \frac{H_1}{H_2} \quad \text{or} \quad \frac{2000}{1000} = \frac{150 + H_2}{H_2}$$

or $2H_2 = 150 + H_2$

$\therefore$ $H_2 = 150$ mm

$H_1 = 150 + H_2 = 150 + 150 = 300$ mm

As the solid is symmetrical about vertical axis,

$\therefore$ $\overline{X} = 200/2 = 100$ mm

To find $\overline{Y}$ proceed as under,

Volume of full cone, ADE

$$v_1 = 1/3 \times \pi \times (100)^2 \times 300 = 1000000\pi \text{ mm}^2$$

Volume of upper cone ABC,

$$v_2 = 1/3 \times \pi \times (50)^2 \times 150 = 125000\pi \text{ mm}^2$$

Volume of hole, $v_3 = \pi r^2 h = \pi \times (25)^2 \times 150 = 93750\pi \text{ mm}^2$

Distance of C. G. from OX axis,

$y_1 = H_1/4 = 300/4 = 75$ mm [for full cone ADE]

$y_2 = 150 + H_2/4 = 150 + 150/4 = 187.5$ mm [for upper cone ABC]

$y_3 = 75$ mm [for cylinder]

Using the relation, $\overline{Y} = \dfrac{v_1 y_1 - v_2 y_2 - v_3 y_3}{v_1 - v_2 - v_3}$

$$\therefore \quad \overline{Y} = \frac{1000000\pi \times 75 - 125000\pi \times 187.5 - 93750\pi \times 75}{100000\pi - 125000\pi - 93750\pi} = \textbf{57 mm} \quad \textbf{[Ans]}$$

$\therefore$ C.G. lies at 57 mm from the base on the axis of symmetry.

Problem 24: Find the C.G. of a uniform square plate ABCD of weight 10 N acting at the centre and weights of 20, 30, 40 and 50 N are placed at its four corners A, B, C and D respectively. Take side of square equal to 2 metre.

Solution: Choose the reference axis AX and AY as shown in Fig. 5.28.

Here $w_a = 20$ N, $w_b = 30$ N,

$w_c = 40$ N, $w_d = 50$ N

$w_o = 10$ N

Distance of each weight from AY axis

$x_a = 0$, $x_b = 2$m, $x_c = 2$m, $x_d = 0$, $x_o = 1$ m

Distance of each weight from AX axis,

$y_a = 0$, $y_b = 0$, $y_c = 2$m,

$y_d = 2$m, $y_o = 1$m

Using the relation,

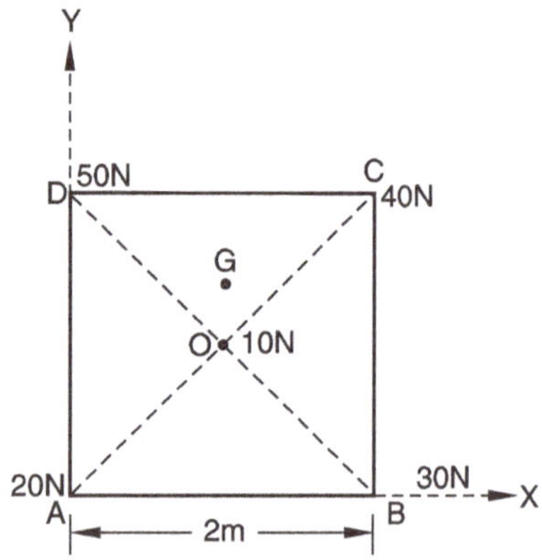

Fig. 5.28

$$\overline{X} = \frac{w_a x_a + w_b x_b + w_c x_c + w_d x_d + w_o x_o}{w_a + w_b + w_c + w_d + w_o}$$

$$\overline{X} = \frac{20 \times 0 + 30 \times 2 + 40 \times 2 + 50 \times 0 + 10 \times 1}{20 + 30 + 40 + 50 + 10} = \textbf{1 m} \quad \textbf{[Ans]}$$

$$\overline{Y} = \frac{w_a y_a + w_b y_b + w_c y_c + w_d y_d + w_o y_o}{w_a + w_b + w_c + w_d + w_o}$$

$$\overline{Y} = \frac{20 \times 0 + 30 \times 0 + 40 \times 2 + 50 \times 2 + 10 \times 1}{20 + 30 + 40 + 50 + 10}$$

$= \textbf{1.26 m [Ans]}$

$\therefore$ The C. G. lies at 1.26 m from AX-axis and 1 m from AY-axis.

5.7. DETERMINING C. G. BY GRAPHICAL MEHTOD

In this method, the C. G. of the body is determined by finding the resultant of parallel forces, graphically. This method is explained below with different steps by taking problems.

Problem 25: Find the position of centre of gravity of unequal angle section 100 × 80 × 20 mm graphically as shown in Fig. 5.29.

Solution: An unequal angle section is shown in Fig. 5.29. The various steps to find C.G. of this section graphically are as under:

1. *Dividing the composite section:*

Divide the given composite section into two rectangles ABCH and DEFH.

Let $a_1 = 20 \times 80 = 1600$ mm^2 area of rectangle ABCH with G_1 as its centroid.

$a_2 = 80 \times 20 = 1600$ mm^2 area of rectangle DEFH with G_2 as its centroid.

2. *Treating the areas as parallel forces:*

Treat these areas a_1 and a_2 as parallel forces (vertical and horizontal) acting through the centroids of the respective rectangles.

3. *Draw force with some scale draw ab = 16 mm = area a_1 and bc = 16 mm = area a_2. Select a pole O and join oa ob and oc as shown in Fig. 5.29 (i). diagram for vertical parallel forces* [see Fig. 5.29 (*i*).

4. *Draw funicular diagram for vertical parallel forces* [see Fig. 5.29 (*ii*)].

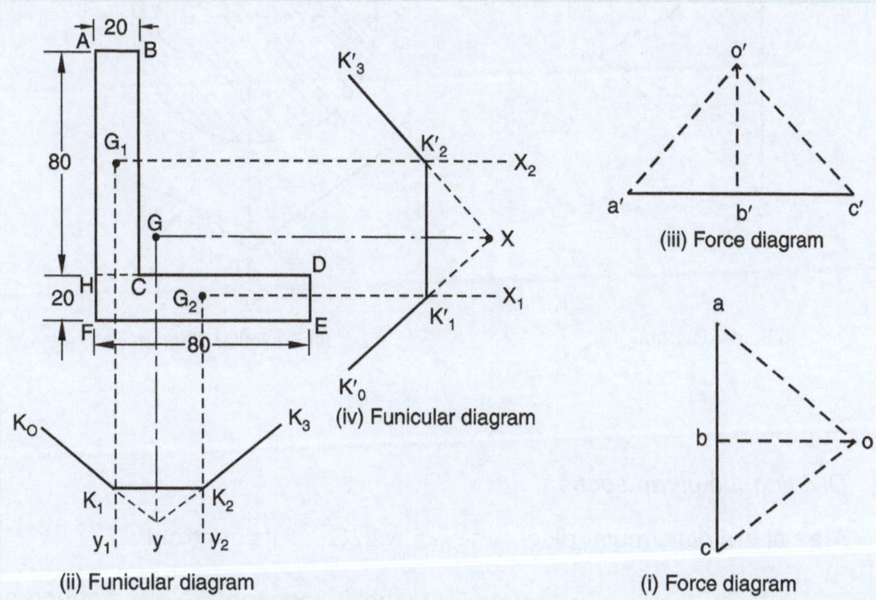

Fig. 5.29.

Draw vertical lines G_1 y_1 and G_2 y_2 through the centroids as shown in fig. 5.29. Choose a point K_1 on G_1 y_1 line and draw K_1 K_0 parallel to oa. Through K_1 draw another line K_1 K_2 parallel to ob meeting the line G_2 y_2 at K_2. Now, from K_2 draw a line K_2 K_3 parallel to oc.

Produce the lines K_3 K_2 and K_0K_1 to meet at y. Through y, draw a vertical line as shown in fig. 5.29 (ii).

5. *Draw force diagram for horizontal parallel forces* [see Fig. 5.29. (iii)].

With same scale draw a′b′ = 16 mm = area a_1 and b′c′ = 16 mm = area a_2. Select a pole o′ and Join o′a′, o′b′ and o′c′ as shown in fig. 5-28 (iii).

6. *Draw funicular diagram for horizontal parallel forces* [see fig. 5.29 (iv)].

Draw horizontal lines G_2 X_2 and G_1 X_2 through the centroids as shown in fig. 5.29 (iv). Choose a point K_1' on the G_2 X_1 line and draw K_1' K_0' parallel to o′a′ Through K_1' draw another line K_1' K_2' parallel to o′b′. meeting in line G_1 X_2, at K_2'. Now from K_2' draw a line K_3' K_2' parallel to oc′. Produce the line K_3' K_2' and K_0' K_1' to meet at x. Through x draw a horizontal line.

7. *Centroid of the angle section.*

The intersection of the vertical and horizontal lines drawn from points y and x of funicular diagrams meet at point G, which is the requried centrid of the section.

Problem 26: From a circular disc of radius 480 mm is cut out circular piece of radius 120 mm. The centre of disc removed is 240 mm from the centre of given disc. Find centre of gravity of the remainder portion graphically.

Solution: The various steps to find the C.G. of the remainder portion graphically are as under:

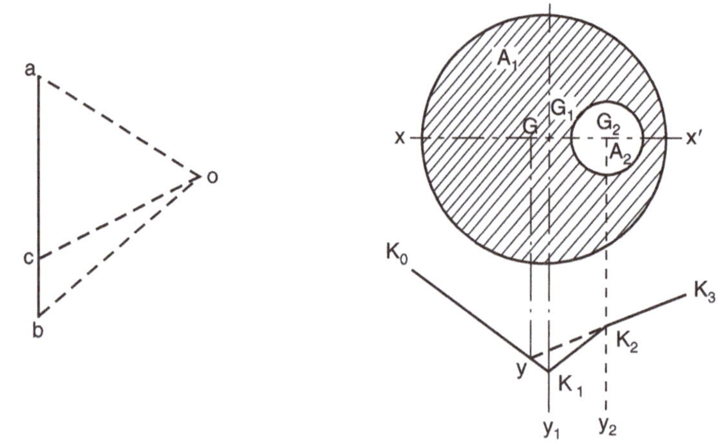

(i) Force diagram

(ii) Funicular diagram

Fig. 5.30

1. *Dividing the given section:*

Area of orginal circular disc, $A_1 = \pi r_1^2$ with G_1 as its centroid

$$= \pi \times (480)^2 = 230400\pi \text{ mm}^2 = 723822.94 \text{ mm}^2$$

Area of cut circular piece, $A_2 = \pi r_2^2$ with G_2 as its centroid

$$= \pi \times (120)^2 = 14400\pi \text{ mm}^2 = 45238.93 \text{ mm}^2$$

2. *Treating the areas as parallel forces:*

Treat the area A_1 as parallel force acting vertically downwards through centroid G_1 and area A_2 as parallel force acting through centroid G_2.

3. *Drawing force diagram* [see Fig . 5.30 (i)]:

With some suitable scale draw ab = 723822.94 mm^2 (area A_1) downwards and bc = 45238.93 mm^2 (area A_2) upwards. Choose a pole O at some suitable place and join oa, oc and ob as shown in Fig. 5.30 (ii).

4. *Drawing funicular diagram* [see Fig. 5.30 (ii)].

Draw vertical lines $G_1 y_1$ and $G_2 y_2$ through the centroids G_1 and G_2 as shown in Fig. 5.30 (i). Choose a point K_1 on the line $G_1 y_1$ line and draw a line $K_1 K_0$ parallel to oa. Through K_1 draw $K_1 K_2$ parallel to ob meeting the line $G_2 y_2$ at K_2. Now, from K_2 draw a line $K_2 K_3$ parallel to oc. Produce the line $K_2 K_3$ and $K_0 K_1$ so to meet at y.

5. *Centroid of the section:*

Through y, draw a vertical line to cut the horizontal axis line x-x′ at G. Then G is the required position of C.G. of the remainder portion.

By measuring **$G G_1$ = 16 mm [Ans]**

MODEL TEST QUESTIONS

1. Define centroid and centre of gravity.
2. What is difference between force of gravity and qravitation?
3. State the various methods of finding C.G. of a body.
4. What do you understand by axis of symmetrical?
5. Derive an expession for the centre of gravity of a plane area using method of moments.
6. Where does the centroid of the following plane figures lie?
   rectangle, semicircle, triangle, quadrant and trapezium.
7. Where does the centre of gravity of the following bodies lie?
   sphere, hemisphere, right circular cone and right circular cylinder.

OBJECTIVE TYPES QUESTIONS

1. **Fill in the blanks:**
   (*i*) The force with which a body is attracted towards the centre of earth is called.......
   (*ii*) The point at which the whole weight of a body is supposed to act is called.....
   (*iii*) The point at which the whole area of a lamina is supposed to act is called.....
   (*iv*) The centroid of a triangle is at the intersection of......
   (*v*) The centroid of a semicircle is .... from the centre on a line perpendicular to the diameter through the centre.

Ans. 1. (i) Weight, (ii) centre of gravity, (iii) centroid, (iv) medians, (v) $4r/3\pi$.

2. **Pick up the correct words:**
   (*i*) There *can/cannot* be more than one C.G. of a body.
   (*ii*) A body *has/does not have* centre of gravity if it is moved very near to the centre of the earth.
   (*iii*) The weight of a body on moon is *1/10th, 1/6th/2* times its weight on earth.
   (*iv*) C.G. of a ball of wood and steel of the same size will be in the same position *(Yes/No)*.

Ans. 2. *(i) cannot, (ii) does not have, (iii) 1/6th, (iv) yes.*

PROBLEMS FOR PRACTICE

1. Locate the position of centroid of a T-section 100 mm × 100 mm × 20 mm.
   [**Ans.** 67.8 mm]
2. Find the poisition of centroid of the channel section 100 × 50 × 15 mm
   [**Ans.** $\bar{x}$ = 17.8 mm, $\bar{y}$ = 50 mm]
3. Find the centroid of an I-section whose dimensions are:
   Top flange = 100 mm × 25 mm ; Web = 100 mm × 25 mm and bottom flange 150 mm × 25 mm [**Ans.** 75 mm, 66.1 mm]

4. Find the position of the centrold of channel section shown in fig 5.31.
[**Ans.** 26.4 mm, 127.9 mm]

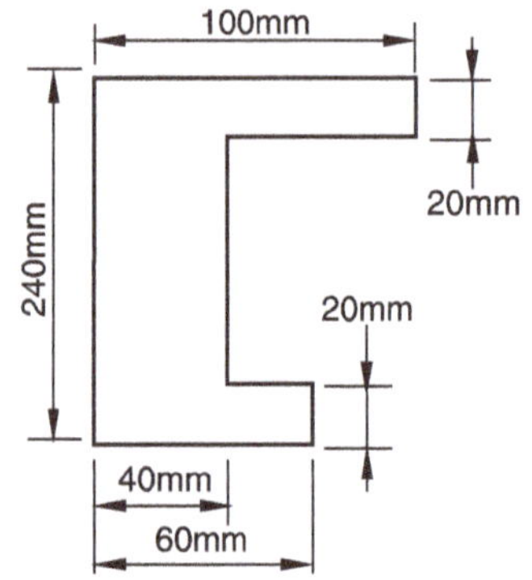

Fig. 5.31

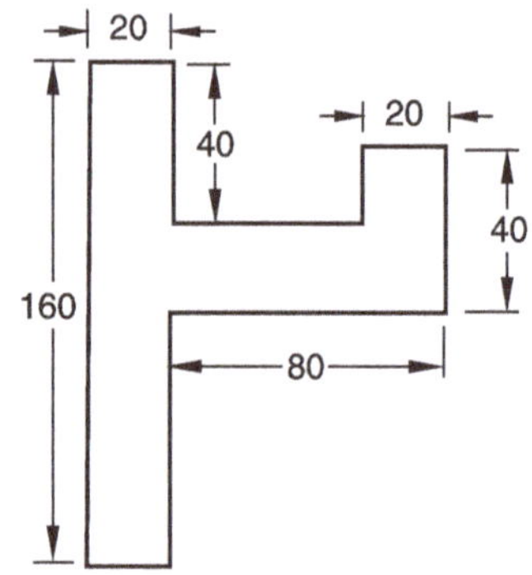

Fig. 5.32

5. Find the centroid of a section shown in fig. 5.32 with respect to two reference axes. [**Ans.** 31.5 mm, 93.1 mm]
6. Locate the centroid of the section shown in fig. 5.33 with respect to the AB axis. [**Ans.** 57.3 cm]
7. Find the position of centre of gravity of a section wall as shown in fig. 5.34

[**Ans.** $\bar{x}$ = 3.68 m, $\bar{y}$ = 2.82 m]

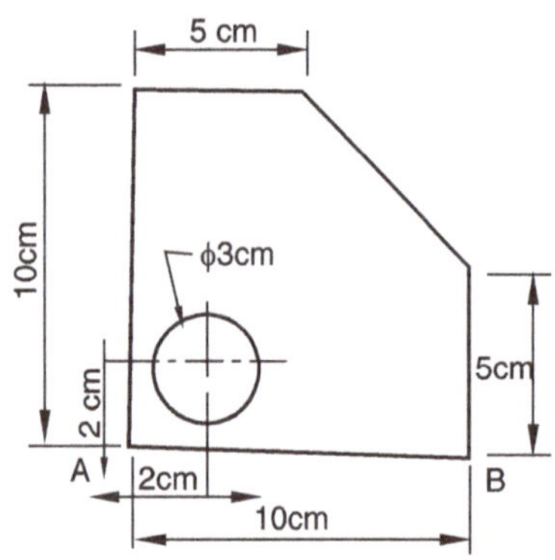

Fig. 5.33

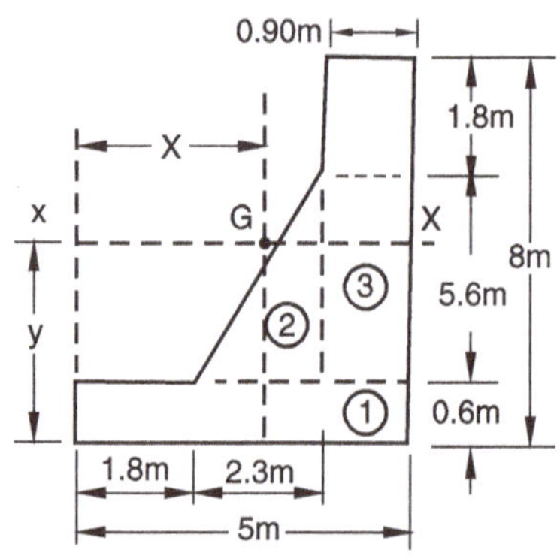

Fig. 5.34

8. From a rectangular sheet of metal ABCD, in which AB = 400 mm and BC = 600 mm, a triangular piece ABX is removed, such that AX = BX = 250, mm. Calculate the distance of C.G. of the remainder. [**Ans.** 335.7 mm]
9. Where must a hole of radius 250 mm be punched out of a circular disc of radius 1000 mm, if centre of gravity of the remainder is 50 mmm from centre of the disc. [**Ans.** 625 mm]

10. Form a rectangular lamina ABCD 100 mm × 120 mm, a rectangular hole of 30 × 40 is cut as shown in Fig. 5.35. Find the C.G. of the remainder from AB and AD.
[**Ans.** 62.2 mm. 47.2 mm]

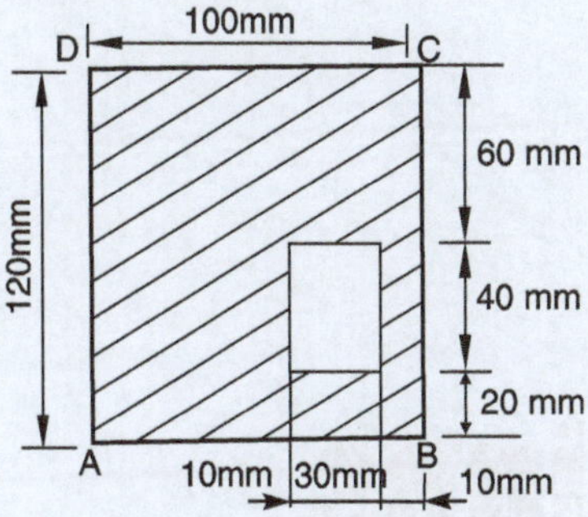

Fig. 5.34

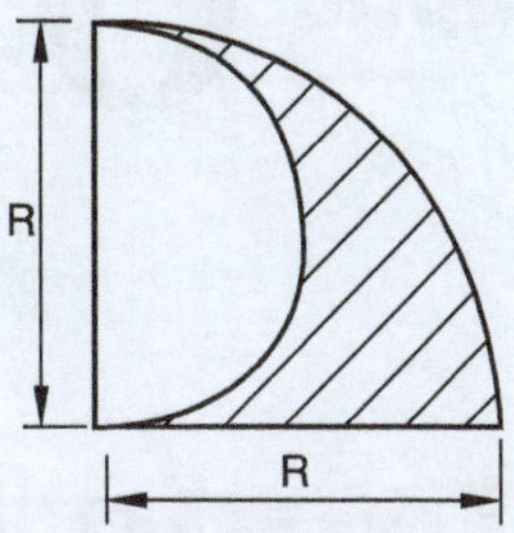

Fig. 5.35

11. Determine the centroid of the shaded area in fig. 5.36 created by cutting a semicircle of diameter R from a quarter circle of ratius R.

[**Ans.** $\overline{X}$ = 0.636 R, $\overline{Y}$ = 0.348 R]

12. A square sheet of metal has a square of one quarter of the original area cut from one corner. Calculate the position of centre of gravity of the remaining position of the sheet. [**Ans.** 5/12 of side of the original square from each uncut side]

13. A solid right circular cylinder has its base scooped out so that the hollow is a right circular cone on the same base and having the same height as the cylinder. Find the C.G. of the remainder. [**Ans.** $\overline{Y} = \frac{5}{8}h$]

14. A body consists of a solid hemisphere of radius 100 mm and a right circular solid cone of height 120 mm. The hemisphere and cone have a common base and are made of the same of the same material. Find the position of the C. G. of the the compound body. [**Ans.** 132.2 mm]

15. The frustum of a right circular cone has an altitude of 600 mm. The radii of the round faces are 600 mm and 300 mm respectively. Find the position of C.G. from larger face. [**Ans.** 235.7 mm]

Chapter 6

Simple Machines

INTRODUCTION

In olden days, various operations such as lifting, draging, etc. were performed manually. This not only involved inconvenience and lot of labour but also results in wastage of time and money. As science and technology progressed, machines were invented which have made it possible to do any operation with amazing degree of ease and convenience. In fact modern age is the machine age and even a minute without machines hangs heavy on us.

6.1. MACHINE

A device which transfers energy put into it in a more useful or convenient form is called a machine.

In other words, a machine is a device which enables us to do some useful work at some point, when an effort is applied to it at some other convenient point.

Examples: (i) Consider a car is too heavy to be lifted manually. However it can be lifted by screw jack by small force, Here, screw jack is a machine for it has helped in applying the effort in a convenient way (see Fig. 6.1).

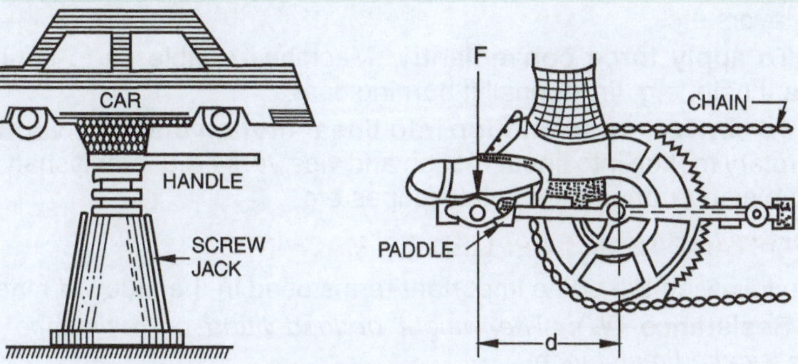

Car lifted by screw jack *Force applied on paddle of bicycle*

Fig. 6.1.

(ii) In case of a bicycle, the force is applied on the paddles by the cycilst to obtain useful work in order to gain in speed (see Fig. 7-2).

From the above discussion, it follows that a machine simply transfers work or enregy put into it in an easy and useful form.

A machine has the following three features:—

(i) The point of application of efforts.

(ii) Mechanism – the working part that makes up a machine.

(iii) The point where useful work is done or load lifted.

6.2. CLASSIFICATION OF MACHINES

The machines are broadly classified into the following two categories: —

1. Simple machines and

2. Compound machines.

Simple machines : *The machines which have one point for application of effort and one point for load to be lifted are called simple machines.*

Examples of simple machines are wheel and axle, screw jack, pulleys etc.

Compound machines : *The machines which have more than one point of application of effort and for the load are called compound machines.*

Examples for compound machines are lathe machine, printing machine, grinding machine etc.

6.3. USES OF MACHINE

Machines have enabled us to transfer work or energy in a most convenient and useful form. Some of the uses of machines are:

1. To lift heavy load by small force: Heavy loads can be lifted with a small force e.g. lifting of car with a screw jack.

2. To gain in speed: Some machines can be used to change slow motion at one point into a rapid motion at the other point. For example, bicycle is a machine in which we gain in speed.

3. To transfer energy: Machines are used to transfer energy e.g. a generator transfers mechanical energy into electrical energy.

4. To change the direction of force: Some machines can help to apply force in a convenient direction and lift heavy loads in some other direction e.g. load lifted by pulleys, levers etc.

5. To apply force conveniantly: Machines enable us to apply the force at a convenient point e.g. fire-tongs lift burning coal.

6. To convert rotary motion into linear motion and vice versa: Machines can convert rotary motion into linear motion and vice versa e.g. crank shaft in petrol engine, rack and pinion arrangement in telescopes etc.

6.4. IMPORTANT TERMS APPLIED TO MACHINES

The following are some important terms used in the study of machines:

1. Resistance (W): *The weight or load lifted or resistance overcome by a machine is called resistance.*

In certain cases a machine lifts load e.g. lifting of car by a screw jack, whereas in some cases a machine has to overcome resistance e.g. bicycle which has to overcome road friction and air resistance etc.

2. Effort (P): *The force applied on machine to overcome resistance (load or weight etc). is called effort.*

For example, force applied on the paddle of a bicycle is the effort.

3. Input : *The work done by the effort on the machine is called input.*

It is measured by the product of effort and the distance through which the effort acts i.e.

$$\text{Input} = P \times y$$

where, P = Effort applied

y = Distance through which effort acts.

4. Output: *The useful work done by machine is called output.*

It is measured by the product of weight i.e. resistance and the distance through which it is overcome i.e.

$$\text{Output} = W \times x$$

Where, W = Resistance overcome

x = Distance through which the resistance is overcome.

5. Mechanical Advantage (M.A.) : *The ratio of weight lifted (W) to the effort (P) applied by the machine is called mechanical advantage.*

$$\therefore \text{ Machanical Advantage, (M.A.)} = \frac{W}{P}$$

The mechanical advantage of a machine depends upon friction present in the machine as a part of the effort applied is used to overcome it.

Explanation: In fact, mechanical advantage represents the advantage gained mechanically (not in energy) from the machines. Thus if with a certain machine, a load of 150 N can be lifted by exerting an effort of 15 N, then mechanical advantage of machine is $\frac{150}{15} = 10$. It simply, represents that by using this machine, we have the advantage to extent of 10 times the effort.

The reader should note that advantage has been gained in weight lifted but we have equally (probably more) lost in the distance through which the weight is lifted so that output is in no case more than input. Inspite of the fact that machine has provided no energy gain, yet it has helped to transfer applied effort into a useful and convenient form. It is due to the ease and convenience with which the machines can transfer energy that mostly machines are designed to give mechanical advantage more than unity.

6. Velocity Ratio (V.R.) : *The ratio of distance (x) moved by the effort to the distance (y) moved by weight is called velocity ratio of the machine.*

$\therefore$ Velocity ratio V.R. = $\frac{y}{x}$

where, y = Distance through which effort is applied

x = Distance through which the weight is lifted.

Thus, in a machine if effort moves through a distance of 100 mm and the weight lifted (resistance overcome) through 10 mm.

Then, Velocity Ratio = $\frac{100}{10} = 10$

It should be noted that here the effort moves through a distance 10 times as large as the weight in the same time i.e. effort moves 10 times faster than weight. In other words, the ratio of their velocities is 10 and hence the name velocity ratio.

7. Efficiency (η) **:** *The ratio of the output to input of a machine is called efficiency of the machine.*

$$\text{Efficiency } (\eta) = \frac{\text{Output}}{\text{Input}}$$

For a perfect machine, output and input are equal so that $\eta = 1$. However, in practice, no machine is perfrect, as a part of input is used to overcome friction between the machine parts. Therefore, out-put from a machine is less than input i.e. $\eta < 1$.

Generally, efficiency is expressed in percentage and is called percentage efficiency i.e.

$$\%\eta = \frac{\text{Output}}{\text{Input}} \times 100$$

If a machine has an efficiency of 80%, it means that only 80% of the input to machine has been converted into useful work (output). The remaining 20% is lost in friction in the machine. While designing a machine, efforts should be made to have as good efficiency as possible.

6.5. RELATION BETWEEN M.A., V.R. AND EFFICIENCY

Let us consider a machine in which

W = Weight lifted

x = Distance through which weight lifted

P = Effort applied

y = distance through which effort applied.

$$\text{We know, } \eta = \frac{\text{Output}}{\text{Input}}$$

$$= \frac{W \times x}{P \times y}$$

$$= \frac{W}{P} \times \frac{x}{y}$$

$$\therefore \quad \eta = \frac{M.A.}{V.R.} \qquad [\because M.A. = \frac{W}{P}, V.R. = \frac{y}{x}]$$

$$\text{i.e. Efficiency} = \frac{\text{Mechanical advantage}}{\text{Velocity ratio}}$$

6.6. REVERSIBLE MACHINE

A machine capable of doing work, in the reversed direction after the effort is removed is called reversible machine.

The action of such a machine is known as reversibility of the machine.

Condition for the reversibiltiy of a machine: Let us consider a reversible machine, on which

P = Effort appied
W = Weight or Load lifted
y = Distance through which the effort is applied
x = Distance through which the weight is lifted

Then, input of the machine = $P \times y$
and output of the machine = $W \times x$
Work lost due to machine firction = Input – output
$= P \times y - W \times x$

As a matter of fact, when a machine has to work in the reversed direction, the output should be more than the machine friction. Thus

$$W \times x > (P \times y - W \times x)$$

$$\text{or} \quad 2(W \times x) > P \times y$$

$$\text{or} \quad \frac{W/p}{y/x} > \frac{1}{2}$$

$$\text{or} \quad \frac{M.A.}{V.R.} > \frac{1}{2}$$

$$\therefore \quad \eta > \frac{1}{2} \quad \text{or} \quad 50\%$$

Hence the condition for a reversible machines is that its efficiency should be more than 50%.

6.7. IRREVERSIBLE MACHINE (SELF LOCKING MACHINE)

A machine which is not capable of doing any work in the reversed direction after the effort is removed, is called irreversible machine or self locking machine.

Thus for the self locking machine, with the notations used in the previous article, the out put of the machine should be less than machine friction.

$$\text{i.e.} \quad W \times x < (P \times y - W \times x)$$

or $2(W \times x) < P \times y$

or $\frac{W/p}{y/x} < \frac{1}{2}$

or $\frac{M.A.}{V.R.} < \frac{1}{2}$

$\therefore$ $\eta < \frac{1}{2}$ or 50%

Hence the condition for an irreversible machine is hat its efficiency should be less than 50%.

6.8. IDEAL MACHINE

A machine which is absolutely free from frictional resistance is known as ideal machine.

The condition of the ideal machine is that its efficiency should be 100%.

6.9. EFFECT OF FRICTION IN MACHINES

In every machine, a part of the energy supplied (input) is wasted in overcoming frictional losses. These losses are converted into heat and cannot be recovered as a useful work. The result is that output from the machine is always less than the input.

Frictional losses may be expressed in terms of effort P or in terms of load W as discussed below:

Let P = Actual effort required to lift the load W.

W = Actual load lifted by effort P.

P_i = Ideal effort i.e. effort required when there is not machine friction.

W_i = ideal load lifted i.e. resistances overcom when there is no machine friction.

It is obvious $P > P_i$ and $W_i > W$

Loss of effort due to friction, $P_F = P - P_i$ ...(i)

Loss of load due to friction, $W_F = W_i - W$ ...(ii)

$\therefore$ We know, $\eta = \frac{M.A.}{V.R.}$

or M.A. = $\eta \times$ V.R

For an ideal machine $\eta = 100\%$ or and effort required to lift load W is equal to ideal effort P_i.

$$\therefore \quad \frac{W}{P_i} = 1 \times V.R.$$

$$\text{or} \quad P_i = \frac{W}{V.R.}$$

Putting the value of P_i in equation (i), we get

$$\therefore \quad \mathbf{P_F = P - \frac{W}{V.R.}}$$

To find the value of W_i again consider an ideal machine.

We know $\eta = \frac{M.A.}{V.R.}$

For ideal machine, $\eta = 100\%$ and ideal load W_i lifted by effort P

$$\therefore \quad \frac{W_i}{P} = 1 \times V.R.$$

$$\text{or} \quad W_i = P \times V.R.$$

Putting the value of W_i in equation (ii), we get,

$$\therefore \quad \mathbf{W_F = P \times V.R. - W}$$

Now, from equtions (iii), $P_F = P - \frac{W}{V.R.}$

$$\text{or} \quad P_F \times V.R. = P \times V.R. - W$$

$$\text{or} \quad P_F \times V.R. = W_F \qquad [\because W_F = P \times V.R. - W]$$

$$\text{or} \quad \mathbf{W_F = P_F \times V.R.}$$

When, for a machine, P, P_i, W and W_i are known,

$$\text{then} \quad \eta = \frac{P_i}{P} = \frac{W}{W_i}$$

6.10. LAW OF MACHINE

The relation between load lifted and effort applied is called the law of machine.

Generally, effort is plotted on y-axis and load on x-axis. To establish this relation, efforts are applied on a machine and corresponding values of loads lifted are noted. Then a graph is drawn between the efforts applied and loads lifted as shown in fig. 7-3. It will be seen that the graph is a straight line AB having an intercept OA. This intercept OA represents the amount of friction offered by the machine i.e. OA is the effort required by the machine to overcome friction before it can lift any load.

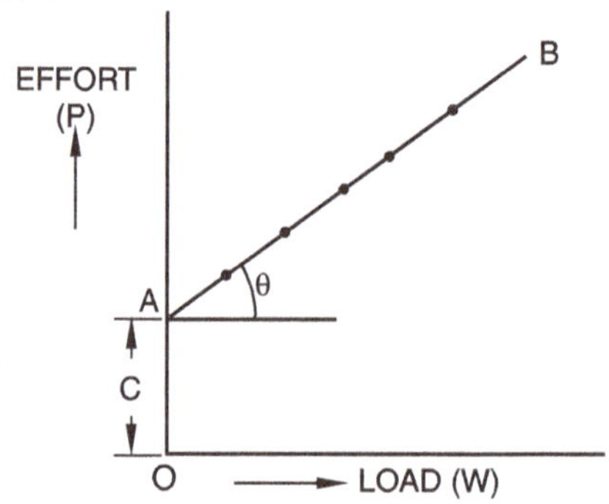

Fig. 6.3. *Concept of law of machine*

The equation of the straight line can be expressed as,

$$P = mW + C \qquad \text{[comparing with equation } y = mx + c\text{]}$$

where, P = Effort applied

W = Load lifted

$$m = \text{Solpe of the line} = \tan\theta = \frac{P - C}{W}$$

C = Constant which represents the machine friction

The equation P = mW + C is known as the law of machine.

Determination of m and C: The constants m and C can be determine as under:

Consider any two points A and B on the graph as shown in fig. 7-4. Let point A corresponds to effort P_1 and load W_1 and B corresponds to efffort P_2 and load W_2.

$$\therefore \quad P_1 = mW_1 + C \qquad \text{...(i)}$$

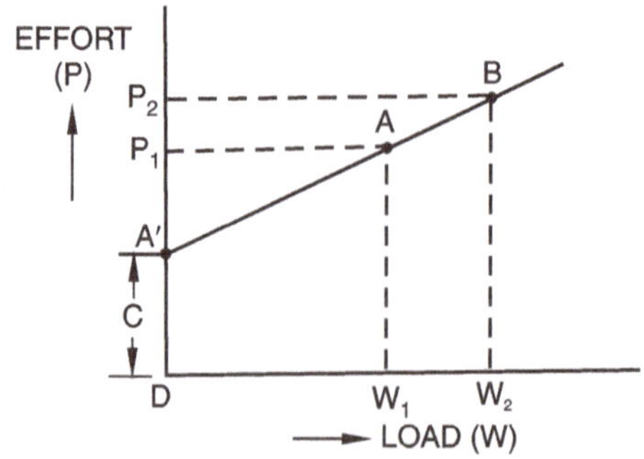

Fig. 6.4

$$P_2 = mW_2 + C \quad ...(ii)$$

Subtracting equation (i) from (ii),

$$P_1 - P_2 = m(W_1 - W_2)$$

$$\therefore \quad m = \frac{P_1 - P_2}{W_1 - W_2}$$

Substituting the value of m in equation (i),

$$P_1 = \frac{P_1 - P_2}{W_1 - W_2} \times W_1 + C$$

$$\therefore \quad C = P_1 - \frac{P_1 - P_2}{W_1 - W_2} \times W_1$$

6.11. MAX. MECHANICAL ADVANTAGE AND MAX. EFFICIENCY

We know that,

$$\text{Mechanical advantage} = \frac{W}{P}$$

$$\therefore \text{Mechanical advantange} = \frac{W}{mW + C} \quad [\because P = mW + C]$$

$$= \frac{1}{m + \frac{C}{W}}$$

The smaller the denominator, greater will be the mechanical advantage. When W is very high the ratio $\frac{C}{W}$ will be very small and can be neglected. Under such conditions machanical advantage will be maximum.

$\therefore$ **Maximum Mechanical advantage =** $\mathbf{\frac{1}{m}}$

For finding the maximum efficiency, we may proceed as:

We know that,

$$\eta = \frac{M.A.}{V.R.} = \frac{W/P}{V.R.} = \frac{W}{P \times V.R.}$$

$$= \frac{W}{(mW + C)} \times \frac{1}{V.R.} \quad [\because P = nW + C]$$

$$= \frac{1}{m + \frac{C}{W}} \times \frac{1}{V.R.}$$

The η will be maximum, if $m + \frac{C}{W}$ and V.R. are minimum. But V.R. for a machine is constant. Thus for

$m + \frac{C}{W}$ to be minimum $\frac{C}{W}$ should be zero.

$$\therefore \quad \eta_{max} = \frac{1}{m} \times \frac{1}{V.R.} \qquad [\because \text{max. M.A.} = \frac{1}{m}]$$

$$\eta_{max} = \frac{\textbf{Max. Mechanical advantage}}{\textbf{V.R.}}$$

SELECTED PROBLEMS

Problem 1: In a machine, effort was found to move a distance of 20 m when the load moved a distance of 1/2 m. With this, a load of 2240 N was raised by application of an effort of 80 N determine :—

(i) Velocity ratio;
(ii) Machanical advantage;
(iii) Efficiency;
(iv) probable effort, has the machine been perfect;
(v) Effort lost in friction;
(vi) Probable load rasied, has the machine been perfect;
(vii) Friction of the machine.

Solution: Load lifted, W = 2240 N

Effort applied, P = 80 N

Distance moved by effort, y = 20 m

Distance moved by load, x = 1/2 m

(i) $V.R. = \frac{y}{x} = \frac{20}{1/2} = \mathbf{40}$ **[Ans]**

(ii) $M.A. = \frac{W}{P} = \frac{2240}{80} = \mathbf{28}$ **[Ans]**

(iii) $\eta = \frac{M.A.}{V.R.} \times 100 = \frac{28}{40} \times 100 = \mathbf{70\%}$ **[Ans]**

(iv) Probable effort, $P_i = \frac{W}{V.R.} = \frac{2240}{40} = \mathbf{56N}$ **[Ans]**

(v) Effort lost in friction, $P_F = P - P_i = 80 - 56 =$ **24 N [Ans]**

(vi) Probable load raised,

$W_i = P \times V.R. = 80 \times 40 =$ **3200 N [Ans]**

(vii) Friction of the machine,

$W_F = W_i - W = 3200 - 2240 =$ **960 N [Ans]**

Problem 2: In a lifting machine, the velocity ratio is 40 and it was found that a load of 800 N could be lifted with an effort of 50 N. Find efficiency of the machine and state whether the machine will reverse if the effort is removed when the load is on the machine. How much of the effort of 50 N go to overcome machine friction?

Solution: Weight lifted, W = 800 N

Effort applied, P = 50 N

Velocity ratio, V.R. = 40

$$\text{Mechanical advantage} = \frac{W}{P} = \frac{800}{50} = 16$$

$$\text{Efficiency, } \eta = \frac{\text{M.A.}}{\text{V.R.}} = \frac{16}{40} \times 100 = \mathbf{40\%} \ \textbf{[Ans]}$$

Since the efficiency is less than 50%, therefore, the machine will not reverse when the effort is removed.

Effort lost in friction, $P_F = P - Pi$

$$= P - \frac{W}{\text{V.R.}} = 50 - \frac{800}{40} = 50 - 20 = \mathbf{30N} \ \textbf{[Ans]}$$

Problem 3: For a certain machine, the velocity ratio is 160 and law of machine is $P = 1/60\,W + 20N$. Find the maximum mechanical advantage and maximum percentage efficiency of the machine. State, wether the machine is reverseible or otherwise.

Solution: The given law of machine is, $P = \frac{1}{60}W + 20$

Now comparing it with $P = mW + C$, we get

$$m = \frac{1}{60}, \text{ and } C = 20$$

$$\text{Then maximum M.A.} = \frac{1}{m} = \frac{1}{1/60} = \mathbf{60} \ \textbf{[Ans]}$$

$$\text{Maximum } \eta = \frac{\text{Maximum M.A.}}{\text{V.R.}} \times 100$$

$$= \frac{60}{100} \times 100 = \mathbf{60\%} \ \textbf{[Ans]}$$

The machine is **reversible** as the efficiency is more than 50%.

Problem 4: The intial friction resistance of a unloaded pulley block machine is 1.5 N and the friction increases @ 1.25 N per 100 N lifted by the block. If the velocity ratio is 15, find the effort required to lift a load of 800 N.

Solution: Weight lifted, W = 800 N

Velocity ratio, V.R. = 15

Now we know that effort lost in friction (P_F) or

Total force of friction P_F = Initial friction + Working friction

$$= 15 + (1.25 + \frac{800}{100}) = 11.5\text{N}$$

$$\text{Then ideal effort, } P_i = \frac{W}{\text{V.R.}} = \frac{800}{15} = 53.33 \text{ N.}$$

Effort lost in friction (P_F) = Actual effort (P) – Ideal effort (P_i)

or $11.5 = P \times 53.33$

$\therefore$ $P = \mathbf{64.83\ N}$ **[Ans]**

Problem 5: The velocity ratio of a machine is 10. An effort of 25 N is applied and it is found that 20% of the effort is lost in overcoming the friction. Find the load lifted. Also find the efficiency at this load.

Solution: Velocity ratio V.R. = 10, Effort P = 25 N

Percentage loss of effort = 20%

$\therefore$ Effort lost in firction, P_F = P × Percentage loss of effort

$$= 25 \times \frac{20}{100} = 5\text{N}$$

Useful effort P_i = Actual effort P – Ideal effort P_F

= 25 – 5 = 20 N

Load lifted, W = Useful effort × V. R. = 20 × 10 $[\because P_1 = \frac{W}{V.R.}]$

$\therefore$ W = **200 N [Ans]**

$$\text{Also M.A.} = \frac{W}{P} = \frac{200}{25} = 8$$

$$\therefore \eta = \frac{M.A.}{V.R.} = \frac{8}{10} \times 100 = \mathbf{80\%} \ \textbf{[Ans]}$$

Problem 6: In a lifting machine, an effort of 100 N lifts a load of 2500 N and an effort of 130 N lifts a load of 4000 N. Establish the law of the machine and calculate the effort required to lift a load of 6000 N. What is the maximum efficiency ? Assume the velocity ratio of the machine is 75.

Solution: We known that the law of machine is P = mW + C

When W = 2500 N, P = 100 N

and when W = 4000 N, P = 130 N

We know that law of a machine is P = mW + C

$\therefore$ 100 = m × 2500 + C ...(i)

and 130 = m × 4000 + C ...(ii)

Now, subtracting equation (i) from (ii), we get

30 = m × 1500

$$\therefore m = \frac{1}{50}$$

and substituting the value of m in equation (i), we get

$$100 = \frac{1}{50} \times 2500 + C$$

or C = 100 – 50 = 50

$\therefore$ The law of the machine $\mathbf{P = \frac{1}{50} W + 50}$ **[Ans]**

Now to find the effort required to lift a load of 6000 N, we have

$m = \frac{1}{50}$, C = 50 and W = 6000 N

From the law of machine,

$$P = mW + C = \frac{1}{50} \times 6000 + 50 = \mathbf{170\ N}\ \textbf{[Ans]}$$

and maximum $\eta = \frac{1}{m} \times \frac{100}{V.R.} = \frac{1}{(1/50)} \times \frac{100}{75}$

$$= 50 \times \frac{100}{5} = \mathbf{66.7\%}\ \textbf{[Ans]}$$

Problem 7: The velocity ratio of a machine is 15 and the effieiency is 60%. Determine the efffort required to lift a load of 1800 N. What is the law of the machine if the frictional resistance of the machine is constant? Determine the effort required to run this machine at (i) no load and (ii) a load of 1000 N.

Solution: Veloctiy ratio, V.R. = 15

Effeclency, η = 60%

Load lifted, W = 1800 N

Using the relation $\eta = \frac{M.A.}{V.R.} = \frac{W}{P \times V.R.}$

$$\therefore \quad P = \frac{W}{\eta \times V.R.}$$

$$= \frac{1800}{0.60 \times 15} = \mathbf{200N}\ \textbf{[Ans]}$$

Effort lost in ffriction $P_F = P - P_i$ $\quad [\because P_i = \frac{W}{V.R.}]$

$$= P - \frac{W}{V.R.}$$

$$= 200 - \frac{1800}{15} = 80\ N$$

In the law of machine, C is a constant which is equal to the effort required to overcome the friction of machine i.e. P_F

$$\therefore C = P_F = 80\ N$$

$\therefore$ The law of machine is $P = mW + C$

i.e. $P = mW + 80$ ...(i)

By putting the values of W = 1800 N and P = 200 N, we get

$$200 = m \times 1800 + 80$$

$$\therefore \quad m = \frac{120}{1800} = 0.067$$

By substituting the value of m in equation (i), the law of the machine becomes as

$$\mathbf{P = 0.067\ W + 80}\ \textbf{[Ans]}$$

(i) Effort required to run the machine at no load i.e. W = 0

$$P = 0.067 \times 0 + 80$$

$$= \mathbf{80\ N}\ \textbf{[Ans]}$$

(ii) Effort required to run the machine at a load of W = 1000 N

$$\therefore \quad P = 0.067\ W + 80$$

$$= 0.067 \times 1000 + 80$$

$$= \mathbf{147\ N}\ \textbf{[Ans]}$$

6.12. LIFTING MACHINES

Those machines which are used for lifting loads are called lifting machines.

The force is applied at one point of the machine and load is lifted at the other point of the machine.

Some important lifting machines are given below:

1. Levers (Already discussed in the chapter of Moments)
2. Inclined plane
3. Simple wheel and axle
4. Wheel and differential axle
5. Worm and worm wheel
6. Pulleys (Systems of pulleys)
7. Weston differential pulley block
8. Screw jacks
9. Winch crabs.

6.13. INCLINED PLANE

Inclined plane is a simple machine used extensively in everyday life. A wooden plank, placed at the back side of a truck, in an inclined position, for loading oil drums by rolling on it is a familiar example.

Fig. 6.5. shows an inclined plane at an angle θ to the horizontal. Let a small effort P slides a load W on the surface of the inclined plane. If the effort P moves a vertical distance y, the load W also moves a distance y along the surface but only a distance x vertically as shown in the figure.

∴ Mechanical advantage

$$\text{M.A.} = W / P$$

Velocity ratio, $\text{V.R.} = Y / X$

$$\therefore \quad \text{V.R.} = \frac{y}{x} = \frac{y}{y\sin\theta}$$

[∵ in Δ ABC, $x = y \sin\theta$]

$$\therefore \quad \text{V.R.} = \frac{1}{\sin\theta}$$

$$\text{Efficiency,} \quad \eta = \frac{\text{M.A.}}{\text{V.R.}} = \frac{W/P}{1/\sin\theta} = \frac{W\sin\theta}{P}$$

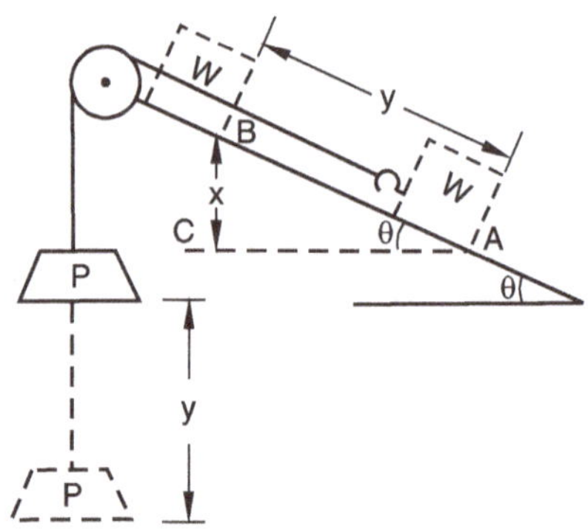

Fig. 6.5. *Inclined plane*

Problem 8: The length of an inclined plane is 5 m and its height is 2.5 m. If a force 400 N pulls a load of 620 N up the inclined plane, find (i) Machanical adventage, (ii) Velocity ratio and (iii) Efficiency of the machine.

Solution: $\sin\theta = x/y = 2.5/5 = 0.5$

Force, P = 400 N

Load, W = 620 N

(i) Machanical adventage (M.A) $= \frac{W}{P} = \frac{620}{400} = \mathbf{1.55}$ **[Ans]**

(ii) Velocity ratio, (V.R.) $= \frac{1}{\sin\theta} = \frac{1}{1.5} = \mathbf{2}$ **[Ans]**

(iii) Efficiency of machine, $(\eta) = \dfrac{\text{M.A.}}{\text{V.R.}} \times 100$

$$= \frac{1.55}{2} \times 100 = \mathbf{77.5\%\ [Ans]}$$

6.14. SIMPLE WHEEL AND AXLE

This machine consists of a simple wheel and axle fixed to the same shaft as shown in Fig. 6.6. The shaft is supported on the ball bearings in order to reduce the frictional resistance to a minimum. A rope is wound round the axle, carrying a load W at its free end and the other end is fixed to the axle. A second rope is wound round the wheel in opposite direction to that of the axle. One end of second rope is fixed to the wheel, while the other end is free and effort is applied to this end.

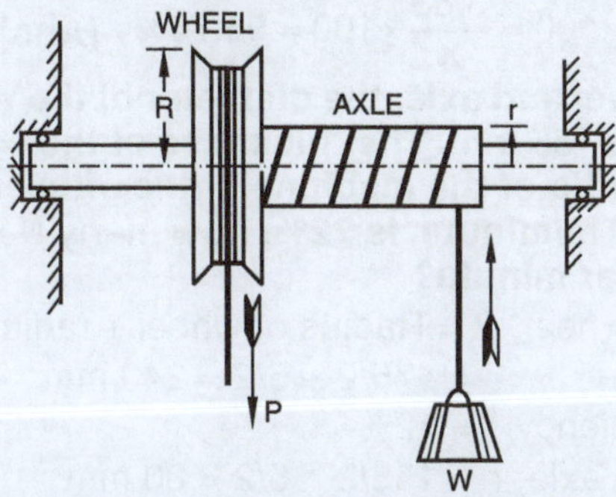

Fig. 6.6. *Simple wheel and axle*

Since, the two ropes are wound in opposite direction, therefore, a downward motion of effort P will raise the load W.

Let W = Load lifted

P = Effort applied to lift the load

R = Radius of effort wheel

r = Radius of the axle.

As the wheel and axle are fixed to the same shaft and when one revolution is given to the wheel, the axle will also make one revolution.

$\therefore$ Distance moved by effort (P) in one revvolution = $2\pi R$ and

Distance moved by load (W) in one revolutin = $2\pi r$

$$\therefore \quad \text{Velocity ratio} = \frac{\text{Distance moved by effort}}{\text{Distance moved by load}}$$

$$\text{V.R.} = \frac{2\pi R}{2\pi r} = \frac{R}{r}$$

Mechanical advantage,

$$\text{M.A.} = \frac{\text{Load lifted}}{\text{Effort applied}} = \frac{W}{P}$$

$$\text{Efficiency, } \eta = \frac{\text{M.A.}}{\text{V.A.}}$$

Problem 9: A drum weighing 60 N and holding 400 N water is to be raised from a well by means of a wheel and axle. The axle is of 100 mm diameter and the wheel is of 400 mm diameter. If the force of 120 N had to be applied to the wheel, find (i) M.A. ; (ii) V.R. and (iii) % η of the machine.

Solution: Radius of wheel, R = 400/2 = 200 mm

Radius of axle, r = 100/2 = 50 mm

Total weight lifted, W = Wt. of drum + Wt. of water

= 60 + 400 = 460 N

Force applied, P = 120 N

$$\therefore \text{ Machanical advantage, M.A.} = \frac{W}{P} = \frac{460}{120} = \mathbf{3.83} \ \textbf{[Ans]}$$

$$\text{Velocity ratio, V.R.} = \frac{R}{r} = \frac{200}{50} = \mathbf{4} \ \textbf{[Ans]}$$

$$\text{Efficiency, } \eta = \frac{\text{M.A.}}{\text{V.R.}} \times 100 = \frac{3.83}{4} \times 100 = \mathbf{95.75\%} \ \textbf{[Ans]}$$

Problem 10: In a wheel and axle, the diameter of the wheel is 475 mm and the diameter of the drum axle 155 mm. The thickness of the cord on the wheel axle is 5 mm. Find the velocity ratio of the machine. If the efficiency, when lifting a load of 1800 N at the rate of 20 m/minute, is 72%, how many N metres of work must be applied to the machine per minute?

Solution: Radius of wheel, R = Radius of wheel + radius off cord

= 475/2 + 5/2 = 240 mm

Efficiency = 72%

Radius of the axle, r = 155/2 + 5/2 = 80 mm

Velocity ratio, V.R. = R/r = 240/80 = **3 [Ans]**

The work done by the machine/minute in lifting the load is the output of the machine = 1800 × 20 = 36000 Nm

$$\text{Also efficency, } \eta = \frac{\text{output}}{\text{input}}$$

$$\text{or} \quad \frac{72}{100} = \frac{36000}{\text{input}}$$

$$\therefore \text{ Input of work applied/minute} = 36000 \times \frac{100}{72} = \mathbf{50000 \ Nm} \ \textbf{[Ans]}$$

6.15. WHEEL AND DIFFERENTIAL AXLE

This machine is an improvement over the simple wheel and axle. In this, the differential axle is made up of two parts of different diameters. The wheel and differential axle are fixed to the same shaft which is supported on ball bearings as shown in Fig. 6.7. One rope is wound round the wheel and at the free end of which, effort P is applied. A second rope is wound round the smaller axle and then after passing round the pulley it is wound round the larger axle but in the opposite direction. Thus the rope is so arranged that when it is wrapped round the larger axle, it is unwarped round the smaller axle. The load lifted is attached with the pulley as shown,

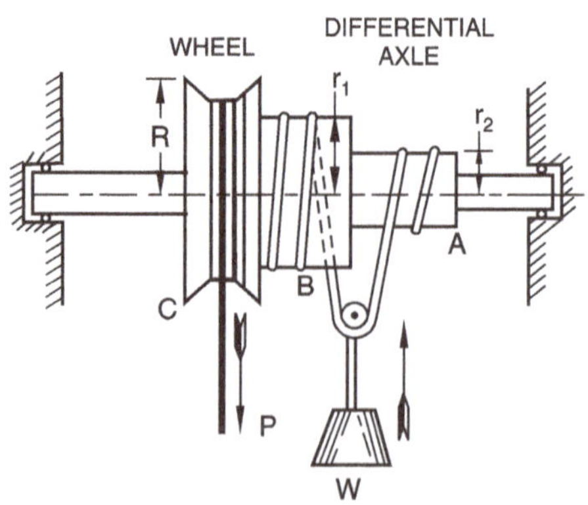

Fig. 6.7. *Wheel and differential axle*

Let W = Load lifted
P = Effort applied to lift the load
R = Radius of the effort wheel C
r_1 = Radius of larger axle B.
r_2 = Radius of smaller axle A.

When one revolution is given to the wheel, the differential axle also moves through one revolution.

$\therefore$ Distance moved by effort = $2\pi R$

Length of rope wrapped round the large axle = $2\pi r_1$

Length of the rope unwarped round the smaller axle = $2\pi r_2$

$\therefore$ Net shortening in lenght = $2\pi r_1 - 2\pi r_2 = 2\pi(r_1 - r_2)$

But the distance moved by load W is half, as the rope is passing over the pulley.

$$\therefore \text{ Distance moved by load, W} = \frac{2\pi(r_1 - r_2)}{2} = \pi(r_1 - r_2)$$

$$\therefore \text{ Velocity ratio, V.R.} = \frac{\text{Distance moved by effort}}{\text{Distance moved by load}}$$

$$= \frac{2\pi R}{\pi(r_1 - r_2)} = \frac{2R}{r_1 - r_2}$$

If diameter of the rope = d then

$$\text{V.R.} = \frac{2(R + d/2)}{(r_1 + d/2) - (r_2 + d/2)}$$

Mechanical advantage, M.A. = $\frac{W}{P}$ and Efficiency, $\eta = \frac{\text{M.A.}}{\text{V.R.}}$

Problem 11: In a differential wheel and axle, the diameter of the wheel is 400 mm and diameters of the axles are 20 mm and 160 mm. If an effort of 100 netwon can lift a load of 1500 netwon, find the efficiency of the machine. Also find the effort lost in friction.

Solution: Radius of wheel, R = 400/2 mm = 200 mm
Radius of larger axle, r_1 = 200/2 mm = 100 mm
Radius of smaller axle, r_2 = 160/2 mm = 80 mm
Effort applied, P = 100 N and Load lifted, W = 1500 N

$$\text{Velocity ratio, V.R.} = \frac{2R}{r_1 - r_2} = \frac{2 \times 200}{100 - 80} = 20$$

$$\text{Mechanical adventage, M.A.} = \frac{W}{P} = \frac{1500}{100} = 15$$

$$\text{Efficiency, } \eta = \frac{\text{M.A.}}{\text{V.R.}} \times 100 = \frac{15}{20} \times 100 = \mathbf{75\%} \textbf{ [Ans]}$$

$$\text{Effort lost in friction, } P_F = P - P_i = P - \frac{W}{\text{V.R.}}$$

$$= 100 - \frac{1500}{20} = 100 - 75$$

$$= \mathbf{25\ N\ [Ans]}$$

6.16. WORM AND WORM WHEEL

It consists of a toothed wheel known as worm wheel and a square threaded screw known a worm geared with each other. A load drum is securely attached to the worm wheel. The worm is rotated by using a hanlde as shown in Fig. 6.8. The worm may be a single start threaded or multi-start threaded.

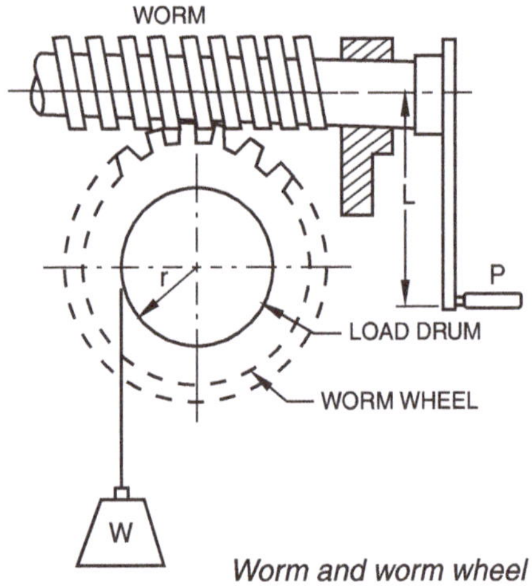

Worm and worm wheel

Fig. 6.8

Let W = Load lifted

P = Effort applied

L = Lenght of the handle

r = Radius of the load drum

T = No. of teeth on the worm wheel

If the worm is single threaded, then for one revolution of the handle the distance moved by effort, P = $2\pi L$

The load drum will move through = $\frac{1}{T}$ revolution

$\therefore$ Distance moved by the load = $\frac{2\pi r}{T}$

$$\therefore \text{ Velocity ratio, V.R.} = \frac{\text{Distance moved by effort}}{\text{Distance moved by load}}$$

$$= \frac{2\pi L}{2\pi r / T} = \frac{LT}{r}$$

Mechanical advatage, M.A. = $\frac{W}{P}$ and Efficiency, $\eta = \frac{\text{M.A.}}{\text{V.R.}}$

Problem 12: **In a single threaded worm and worm wheel, the worm wheel has 100 teeth. The crank handle is 600 mm long and the load drum has a radius of 400 mm. Calculate the velocity ratio of the machine. If it has an dfficiency of 80%, calculate the effort required to lift a load of 8000 N.**

Solution: Load lifted, W = 8000 N

Lenght of handle, L = 600 mm

Radius of load drum, r = 400 mm

No. of teeth, T = 100 and Efficiency, η = 80%

$$\text{Velocity ratio V.R.} = \frac{L \times T}{r} = \frac{600 \times 100}{400} = \mathbf{150}\ \textbf{[Ans]}$$

$$\text{Mechanical advantage, M.A.} = \frac{W}{P} = \frac{8000}{P}$$

$$\text{Also \% efficiency, } \eta = \frac{\text{M.A.}}{\text{V.R.}} \times 100$$

$$\text{or} \quad \frac{80}{100} = \frac{80000}{P \times 150}$$

$$\therefore \quad P = \frac{8000 \times 100}{80 \times 150} = \mathbf{66.67\ N}\ \textbf{[Ans]}$$

6.17. PULLEYS

A machine changing the direction of the tension of a string is called a pulley.

It essentially consists of a small wheel of wood or metal with a groove cut along the circumference so as to received a rope or chain. It rotates freely about an axle passing through its hole which is held by a frame called the block (see Fig. 6.9.)

The word pulley is often used to indicate the pulley and the block as a whole. A pulley may be movable or fixed according as the supporting block is movable or fixed.

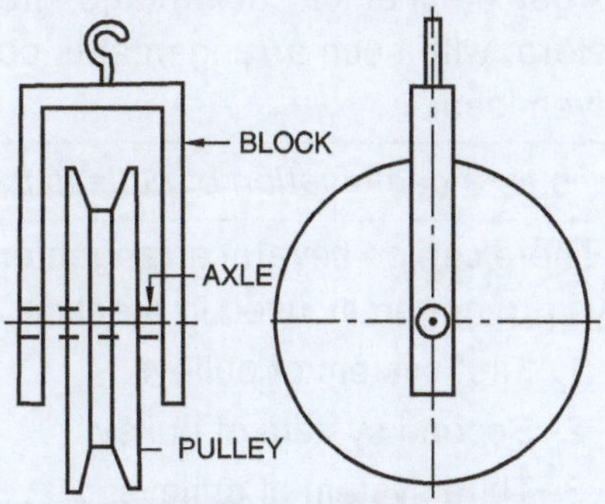

Fig. 6.9. *Pulley*

Single fixed pulley: In this case the load W is attached to one end of the string and effort P is applied at the other end. The two portions of the string may either be parallel or inclined to each other as shown in Fig. 6.9 (i) and (ii).

In each case the weight W and effort P are equal.

$$\therefore \text{ Mechanical advantage} = \frac{W}{P} = 1 \qquad [\because W = P]$$

Thus, in single fixed pulley same effort is required to lift the weight. But the only advantage is that it enables us to apply the effort in a convenient direction.

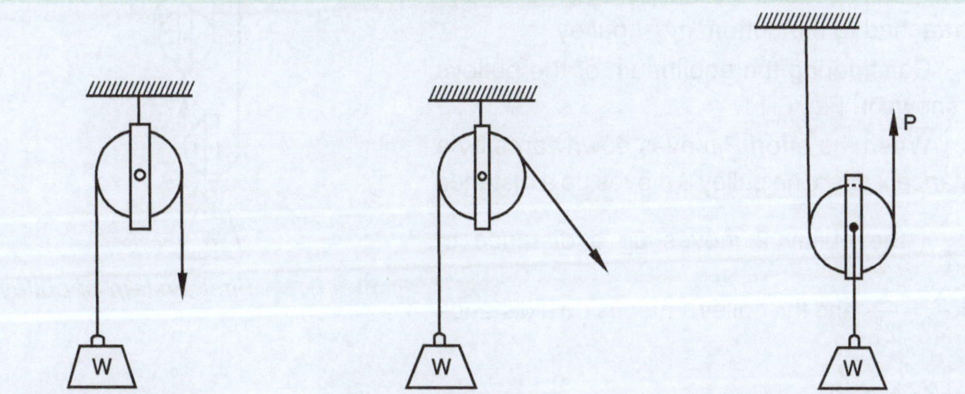

(I) Two ends of string parallel to each other

(II) Two ends of string inclined to each other

(III) Single movable pulley

Fig. 6.10. *Types of pulleys*

Single movable pulley: In this pulley, the load W is suspended from the pulley block and one end of the string passing round the pulley is attached to a fixed support and effort P is applied upward at the other end [see Fig. 6.9. (iii)].

Thus, the weight W is supported by the resultant of the tensions in the two portions of the string, so that if they are parallel, then

$$W = 2\,P$$

$$\therefore \text{ Mechanical advantage, M.A.} = \frac{W}{P} = \frac{2P}{P} = 2$$

6.18. SYSTEM OF PULLEYS

In engineering practice, two or more pulleys may be arranged in different ways so as to get mechanical advantage greater than that obtained with a single pulley. Therefore, with such arrangements comparatively less effort will be required to raise the given load.

> *Thus, a combination of pulleys is called a system of pulleys.*

There can be several arrangements of pulleys, but the following three systems of pulleys are generally used in practice:

1. First system of pulleys
2. Second system of pulley
3. Third system of pulleys.

6.19. FIRST SYSTEM OF PULLEYS

In this system, the pulleys are arranged in such a way that there are as many strings as the no. of pulleys. One end of each string is tied to the rigid support, while the other end passing round the bottom periphery of the pulley. The last topmost pulley which is fixed, is used to change the direction of load.

The effort P is applied at free end of the string passing round the last pulley. The load W is attached to the bottom most pulley.

Considering the equilbrium of the pulleys as shown in Fig. 6.11.

When the effort P moves downwards by a distance x, then the pulley 3 moves up a distance of $\frac{x}{2}$, the pulley 2 moves up a distance of $\frac{1}{2}\times\frac{x}{2}=\frac{x}{2^2}$ and the pulley 1 moves up a distance of $\frac{1}{2}\times\frac{x}{2^2}=\frac{x}{2^3}$.

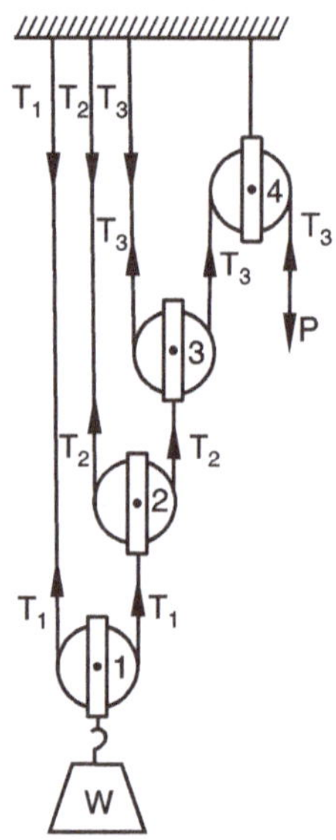

Fig. 6.11. *First system of pulley*

Therefore, the load W attached to the pulley 1 moves a distance = $\frac{x}{2^3}$

If there are n number of movable pulleys, then

Distance moved by the load = $\frac{x}{2^n}$

$$\therefore \text{ Velocity ratio, V.R.} = \frac{\text{Distance moved by effort}}{\text{Distance moved by load}}$$

$$= \frac{x}{x/2^n}$$

Mechanical advantage, M.A. = $\frac{W}{P}$

And Efficiency $\eta = \frac{M.A}{V.R.}$

When weight of the pulley is also taken into account:

Let 'w' be the weight of each pulley. Now considering the equilibrium of pulley 1, pulley 2, pulley 3, we get

$$2T_1 = (W + w); \qquad \text{or} \quad T_1 = \frac{W + w}{2}$$

Now $$2T_2 = (T_1 + W)$$

or $$T_2 = \frac{W}{2^2} + \frac{W}{2^2} + \frac{w}{2} \qquad \left[\because T_1 = \frac{W + w}{2}\right]$$

Also $$2T_3 = T_2 + w$$

or $$T_3 = \frac{W}{2^3} + \frac{w}{2^3} + \frac{w}{2^2} + \frac{w}{2} = P \qquad \left[\because T_2 = \frac{W}{2^2} + \frac{w}{2^2} + \frac{w}{2}\right]$$

Following the same analogy, if there are n pulleys,

$$T_n = P = \frac{W}{2^n} + \frac{w}{2^n} + \frac{w}{2^{n-1}} + \frac{w}{2^{n-2}} + ... + \frac{w}{2}$$

Multiplyingg both sides by 2^n, we get

$$2^nP = W + w + 2w + 2^2w + ... + 2^{n-1}w$$
$$= W + w(1 + 2 + 2^2 + ... + 2^{n-1})$$
$$= W + w(2^n - 1)$$

But Mechanical Advantage, M.A. = $\frac{W}{P}$

[Multiplying both numerator and denominator by 2^n]

or $$M.A. = \frac{2^nW}{2^nP}$$

∴ $$M.A. = \frac{2^nW}{W + w(2^{n-1})} \qquad [\because 2^nP = W + w(2^{n-1})]$$

6.20. SECOND SYSTEM OF PULLEYS

This system of pulleys consists of two blocks, one upper and the other lower. The upper block is fixed while the lower block is movable. The upper and lower blocks either carry equal number of pulleys or the upper block may have one pulley more than pullieys at the lower one as shown in Fig. 6.12. There is only one string which passes round all the pulleys. If there are three pulleys in upper block and two pulleys in lower block and after passing round all the pulleys, effort is applies on the other end (see Fig. 6.12).

Since, only one string passes round all the pulleys, therefore, the tension in the string will be equal to the magnitude of the effort applied.

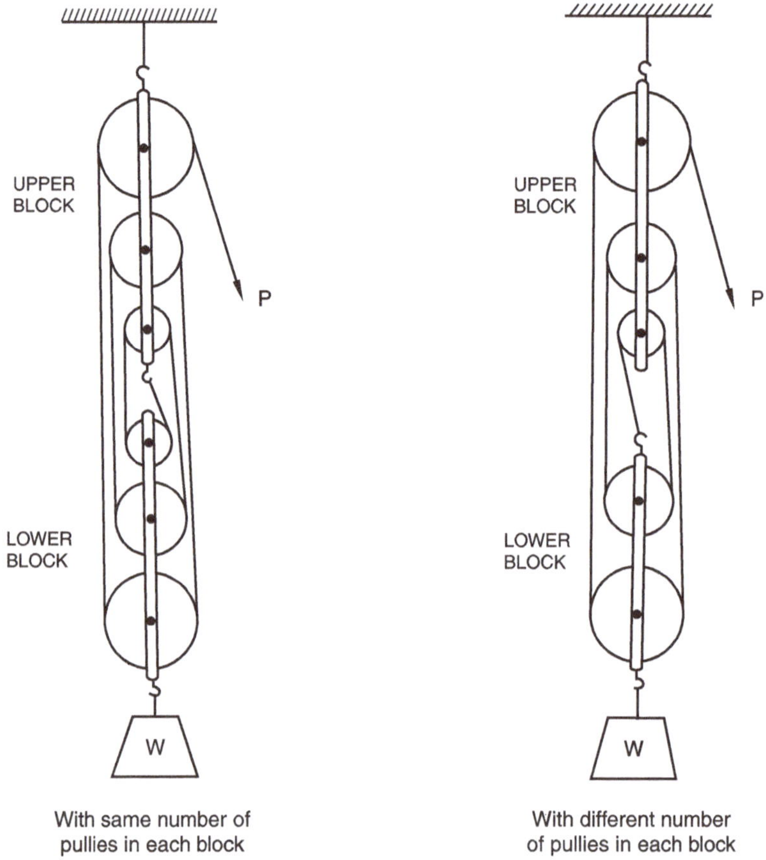

Fig. 6.12. *Second system of pulleys*

If there are n number of pulleys in both the blocks and load is lifted through a distance x, then each string will moves through a distance x, causing a slackness of x in each string.

The total slackness = nx will be taken by the effort to moves downward.

$$\therefore \quad \text{V.R.} = \frac{nx}{x} = n = \text{Total no. of pulleys in two blocks}$$

$$\text{M.A.} = \frac{W}{P} \qquad \text{and} \qquad \eta = \frac{\text{M.A.}}{\text{V.R.}}$$

If w = weight of lower block, then $\text{M.A.} = \dfrac{W + w}{P}$

6.21. THIRD SYSTEM OF PULLEYS

In this system, the pulleys are arranged so that there are as many strings as the total number of pulleys. Each end of the string is fixed to the lower block to which the load is attached. The other end of the string passes round the upper priphary of pulley and is fastened to the next lower pulley as shown in Fig. 6.13.

In this all the pulleys are movable except the top-most pulley which is fixed to rigid support.

When the load W is lifted by distance x, then all the strings supporting the load will be slackened by x.

Now, consider pulley 1, which is fixed to the rigid support.

The slackness in string, x_1 = x and this slackness will be taken up by pulley 2.

Therefore, the pulley 2 will come down through a distance of 2x.

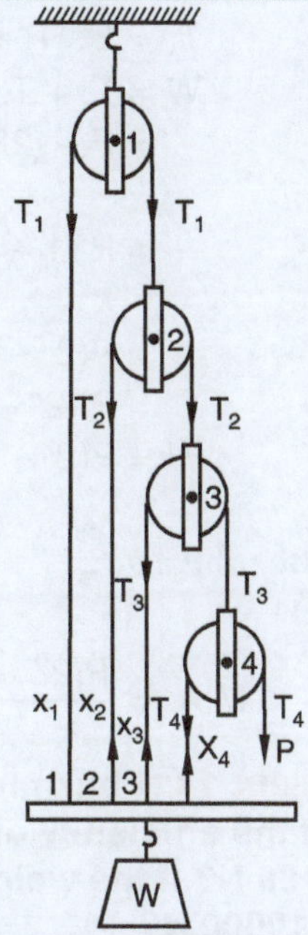

Fig. 6.13. *Third system of pulleys*

The string x_2 will come down through a distance

$$2x + x = 3x = (2^n - 1)\ x.$$

Similarly, the string x_3 will come down $2 \times 3x + x = 7x = (2^3 - 1)x$ and the string x_4 will come down through a distance $2 \times 7x + x = 15x = (2^4 - 1)x$.

Distance moved by effort = $(2^4 - 1)\ x$.

$$\therefore \text{ Velocity ratio, V.R.} = \frac{\text{Distance moved by effort}}{\text{Distance moved by load}}$$

$$= \frac{(2^4 - 1)x}{x} = (2^4 - 1)$$

If there are total n number of pulleys, then velocity ratio,

$$\text{V.R.} = 2^n - 1$$

$$\text{Mechanical advantage, M.A.} = \frac{W}{P} \qquad \text{Efficiency, } \eta = \frac{\text{M.A.}}{\text{V.R.}}$$

When the weight of the pulley is also taken into account:

Let 'w' be the weight of each pulley. Now consider the equallibrium of pulleys 1, 2, 3, 4, etc., we have

$$T_4 = P$$

$$T_3 = T_4 + P + w \qquad \text{[T is tension in the string]}$$

$$= 2P + w \qquad \text{...(i)} \qquad [\because T_4 = P]$$

$$T_2 = 2T_3 + w$$

$$= 2^2P + 2w + w \qquad \text{...(ii)}$$

$$\text{Also,} \quad T_1 = 2T_2 + w$$

$$= 2^3P + 2^2w + 2w + w + \ldots \qquad \text{...(iii)}$$

For 'n' number of pulleys,

$$T_n = 2^{n-1} P + 2^{n-2}w + 2^{n-3}w + 2^{n-4}w + \ldots + 2w + w \quad \ldots(iv)$$

But

$$W = T_1 + T_2 + T_3 + T_4 + \ldots + T_n$$
$$= P + (2P + w) + (2^2P + 2w + w) + (2^3P + 2^2w + 2w + w) + (2^{n-1}P + 2^{n-3} w + \ldots + 2w + w)$$
$$= P (1 + 2^2 + 2^3 + \ldots + 2^{n-1}) + w(1 + 2 + 2^2 \ldots + 2^{n-2}) + w(1 + 2 + \ldots + 2^{n-3}) + w(1 + \ldots + 2^{n-4}) + \ldots + 2w + w$$
$$= P(2^n - 1) + w(2^{n-1} - 1) + w(2^{n-2} - 1) + w(2^{n-3} - 1)$$
$$= P (2^n - 1) + w(2 + 2^2 + \ldots + 2^{n-2} + 2^{n-3} + 2^{n-1}) - w(n - 1)$$

$$\therefore \quad W = P(2^n - 1) + w(2^n - n - 1)$$

But $\text{M.A.} = \dfrac{W}{P}$

$$\therefore \quad \text{M.A.} = \frac{P(2^n - 1) + w(2^n - n - 1)}{P}$$

Problem 13: Four movalbe pulleys are arranged as in the first system of pulleys. If the efficiency of the system is 85%, find what load can be raised by an effort of 180 N? If the weight of each pulley is 10 N. Calculate what effort can lift a load of 10000 N?

Solution: No of pulley, n = 4

Efficiency, $\eta = 85\%$

(i) When the weight of each pulley is not taken into account: —

Velocity ratio, V.R. = $2^n = 2^4 = 16$

Mechanical advantage, M.A. = W/P = W/180 [∵ P = 180 N]

$$\text{Efficiency, } \eta = \frac{\text{M.A.}}{\text{V.R.}} = \frac{W/180}{16}$$

∴ W = 85/100 × 16 × 19 = **2448 N [Ans]**

(ii) When the weight of each pulley is taken into account: —

Here, weight of pulley, w = 10 N

$$\text{We know that} \quad \text{M.A.} = \frac{2^n W}{W + w(2^n - 1)} \qquad \left[\because \text{M.A.} = \frac{W}{P}\right]$$

$$\text{or} \quad \frac{10000}{P} = \frac{2^4 \times 10000}{10000 + 10(2^4 - 1)}$$

$$\therefore \quad P = \frac{10000 \times 1500}{16 + 10000} = \textbf{634.37 N [Ans]}$$

Problem 14: In a second system of pulleys, the same rope goes round all the pulleys. One end of the rope is tied to the upper block and there are six pulley in all. Find the force required to lift a load of 1000 N, if η = 60%. The weight of the lower block is 50 N.

Solution: Load lifted, W = 1000 N,

Efficiency, $\eta = 60\%$

No. of pulleys $n = 6$

Weight of the lower block, $w = 50$ N

Velocity ratio, V.R. = n = 6

$\therefore \quad W = 1000 + 50 = 1050$

$$\text{Also } \eta = \frac{\text{M.A.}}{\text{V.R.}} = \frac{W/P}{\text{V.R.}} = \frac{1050/P}{\text{V.R.}}$$

$$\frac{60}{100} = \frac{1050}{P \times 6}$$

$$\therefore \quad P = \frac{1050 \times 100}{6 \times 60} = \mathbf{291.66\ N\ [Ans]}$$

Problem 15: A person weighing 680 N sits on a plateform attached to the lower block of a system of 5 pulleys arranged in the second system. The plateform and the lower block weigh 110 N. The man supports himself by exerting a down ward pull at the free end of the rope. Neglecting friction, find the maximum pull the man exerts.

Solution: Weight of the man = 680 N

Weight of lower block and plateform = 110 N

Total load = 680 + 110 = 790 N

Due to pull, effective load on the lower plateform will reduce to (790 – P) = W

V.R. = n = 5

$$\eta = \frac{\text{M.A.}}{\text{V.R.}} = \frac{W}{P \times n}$$

$$\frac{100}{100} = \frac{(790 - P)}{P \times 5} \qquad [\because \eta = 100\% \text{ for no friction}]$$

$$5P = (790 - P)$$

$$6P = 790$$

$$\therefore \quad P = 790/6 = \mathbf{131.67\ N\ [Ans]}$$

Problem 16: In third system of pulleys, there are 4 pulleys. Determine the velocity ratio. Also find out the effort required to lift a load of 1800 N, when efficiency of the machine be 75%.

If the weight of each pulley is 5N, how much weight can be supported by effort of 100 N.

Solution: No. of pulleys. n = 4

Weight to be lifted, W = 1800 N

Efficiency, $\eta = 75\%$

Velocity ratio, V.R. = $2^n - 1 = 2^4 - 1 =$ **15 [Ans]**

Mechanical adventage, M.A. = W/P = 1800 /P

$$\text{Efficiency, } \eta = \frac{\text{M.A.}}{\text{V.R.}}$$

$$\text{or} \quad \frac{75}{100} = \frac{1800}{P \times 15}$$

$$\therefore \quad P = \frac{100 \times 1800}{75 \times 75} = \mathbf{160\ N\ [Ans]}$$

When the weight of each pulley is taken into account:—

Here, w = 5N

Now to find the total weight which can be supported by the system we may use the relation,

$$W = P(2^n - 1) + w(2^n - n - 1)$$
$$= 10(2^4 - 1) + 5(2^4 - 4 - 1) = 10 \times 15 + 5 \times 11$$
$$= 150 + 55 = \mathbf{205\ N\ [Ans]}$$

6.22. WESTON DIFFERENTIAL PULLEY BLOCK

This machine consists of two blocks upper and lower. The upper block is fixed and has pulleys of different radii but fixed to a common axle. These two pulleys rotate together. The lower block is movable and consists of a single pulley through which the load is attached. An endless chain or a rope passes round the smaller pulley of the lower block and finally round the pulley of greater radius as shown over the large pulley. To avoid slipping, the rim of these pulleys are provided with projections or teeth.

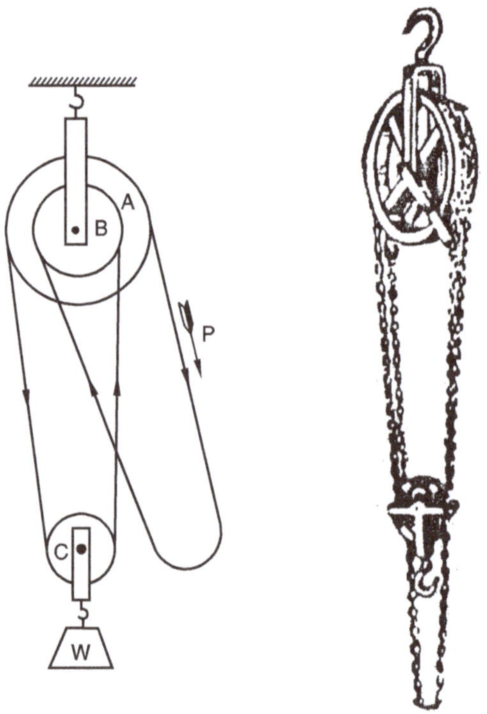

Fig. 6.14. *Weston differential pulley block*

Let W = Load lifted

P = Effort applied to lift load W

R = Radius of the larger pulley A

r = Radius of the smaller pulley B

When one complete revolution is given to the larger pulley A by the application of effort P, as shown.

Displacement of effort P, will be equal to the length of rope or chain pulled over the large pulley, A = $2\pi R$

Length of chain released by the smaller pulley, B = $2\pi r$

Net shortening in length of chain = $2\pi R - 2\pi r = 2\pi(R-r)$

But the distance moved by load W is half that of the shortening in length of chain = $\frac{2\pi(R-r)}{2} = \pi(R-r)$

But the distance moved by the load, W = $\pi(R-r)$

$$\therefore \quad \text{Velocity ratio, V.R.} = \frac{\text{Distance moved by effort}}{\text{Distance moved by load}}$$

$$= \frac{2\pi R}{\pi(R-r)} = \frac{2R}{R-r}$$

Mechanical advantage, M.A. = $\frac{W}{P}$ and $\eta = \frac{\text{M.A.}}{\text{V.R.}}$

> **Note**: *If N_1 and N_2 are the number of recesses in the larger and smaller pulleys of upper block.*

$$\text{Then, V.R.} = \frac{2N_1}{N_1 - N_1}$$

Problem 17: A differential pulley machine is used for lifting a heavy steel slab which measures 600 mm × 250 mm × 200 mm. The upper block has two cogged groves, the diameters are of 500 mm and 550 mm. If the efficiency of the machine is 65%, find the effort required to lift the slab. Take density of steel as 14.5 gm per cubic centimetre.

Solution: Radius of larger pulley, R = 550/2 mm = 275 mm

Radius of smaller pulley, r = 500/2 = 250 mm

Efficiency, η = 65%

Weight of the slab = $600 \times 250 \times 200 \times 145 \times 10^{-6} = 4350$ N

$$\text{Velocity ratio, V.R.} = \frac{2R}{R-r} = \frac{2 \times 275}{275 - 26} = 22$$

$$\text{Mechanical advantage, M.A.} = \frac{W}{P} = \frac{435}{P}$$

$$\text{Also efficiency } \eta = \frac{\text{M.A.}}{\text{V.R.}}$$

$$\text{or} \quad \frac{65}{100} = \frac{435/P}{22}$$

$$\therefore P = \frac{435}{22} \times \frac{10}{65} = \mathbf{30.42\ [Ans]}$$

Problem 18: In a Weston differential pulley bolck, the number of recesses in the smaller wheel is 9/10 of the large wheel. If efficiency of the machine is 50%, find the load lifted an effort of 25 newton.

Solution: Let the recesses in the larger wheel, $N_1 = 10$

$\therefore$ Recesses in the smaller wheel, $N_2 = 9/10 \times 10 = 9$

Velocity ratio, $\text{V.R.} = \dfrac{2N_1}{N_1 - N_2} = \dfrac{2 \times 10}{10 - 9} = 20$ and

Mechanical advantage, M.A. = W/P = W/ 25 [$\because$ P = 25N]

Also efficiency $\eta = \dfrac{\text{M.A.}}{\text{V.R.}} = \dfrac{W/25}{20}$

or $\dfrac{50}{100} = \dfrac{W/25}{20}$

$\therefore$ $W = \dfrac{50}{100} \times 20 \times 25 =$ **250N [Ans.]**

6.23. SCREW JACK

A screw jack is used for lifting heavy loads by applying comparatively smaller effort at the end of the lever. For example to lift cars, buses, trucks etc. while reparing or replacing their punchered wheels.

The screw jack consists of a body which is threaded at the top in which a threaded spindle is rotated by the application of effort P to lift the load W (see Fig. 6.15). The principle, on which a screw jack works, is similar to that of an inclined plane on which a load is dragged by applying horizontal effort.

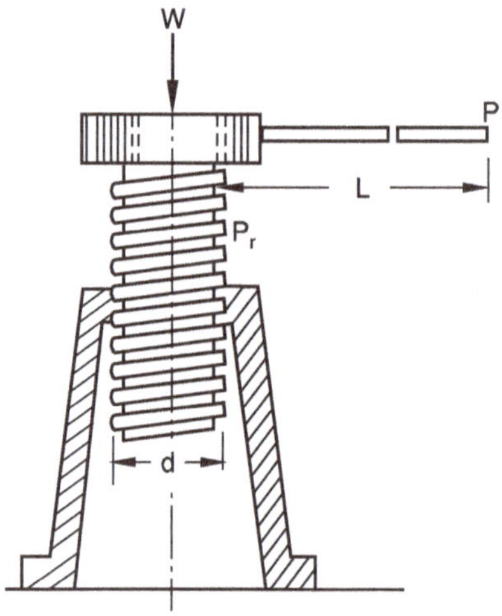

Fig. 6.15. *Screw jack*

Let W = load lifted

P = Effort applied at the end of lever

L = length of the arm (lever)

p = *pitch of the threads

For one revolution of the lever, distance moved by effort = $2\pi L$

Distance moved by the load = pitch = p

$\therefore$ Velocity ratio, $\text{V.R.} = \dfrac{\text{Distance moved by effort}}{\text{Distance moved by load}} = \dfrac{2\pi L}{p}$

Mechanical advantage, $\text{M.A.} = \dfrac{W}{P}$ and

Efficiency, $\eta = \dfrac{\text{M.A.}}{\text{V.R.}}$

Force required to lift the load

If one complete revolution of a screw thread be imagined to be unwound from the body of screw [see Fig. 6.16 (i)] and developed, it will form an inclined plane as shown in Fig 6.16 (ii).

*pitch is the axial distance between a point on one thread to a similar point on the next thread.

Let d = Mean diameter of screw

p = Pitch of the screw

θ = Angle of inclination

Therefore, the angle of inclination can be calculated by,

$$\tan\theta = \frac{p}{\pi d}$$

Since the principle on which a screw jack works is similar to that of an inclined plane to which a horizontal force is applied to a load W (refer to chapter friction).

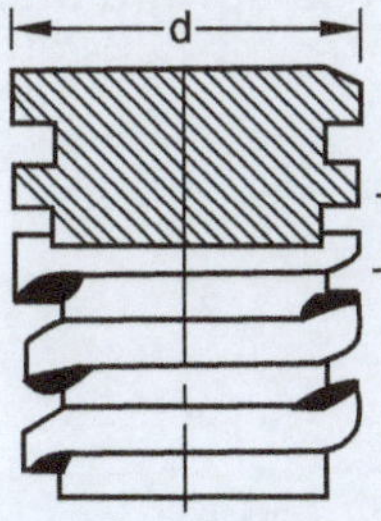

Fig. 6.16. (i)

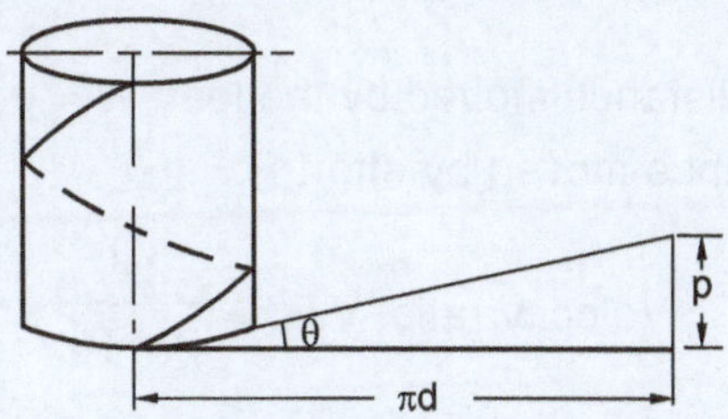

Fig. 6.16. (ii)

$\therefore$ Effort required to lift the load, $P_r = W\tan(\theta + \phi)$.

If P is the effort applied at the end of the handle of length L, then its moment about the axis of the screw must be equal to the moment of P_r acting at the thread about the axis of the screw.

$$\therefore \qquad P \times L = P_r \times \frac{d}{2}$$

$$\text{or} \qquad P = \frac{P_r \times d}{L \times 2} = W\tan(\theta + \phi) \times \frac{d}{2L}$$

$$\therefore \qquad P = \frac{Wd}{2L}\tan(\theta + \phi)$$

When the load is lowered, then $P_r = W\tan(\theta - \phi)$

$\therefore$ Effort required to lower the load, $P = \dfrac{Wd}{2L}\tan(\theta - \phi)$

$\text{M.A.} = \dfrac{W}{P}$ and $\eta = \dfrac{\text{M.A.}}{\text{V.R.}}$ also

$$\eta = \frac{\tan\theta}{\tan(\theta + \phi)}$$

The maximum efficiency i.e.,

$$\theta = 45° - 1/2.$$

6.21. DIFFERENTIAL SCREW JACK

This jack is an improvement of simple jack in order to obtain high velocity ratio. In this, the differential screw is made in two parts B and C as shown in Fig. 6.17. The screw C which is threaded from outside only rotates in screw B and carries load W on its top. Screw B is threaded from both inside, and outside and rotates in the body A. Thus the screw B

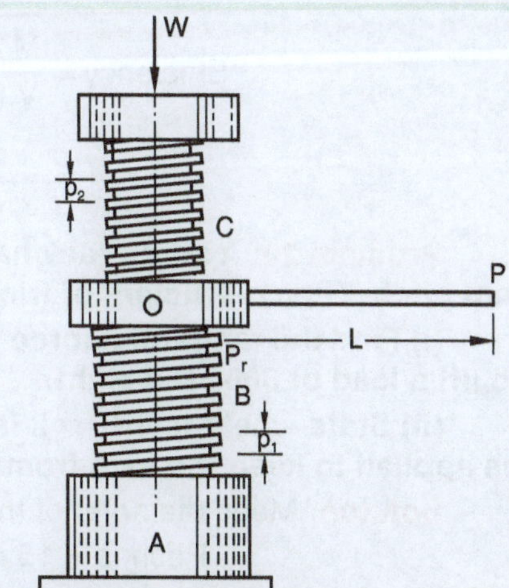

Fig. 6.17. *Differential screw jack*

acts like a nut to screw A. In this type of screw jack, the screw does not rotate but moves upward and downward only. The effort P is applied at the end of lever, which is fixed with screw B.

Let W = Load lifted

P = Effort applied

p_1 = Pitch of screw B

p_2 = Pitch of screw C

L = Lenght of the lever

When one revolution is given to lever, screw B moves upward and the screw C downward.

∴ Net distance moved by the load W = $p_1 - p_2$

Distance moved by effort P = $2\pi L$

$$\therefore \quad \text{Velocity ratio, V.R.} = \frac{2\pi L}{p_1 - p_2}$$

$$\text{Mechanical advantage, M.A.} = \frac{W}{P} \text{ and Efficiency, } \eta = \frac{\text{M.A.}}{\text{V.R.}}$$

Problem 19: In a simple screw jack, the pitch of the screw is 10 mm and the length of the lever is 500 mm. Find the velocity ratio. If an effort of 100 N is applied at the end of the lever to lift a load of 12000 N, what is the efficiency of the screw jack?

Solution: Pitch, p = 10 mm

Length, L= 500 mm

Effort, P = 100 N

Load, W = 12000 N

$$\text{(i) Velocity ratio, V.R.} = \frac{2\pi L}{p} = \frac{2\pi \times 500}{10} = \mathbf{314.16} \quad \textbf{[Ans]}$$

(ii) Efficiency of the screw jack (η)

$$\text{M.A.} = W/P = 1200/100 = 12$$

$$\text{Efficiency} = \frac{\text{M.A.}}{\text{V.R.}} \times 100$$

$$\eta = \frac{12}{314.16} \times 100 = \mathbf{38.9\%} \quad \textbf{[Ans]}$$

Problem 20: A screw jack has square threads, 75 mm mean diameter and 15 mm pitch. The co-efficient of friction at the screw threads is 0.05.

(i) Find the tangential force to be applied to the jack at 360 mm radius so as to lift a load of 6000 N weight.

(ii) State whether the jack is self locking. Also find the torque, which must be applied to keep the load from decending.

Soluton: Mean diameter of threads, d = 75 mm

Pitch, p = 15 mm

Coefficiency of friction, $\mu = 0.05$

(i) Tangentail force (P) applied at L = 360 mm radius to lift a load of 6000N,

$$\tan\theta = \frac{p}{\pi d} = \frac{15}{\pi \times 75} = 0.0636$$

$$\tan\phi = \mu = 0.05 \qquad [\because \mu = \tan\phi]$$

and $$P = W\tan(\theta+\phi) = W \times \frac{\tan\theta + \tan\phi}{1-\tan\theta\tan\phi}$$

$$\therefore \quad P_r = 6000 \times \frac{0.0636+0.05}{1-0.0636\times 0.05} = \mathbf{683.77\ N\ [Ans]}$$

Also $P \times 360 = P_r \times d/2$

$$\therefore \quad P = \frac{P_r}{360} \times \frac{d}{2} = \frac{683.77}{360} \times \frac{75}{2} = \mathbf{71.22\ [Ans]}$$

(ii) For self looking

$$\eta = \frac{M.A.}{V.R.} = \frac{W/P}{2\pi L/p} = \frac{6000/71.22}{2\pi \times 360/15} \times 100 = 55.84\%$$

Since the efficiency of the screw lack is more tan 50%, therefore **the screw jack is not self looking [Ans]**

(iii)Torque (T) applied to keep the load from deceding

We know that $P = W\tan(\theta - \phi)$

$$= 600 \times \frac{0.636-0.05}{1+0.0636\times 0.05} = \mathbf{81.34\ N\ [Ans]}$$

$$T = P \times r = 81.34 \times 75/2 = \mathbf{3050\ N\ mm\ [Ans]}$$

Problem 21: In a differential screw jack, the pitch of the upper screw is 6 mm and that of the lower screw is 12 mm. It the efficiency of the machine is 60%, find the effort required at the end of lever 0.6 m long, to lift a load of 1200 N.

Solution: Pitch, $p_1 = 12$ mm

Pitch, $p_2 = 6$ mm

Efficiency, $\eta = 60\%$

Lenght of level, L = 0.6 m = 600 mm

Load lifted W = 1200 N

We know that, $$V.R. = \frac{2\pi L}{p_1 - p_2} = \frac{2\pi \times 600}{12-6} = 628.32$$

and $$M.A. = \frac{W}{P} = \frac{12000}{P}$$

$$\eta = \frac{M.A.}{V.R}$$

$$\frac{60}{100} = \frac{12000/P}{628.32}$$

$$\therefore \quad P = \frac{12000}{628.32} \times \frac{100}{60} = \mathbf{31.83\ N\ [Ans]}$$

6.24. SINGLE PURCHASE WINCH CRAB

It consists of two-toothed wheels, smaller and larger. The smaller wheel A is called the pinion and the larger wheel B is called the gear as shown in Fig. 6.18. The gear and the load drum are fixed on shaft S_2. A rope is wound round the drum which carries the load W to be lifted. Effort P is applied at end of the lever handle which is fixed to the shaft S_1 on which the pinion is mounted and meshed with gear.

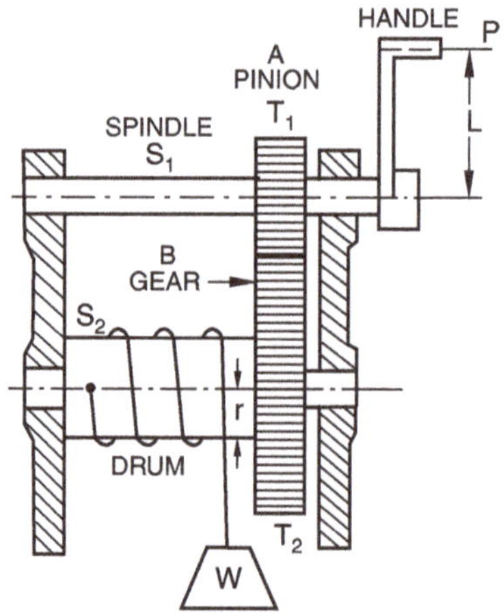

Fig. 6.18. *Single purchase winch crab*

Let W = Load lifted

P = Effort applied to lift the load W

T_1 = No. of teeth on the pinion A

T_2 = No. of teeth on the gear B

L = Length of the handle

r = Radius of the load drum

For one revolution of the handle,

Distance moved by effort, P = $2\pi L$

Number of revolution made by pinion, A = 1

$$\text{Number of revolutions made by gear, B} = \frac{T_1}{T_2}$$

As the spur gear is fixed with load drum, therefore

$$\text{Number of revolutions made by load drum} = \frac{T_1}{T_2}$$

$$\text{Now, the distance moved by load W} = 2\pi r \times \frac{T_1}{T_2}$$

$$\therefore \quad \text{Velocity ratio, V.R.} = \frac{\text{Distance moved by effort}}{\text{Distance moved by load}}$$

$$= \frac{2\pi L}{2\pi r \times \frac{T_1}{T_2}}$$

$$= \frac{L}{r} \times \frac{T_2}{T_1}$$

If the thickness, t of the lifting rope is taken into consideration, then,

$$\text{V.R.} = \frac{L}{r + t/2} \times \frac{T_2}{T_1}$$

$$\text{Mechanical advantage, M.A.} = \frac{W}{P} \quad \text{and Efficiency, } \eta = \frac{\text{M.A.}}{\text{V.R.}}$$

6.25. DOUBLE PURCHASE WINCH CRAB

This winch crab is improvement over the single purchase winch crab in order to obtain higher velocity ratio.

In this machine, there are two pinions with teeth T_1 and T_3 fixed to shafts S_1 and S_2 as shown in Fig. 6.19.

The gear with teeth T_2 is fixed on the shaft S_2 and geared with pinion having teeth T_1. The gear with teeth T_4. Is fixed to shaft S_3 and geared with pinion having teeth T_3. The gear teeth T_4 is attached with the load drum. A rope is would around the load drum and carries load on its free end.

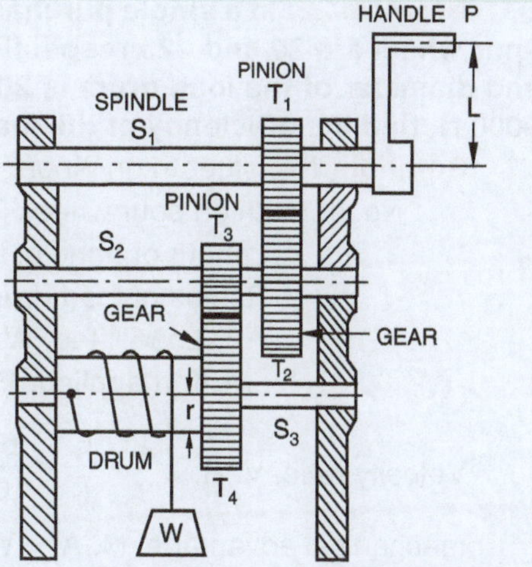

Fig. 6.19. *Double purchase which crab*

Let W = Weight lifted

P = Effort applied at the end of handle

T_1 and T_3 = Number of teeth on the pinions

T_2 and T_4 = Number of teeth on the gears

L = Lenght of the handle

r = Radius of the load drum

For one revolution of the handle,

Distance moved by effort P = $2\pi L$

For one revolution of the handle, the shaft S_1 and pinion having teeth T_1 also rotates through one revolution.

Pinion having teeth T_1 rotates the gear have teeth T_3 on shaft S_2.

$\therefore$ Number of revolutions made by shaft $S_2 = \dfrac{T_1}{T_2}$

Number of revolutions made by pinion on shaft $S_3 = \dfrac{T_1}{T_2} \times T_3$

This pinion rotates the gear fixed on shaft S_3.

$\therefore$ Number of revolutions made by gear on shaft $S_3 = \dfrac{T_1}{T_2} \times \dfrac{T_3}{T_4}$

$\therefore$ Distance moved by the load, W = $2\pi r \times \dfrac{T_1}{T_2} \times \dfrac{T_3}{T_4}$

$$\text{Velocity ratio, V.R.} = \frac{\text{Distance moved by effort}}{\text{Distance moved by load}}$$

$$= \frac{2\pi L}{2\pi r \times \dfrac{T_1}{T_2} \times \dfrac{T_3}{T_4}}$$

$$= \frac{L}{r} \times \frac{T_2}{T_1} \times \frac{T_4}{T_3}$$

Mechanical advange, M.A. $= \dfrac{W}{P}$ and Efficiency, $\eta = \dfrac{\text{M.A.}}{\text{V.R.}}$

Problem 22: **In a single purchase which crab, number of teeth on pinion and spur wheel are 20 and 125 respectivley. The length of the effort handle is 0.6 m and diameter of the load drum is 200 mm. If an effort of 400 N just lifts a load of 9000 N, find the efficiency of the machine.**

Solution: No. of teeth on pinon, $T_1 = 20$

No. of teeth on spur wheel, $T_2 = 125$

Length of handle, L = 0.6 m = 600 m

Radius of load drum = 200/2 = 100 mm

Load lifted, W = 9000 N

Effort applied, P = 400 N

$$\text{Velocity ratio, V. R.} = \frac{L}{r}\times\frac{T_2}{T_1} = \frac{60}{100}\times\frac{120}{20} = 37.5$$

mechanical advantage, M. A. = W/P = 9000/400 = 22.5

$$\therefore \text{ Efficiency, } \eta = \frac{\text{M.A.}}{\text{V.R.}} = \frac{22.5}{37.5}\times 100 = \mathbf{60\%} \text{ [Ans]}$$

Problem 23: **The following data related to a single purchase winch crab:—**

Lenght of handle = 400 mm

Diameter of drum = 120 mm

Diameter of lifting rope = 10 mm

No. of teeth on pinion = 25

Effort applied = 20 N

No. of teeth of wheel = 130

Load raised = 180 N

Find (i) V. R. ; (ii) M. A.; (iii) η

Solution: $T_1 = 25$, $T_2 = 130$, L = 400 mm, r = 120/2 = 60 mm

P = 10 N, W = 180 N and t = 10 mm

$$\text{(i) Velocity ratio, V. R.} = \frac{L}{(r+\frac{t}{2})}\times\frac{T_2}{T_1}$$

$$\therefore \quad \text{V. R.} = \frac{400}{60+10/2}\times\frac{130}{25} = \mathbf{32} \text{ [Ans]}$$

$$\text{(ii) Mechanical advantage, M.A.} = \frac{W}{P} = \frac{180}{20} = \mathbf{9} \text{ [Ans]}$$

$$\text{(iii) Efficiency, } \eta = \frac{\text{M.A.}}{\text{V.R.}} = \frac{9}{32}\times 100 = \mathbf{28.12\%} \text{ [Ans]}$$

Problem 24: **In a double purchase which carb, teeth of pinions are 20 and 25 those of gears are 50 and 60 respectively. Length of the handle is 2 metres and effective diameter of load drum is 0.5 metre. If the efficiency of machine is 60%, find the effort required to lift a load of 4800 N.**

Solution: No of teeth in first pinion $T_1 = 20$

No. of teeth in second pinion, $T_3 = 25$

No. of teeth on first gear, T_2 = 50

No. of teeth on second gear, T_4 = 60

Lenght of handle, L = 2 m

Radius of load drum, r = 0.5/2 m = 0.25 m

Efficiency, η = 60%

Load lifted, W = 4800 N

Velocity ratio, $V.R. = \frac{L}{r} \times \frac{T_2}{T_1} \times \frac{T_4}{T_3} = \frac{2}{0.25} \times \frac{50 \times 60}{20 \times 25} = 48$

Mechanical advantage, M.A. = W/P = 4800/P

Efficiency, $\eta = \frac{M.A.}{V.R.} = \frac{4800}{P \times 48}$ or

$$\frac{60}{100} = \frac{4800}{P \times 48}$$

$$\therefore \quad P = \frac{4800 \times 100}{60 \times 48} = \mathbf{166.7\ N\ [Ans]}$$

MODEL TEST QUESTIONS

1. Define a machine? What are the uses of machines?
2. Define (a) Mechanical advantage (b) velocity ratio (c) Input of a machine (d) efficiency of a machine.
3. Establish relation between efficiency, mechanical advantage and velocity ratio of a machine.
4. Explain the difference between reversible machine and self looking machine.
5. Prove that the condition for a self looking machine is that its efficiency is less than 50%.
6. What is the law of a machine?
7. Define the expression for maximum mechanical advantage and maximum efficiency in case of a machine.
8. Discuss briefly with neat diagrams the following lifting machines:
   (a) Wheel and differential axle;
   (b) Simple screw-jack
   (c) Weston differential pulley block;
   (d) Single purchase winch crab;
   (e) Worm and worm wheel.

QUESTIONS FOR SELF EXAMINATION

1. **Fill in the blanks:**
   (i) Working parts make up a machine constitute ......
   (ii) Input to a machine = ... × ..... moved.
   (iii) Output of a machine = ... lifted × .....
   (iv) Ratio between ..... and ..... is known as machanical advantage.
   (v) Efficiency of a machine is the ratio its ..... and .....

(vi) An ideal machine is one in which efficiency is .....

(vii) M. A. ..... as the load increases.

(viii) Efficiency ..... as the load, the decreases.

(ix) At very high load, the efficiency will be .....

(x) Machine will work in the reversed direction if its efficilency is .... then 50%.

Ans. 1. (i) mechanism; (ii) effort, distance; (iii) load, height; (iv) load lifted, effort applied; (v) M. A., V. A.; (vi) 100%; (vii) Increases; (viii) decreases; (ix) maximum; (x) more.

2. **State whether the following statements are true or false:**

(i) Machine is a device by which the resistance is overcome.

(ii) Machine helps in lifting heavy load with small effort.

(iii) Resistance to be overcome by a machine is called effort.

(iv) The work done by the effort on the machines is known as output of the machine.

(v) For an ideal machine, Input is equal to output.

(vi) In a self looking machine, efficiency is more than 50%.

(vii) The law of machine is P = W + mC

(viii) In second system of pulleys, the velocity ratio is always equal to number of pulleys in two blocks.

(ix) In third system of pulleys, welght of the moveable pulleys add to the machanical advantage of the system.

(x) If the diameter of load drum in a single purchase crab is increased, the velocity ratio will decrease.

Ans. 2. (i) False ; (ii) True; (iii) False; (iv) False; (v) True; (vi) False; (vii) False ; (viii) True; (ix) False; (x) False.

PROBLEMS FOR PRACTICE

1. In a lifting machine, a weight of 1000 N is lifted by effort of 25N. If the distance moved by the weight is 100 mm and the point of application of effort by 8m. Find:

(i) Mechanical advantage;

(ii) Velocity;

(iii) Efficiency of the machine.

[**Ans.** (i) 40, (ii) 80, (iii) 50%]

2. The efficiency of a machine is 80% when an effort of 150 N is required to lift of 1300 N. Calculate the velocity ratio and the frictional force of the machine in terms of effort and load.

[**Ans.** 10.833, 30 N]

3. The efficiency of a lifting machine is 70%. An effort of 100 N is applied through a distance of 3 m to the machine to raise a load through a distance of 300 mm. Find;

(i) Load lifted by the machine;

(ii) Mechanical advantage;

(iii) Velocity ratio [**Ans.** (i) 700 N, (ii) 7, (iii) 10]

4. An effort of 200 N is applied to a lifting machine to raise a load,out of which 10% is lost in friction. If the velocity ratio of the machine is 10, find ; (i) load lifted and (ii) efficiency. [**Ans.** 1800 N, 90%]

5. In a machine, the efforts required to lift loads of 200 N and 300 N were 50 and 60 N respectively. If the velocity ratio of the machine is 20, find:
   (i) Law of machine.
   (ii) Efficiencies corresponding to loads of 200 N and 300 N.
   (iii) Effort lost in friction in both cases and
   (iv) The maximum efficiency which can be expected from this machine.

   [**Ans.** (i) $P = \frac{1}{10}W + 3.$ (ii) 20%, (iii) 45 N (iv) 50%]

6. The length of an inclined plane is 5 m and its height is 2.5 m. If a force of 400 N pulls a load of 620 N up the inclined plane, find:
   (i) M. A. (ii) V. R. and (iii) η of the machine.

   [**Ans.** (i) 1.55, (ii) 2, (iii) 77.5%]

7. A drum weighing 60 N and holding 400 N of water is to be raised form a well by means of wheel and axle. The axle is 100 mm in diameter and the wheel is 400 mm in diameter. If a force of 120 N has to be applied to the wheel, find
   (i) Mechanical advantage;
   (ii) Velocity ratio;
   (iii) Efficiency of the machine.

   [**Ans.** (i) 3.83 (ii) 4 (iii) 83.25%]

8. In a differential wheel and axle the diameters of the axles are 150 and 200 mm while the diameter of wheel is 500 mm. Find the velocity ratio of the machine. If a effort of 200 N can raise a load of 2500 N, find the mechanical advantage and efficiency of the machine.

   [**Ans.** 20, 12.5 and 62.5%]

9. In a single threaded worm and worm wheel, the number of teeth of the worm wheel is 50. The diameter of the effort wheel, is 200 mm and that of the load drum is 100 mm. It the efficiency of the machine is 20%, find the effort required to lift a load of 300 N.

   [**Ans.** 15 N]

10. Five movable pulleys are arranged as in the first systern of pulleys. If the efficiency of system Is 80% find the effort required to raise a mass of 10 kN.

    [**Ans.** 390.63 N]

11. Five pulleys are arranged In the second system of pulleys. When not loaded, the effort required to raise the movable block is 35 N. Further wastage In friction Increases the pull at the rate of 3% of load lifted. Find :
    (i) The effort required to raise a load of 200 N;
    (ii) Efficiency of the system at this load

    [**Ans.** (i) 495 N, (ii) 80·81%]

12. In a third system of pulleys, there are 4 pulleys. Find out the effort required to lift a load of 1800 N. It the efficiency of the machine be 75%, calculate the amount of effort wasted In friction.

    [**Ans.** 160 N, 40 N]

13. A Weston differential pulley block consists of a lower block and an upper block. The upper block has two cogged grooves, one of which has a radius of 125 mm and the other has a radius of 115 mm. If the efficiency of the machine Is 40%, calculate the effort required to raise a load of 1500 N.

[**Ans.** 150 N]

14. In a Weston differential pulley block, the number of recesses In the smaller wheel is 9/10 of that of the larger wheel. If the efticlency of the machine Is 50%, find the load lifted by an effort of 25 N,

[**Ans.** 250 N]

15. The efficiency of a screw jack Is 35%, when a load of 15 KN. Is lifted by an effort applied at the end of a handle of length 50 cm Determine the effort applied If the pitch of the screw thread Is 10 mm.

[**Ans.** 86.8 N]

16. In a differerntial screw lack, the screw threads have pitch of 10 mm and 7 mm. If the efficiency of the machine Is 28%, find the effort required. at the end of an arm 360 mm long to lift a load of 5 kN.

[**Ans.** 23·7 N]

17. In a single purchase winch crab, the number of tooth on pinion Is 25 and that on the spur wheel Is 250. The diameter of the drum is 300 mm. If the length of the handle Is 300 mm. Find (i) Velocity ratio. (ii) Efficiency of the crab, If an effort of 20 N can lift a load of 360 N.

[**Ans.** (i) 20, (ii) 90%]

18. In a double purchase crab, teeth of pinion are 20 and 25 that of spur wheel are 50 and 60. Length of the handle Is 2 m and diameter of the load drum Is 0.50 m. Determine the efficiency of the crab if 25 N effort is required to lift a load of 720 N.

[**Ans.** 60%]

Chapter 7

Moment of Inertia

INTRODUCTION

As we know, moment of a force about a point, is the product of the force (F) and the perpendicular distance (K) between the point and the line of the force. This moment is called first moment of the force (FK). If this moment is again multiplied by the perpendicular distance (K), then the quantity (FK^2) is called the second moment or moment of inertia.

Similarly, instead of force, if the area of a section (A) is taken into consideration, then the second moment (AK^2) is known as second moment of area or moment of inertia of area. The knowledge of moment of inertia is extremely important for the design of structural members in bridges, trusses and dams.

In this chapter, we shall deal with the study of moment of inertia of plane areas only.

7.1. MOMENT OF INERTIA

The product of the area of a section and the square of the perpendicular distance of the centre of gravity of the area from an axis is called moment of inertia.

In Fig. 7.1, the area of the section is A and the perpendicular distance of the C.G. of the area from the axis AB is K. Then, moment of inertia, $I = AK^2$

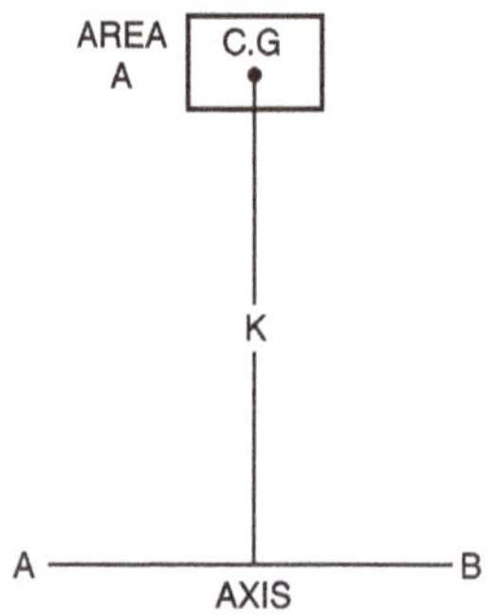

Fig. 7.1

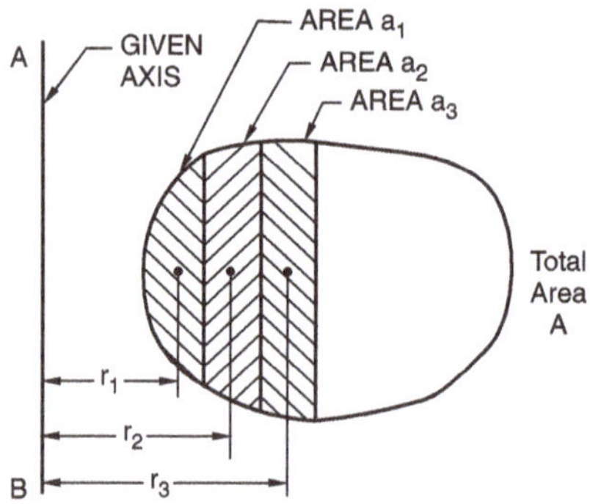

Fig. 7.2

7.2. MOMENT OF INERTIA OF A PLANE SECTION

Consider a plane section of area A as shown in Fig. 7.2. Split up the whole area of section into a number of small areas.

Let $a_1, a_2, a_3 \ldots$ = Small areas of the section.

$r_1, r_2, r_3 \ldots$ = Distances of C.G. of small areas from AB axis.

Moment of inertia of whole area, about AB = Algebraic sum of second moments of small areas about axis AB

or $$I = a_1r_1^2 + a_2r_2^2 + a_3r_3^2 + \ldots\ldots \ldots$$

$\therefore$ Moment of inertia, $I = \Sigma ar^2$

7.3. UNITS OF MOMENT OF INERTIA

The units of moment of inertia of a section depends upon the units of area and the distance from C.G.

If area is in mm^2 and distance is also in mm, then moment of inertia is expressed in mm^4.

Similarly, if area is in m^2 and distance is in m, then moment of inertia is expressed in m^4.

7.4. RADIUS OF GYRATION

The distance of a point where the whole area of a section is assumed to be concentrated (i.e., C.G.) from a given axis is called the radius of gyration.

$\therefore$ Radius of gyration, $K = \sqrt{\dfrac{I}{A}}$

7.5. MODULUS OF SECTION

The ratio of moment of inertia of a section about its centre of gravity and distance of the extreme fibre from centroidal axis is called modulus of section or section modulus.

Consider a rectangle section as shown in Fig. 7.3.

Let I_{xx} = Moment of inertia about x-x axis.

I_{yy} = Moment of inertia about y-y axis.

∴ Modulus of section about x-x axis,

$$Z_{xx} = \frac{I_{xx}}{d/2}$$

Similarly, modulus of section about y-y axis.

$$Z_{yy} = \frac{I_{yy}}{b/2}$$

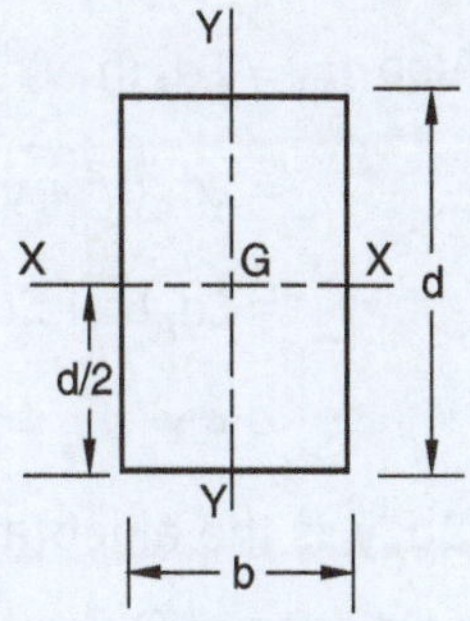

Fig. 7.3

7.6. THEOREM OF PERPENDICULAR AXIS

If I_{xx} and I_{yy} be the moment of inertia of a plane section about two perpendicular axis meeting at O, then the moment of inertia I_{zz} about the axis perpendicular to the plane and passing through the inter-section of O-X axis and O-Y axis is given by,

$$I_{zz} = I_{xx} + I_{yy}$$

Consider a small area A, at a distance x from O-Y axis, y form O-X axis and z from O-Z axis which is perpendicular to O-X and O-Y axes as shown in Fig. 7.3

Then $z^2 = x^2 + y^2$

Now $I_{xx} = d_A y^2$ and $I_{yy} = d_A x^2$ and $I_{zz} = d_A z^2$

$$= \Sigma d_A (x^2 + y^2)$$

$[\because r^2 = x^2 + y^2]$

or $I_{zz} = d_A x^2 + d_A y^2$

$$I_{zz} = I_{yy} + I_{xx}$$

∴ $I_{zz} = I_{xx} + I_{yy}$

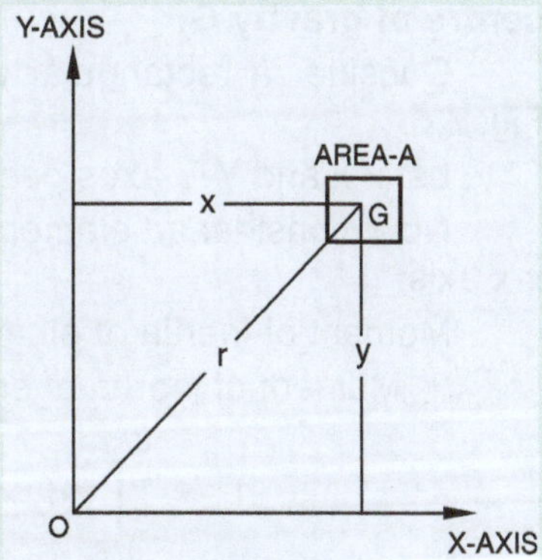

Fig. 7.4

Note: *z – z axis is known as polar axis; I_{zz} is called polar moment of inertia and is used for analysing the torsional stresses.*

7.7. THEOREM OF PARALLEL AXIS

If the moment of inertia of a plane area about an axis passing through its centre of gravity, be denoted by I_G, then moment of inertia of the area about axis A-B parallel to the first and at a distance h from the centre of gravity is given by

$$I_{AB} = I_G + Ah^2$$

Where I_{AB} = moment of inertia about AB

I_G = Moment of inertia about C.G.

A = Area of the section

h = Distance between C.G. of the section and axis AB.

Proof: Let d_A be the area of the strip at a distance y from the C.G. of a section as shown in Fig. 7.5.

Then $I_G = \Sigma d_A y^2$

Also $I_{AB} = \Sigma d_A (h+y)^2$

$$= \Sigma d_A (h^2 + y^2 + 2hy)$$

$$= \Sigma d_A h^2 + \Sigma d_A y^2 + \Sigma d_A 2hy$$

$$= Ah^2 + I_G + 2h\,\Sigma d_A y \qquad \ldots(i)$$

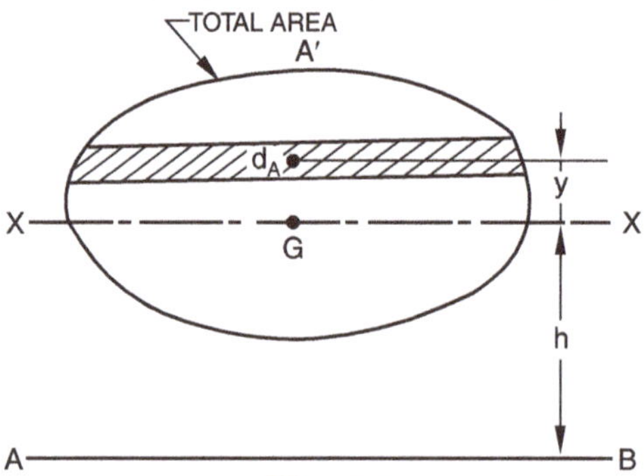

Fig. 7.5

$\Sigma d_A y$ is the algebraic sum of the moments of all the small areas about axis x-x.

As the distance of C.G. of the total area from x-x axis is zero, therefore, $\Sigma d_A y$ will also be equal to zero.

By substituting $\Sigma d_A y = 0$ in equation (i), we get

$$I_{AB} = Ah^2 + I_G + 0$$

$$\therefore \qquad I_{AB} = I_G + Ah^2$$

7.8. MOMENT OF INERTIA OF RECTANGLE

Case I: Moment of inertia of a rectangle about the axis passing through the centre of gravity G.

Consider a rectangular section ABCD of breadth b and depth d as shown in Fig. 7.6.

Let x-x and Y-Y axes passing through its centre of gravity.

Now, consider an elementry strip of width b and thickness dy at a distance y from x-x axis.

Moment of inertia of elementry strip = $(b \times dy)y^2$

$\therefore$ Moment of inertia of entire rectangle about x-x axis

$$I_{xx} = \int_{-d/2}^{d/2} bdy \times y^2 = b \int_{-d/2}^{d/2} y^2 dy$$

$$= b\left[\frac{y^3}{3}\right]_{-d/2}^{d/2}$$

$$= \frac{b}{3}\left[\left(\frac{d}{2}\right)^3 - \left(-\frac{d}{2}\right)^3\right]$$

$$= \frac{b}{3}\left(\frac{d^3}{8} + \frac{d^3}{8}\right)$$

$$\therefore \qquad I_{xx} = \frac{bd^3}{12}$$

Similarly $\qquad I_{yy} = \frac{db^3}{12}$

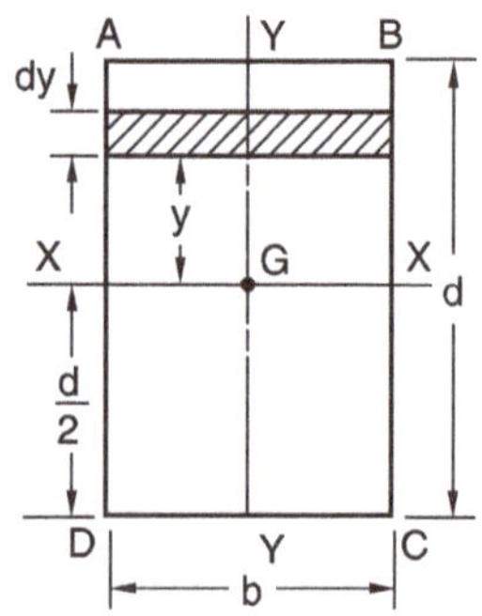

Fig. 7.6

Case II: Moment of inertia of rectangle about a line passing through the base DC.

Consider a rectangular section ABCD having width = b and depth = d (see Fig. 7.7)

By theorem of parallel axis

$$I_{DC} = I_G + Ah^2$$

$$= \frac{bd^3}{12} + bd(d/2)^2$$

$$= \frac{bd^3}{12} + \frac{bd^3}{4}$$

$$= \frac{bd^3 + 3bd^3}{12} = \frac{4bd^3}{12}$$

$$\therefore \ I_{DC} = \frac{bd^3}{3}$$

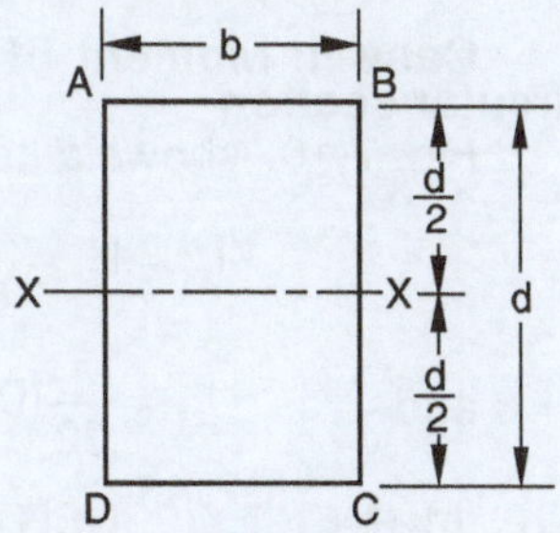

Fig. 7.7

Case III: Moment of inertia of a hollow rectangular section about the axis passing through the C.G.

Fig. 7.8 shows a simple hollow rectangle section, where

$$\therefore \ I_{xx} = \frac{BD^3}{12} - \frac{bd^3}{12} \text{ and}$$

$$\therefore \ I_{yy} = \frac{DB^3}{12} - \frac{db^3}{12}$$

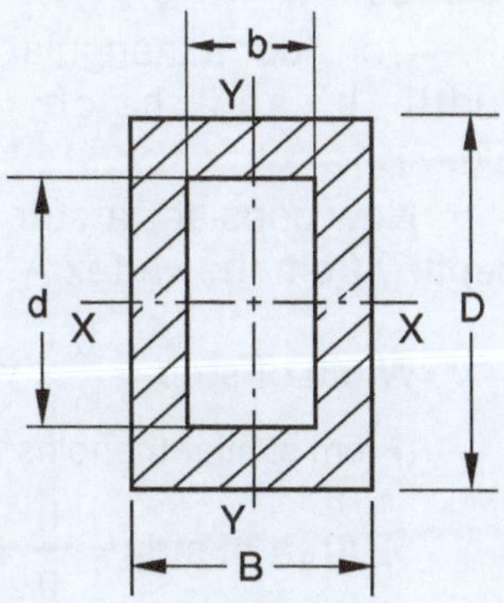

Fig. 7.8

7.9. MOMENT OF INERTIA OF CIRCULAR SECTION

Case I: Moment of inertia of solid circular section about C.G.

Consider a circular section of radius R, as shown n Fig. 7.9. Let I_{zz} be the moment of inertia about an axis perpendicular to the plane of the section at C.G. of the circle.

Now, consider an elementry circular ring of radius r and thickness dr

Area of circular ring = $2\pi r.dr$

Moment of inertia of an elementary ring about Z-Z axis

$$= \text{area} \times (\text{distance})^2$$

$$= 2\pi r.dr.r^2 = 2\pi r^3.dr$$

$\therefore$ Total moment of inertia of the whole figure about Z-Z axis,

$$I_{zz} = \int_0^R 2\pi r^3 dr = 2\pi \int_0^R r^3 dr = 2\pi \left[\frac{r^4}{4}\right]_0^R$$

$$= 2\pi \times \frac{R^4}{4} = \frac{\pi R^4}{2}$$

$$\therefore \ I_{zz} = \frac{\pi}{2}\left(\frac{D}{2}\right)^4 = \frac{\pi D^4}{32}$$

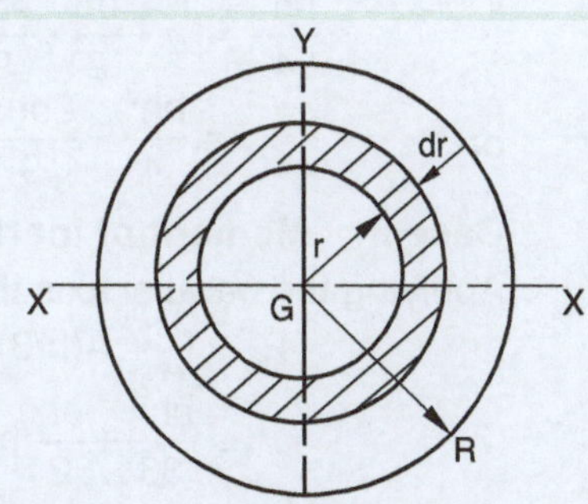

Fig. 7.9

From perpendicular axis theorem,

$$I_{zz} = I_{xx} + I_{yy}$$

$$\therefore \quad I_{xx} = I_{yy} = \frac{I_{zz}}{2} = \frac{\pi D^4}{32} \times \frac{1}{2} = \frac{\pi D^4}{64}$$

Case II: Moment of inertia of a hollow circular section

Fig. 7.10. shows a hollow circular section

$$I_{xx} = I_{yy} = \frac{\pi}{64}(D^4 - d^4)$$

and $$I_{zz} = \frac{\pi}{32}(D^4 - d^4)$$

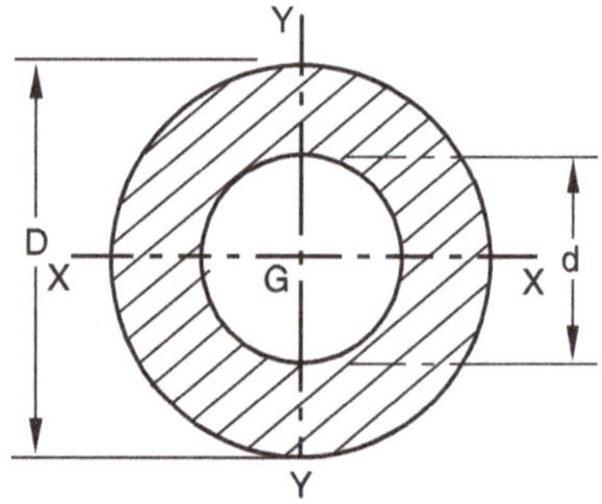

Fig. 7.10

7.10. MOMENT OF INERTIA OF A TRIANGULAR SECTION

Case I: Moment of inertia of triangular section about centroidal axis parallel to base.

Consider a triangular section ABC of base width b and height h as shown in Fig. 7.11

Now consider a strip of thickness dy at a depth y from the vertex A.

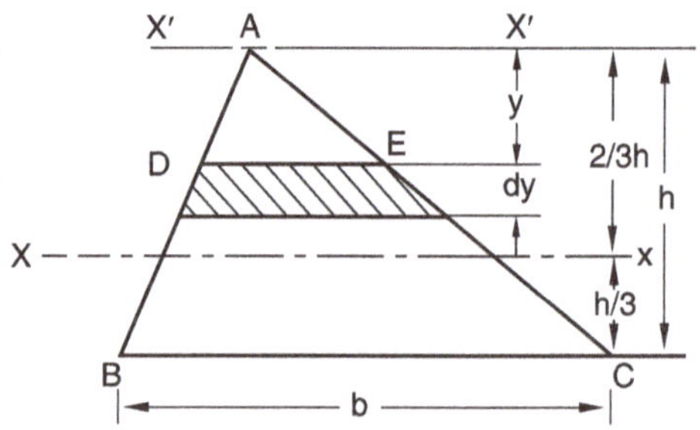

Fig. 7.11

Width of strip = $\frac{b}{h} \times y$

(From similar triangles ADE and ABC)

$\therefore$ Area of strip = $\frac{b}{h} \times y \times dy$

Moment of inertia of the strip about x'-x' axis.

$$= \frac{b}{h} \times y \times dy \times y^2 = \frac{b}{h} \times y^3 \times dy$$

$\therefore$ Moment of inertia of the whole area about x'-x' axis,

$$I_{x'x'} = \int_0^h \frac{b}{h} \times y^3 \times dy$$

$$I_{x'x'} = \frac{b}{h}\left[\frac{y^4}{4}\right]_0^h = \frac{bh^4}{4h} = \frac{bh^3}{4}$$

Applying parallel axis theorem, $I_{x'x'} = I_{xx} + A(2h/3)^2$

$$\therefore \quad \frac{bh^3}{4} = I_{xx} + \frac{bh}{2} \times \frac{4h^2}{9}$$

or $$I_{xx} = \frac{bh^3}{4} - \frac{2bh^3}{9} = \frac{bh^3}{36}$$

Case II: Moment of inertia about base BC

Applying the parallel axis theorem,

$$I_{BC} = I_{xx} + A(h/3)^2$$

$$= \frac{bh^3}{36} + \frac{bh}{2}(h/3)^2 = \frac{bh^3}{36} + \frac{bh^3}{18}$$

$$\therefore \quad I_{BC} = \frac{bh^3}{12}$$

TABLE 7.1

MOMENTS OF INERTIA OF COMMON GEOMETRIC FIGURES

Shape	Moment of Inertia
Rectangle 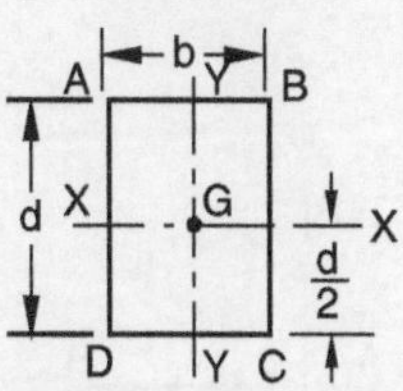	$I_{XX} = \dfrac{bd^3}{12}$ $I_{YY} = \dfrac{db^3}{12}$ $I_{CD} = \dfrac{bd^3}{3}$
Triangle 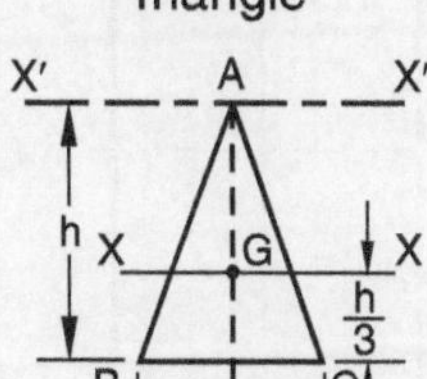	$I_{X'X'} = \dfrac{bh^3}{4}$ $I_{XX} = \dfrac{bh^3}{36}$ $I_{BC} = \dfrac{bh^3}{12}$
Circle 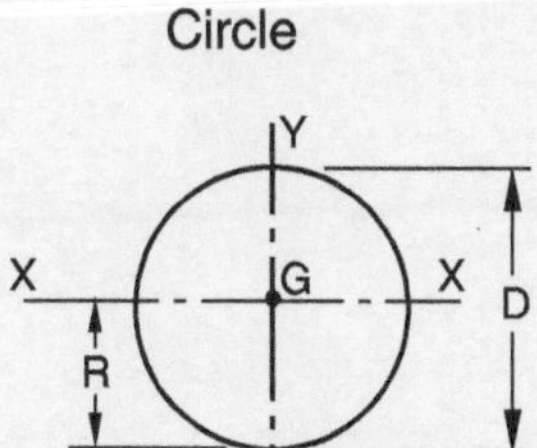	$I_{XX} = I_{YY} = \dfrac{\pi D^4}{64}$
Quadrant 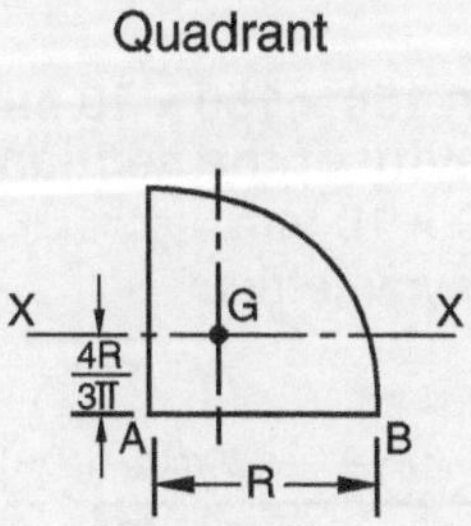	$I_{X'X'} = 0.055R^4$ $I_{AB} = \dfrac{\pi R^4}{16} = 0.196R^4$
Semicircle 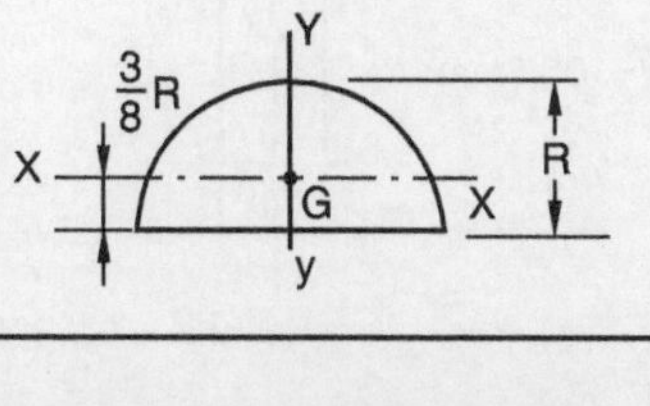	$I_{XX} = 0.11R^4$ $I_{YY} = \dfrac{\pi R^4}{8} = 0.392R^4$

SELECTED PROBLEMS

Problem 1: Find the moment of inertia of a rectangular section of 60 cm × 120 cm about,

(i) horizontal (x-x′) axis passing through the centre of gravity.
(ii) vertical (y-y′) axis passing through the C.G.
(iii) base DC
(iv) face BC

Solution: Here b = 60 mm and d = 120 mm [see Fig. 7.12]

(i) *Moment of inertia about x-x axis*

$$I_{xx} = \frac{bd^3}{12}$$

$$= \frac{60 \times 120^3}{12}$$

$$= \mathbf{864 \times 10^4\ mm^4}$$ **[Ans.]**

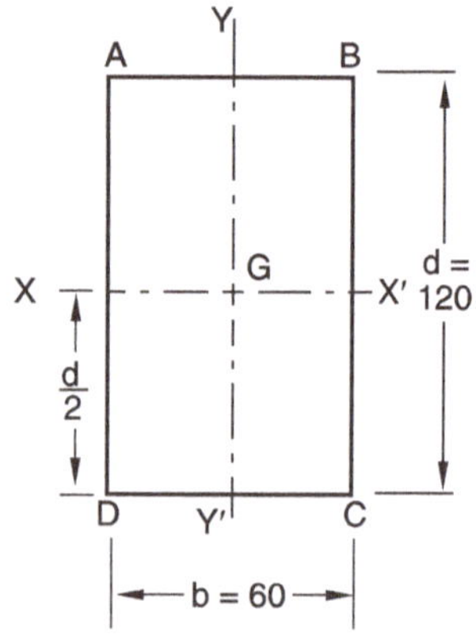

Fig. 7.12

(ii) *Moment of inertia about y-y axis*

$$I_{yy} = \frac{db^3}{12} = \frac{120 \times 60^3}{12}$$

$$= \mathbf{216 \times 10^4\ mm^4}$$ **[Ans.]**

(iii) *Moment of inertia about base DC*

$$I_{DC} = I_G + Ah^2$$

$$= 864 \times 10^4 + 60 \times 120 \times 60^2 \qquad [\because I_G = I_{XX}]$$

$$= (864 + 2592) \times 10^4$$

$$= \mathbf{3456 \times 10^4\ mm^4}$$ **[Ans.]**

(iv) *Moment of inertia about face BC*

Applying parallel axis theorem

$$I_{BC} = I_G + Ah^2$$

$$= I_{yy} + (120 \times 60) \times 30^2$$

$$= 216 \times 10^4 + 648 \times 10^4$$

$$= \mathbf{864 \times 10^4\ mm^4}$$ **[Ans.]**

Problem 2: Find the moment of inertia of T-section 150 × 150 × 10 mm about x-x axis and y-y axis passing through the centre of gravity of the section.

Solution: Fig. 7.13 shows the T-section of 150 × 150 × 10 mm.

Position of centre of gravity (G): As the section is symmetrical

$$\therefore \quad \bar{x} = 150/2 = 75 \text{ mm}$$

To find the position $\bar{y}$ proceed as under:

Area, $a_1 = 10 \times 140 = 1400$ mm^2

Area, $a_2 = 150 \times 10 = 1500$ mm^2

Distance of centroids of two rectangles:

$y_1 = 140/2 = 70$ mm

and $y_2 = 140 - 10/2 = 145$ mm

$$\therefore \quad \bar{y} = \frac{a_1y_1 + a_2y_2}{a_1 + a_2}$$

$$= \frac{1400 \times 70 + 1500 \times 145}{1400 + 1500} = 108.9 \text{ mm}$$

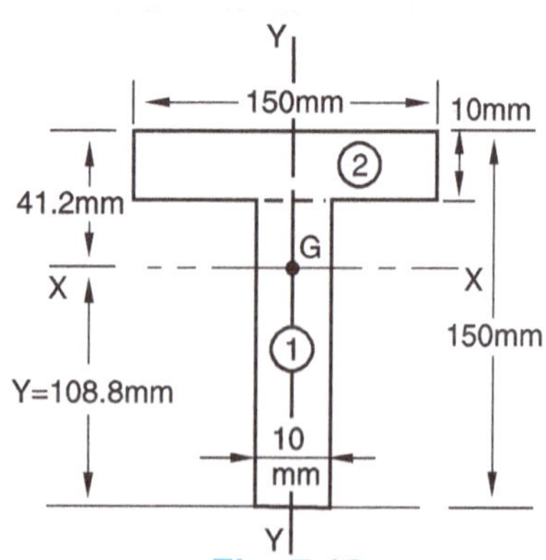

Fig. 7.13

Moment of inertia about x-x axis passing through centre of gravity (G) of section

$$I_{xx} = I_{xx1} + I_{xx2}$$

$$= \left[\frac{b_1 d_1^3}{12} + a_1 h_1^2\right] + \left[\frac{b_2 d_2^3}{12} + a_2 h_2^2\right]$$

$$= \frac{10 \times 140^3}{12} + 10 \times 140 \times 38.8^2 + \frac{150 \times 10^3}{12} + 150 \times 10 \times 36.2^2 .$$

$= 4394283 + 1978160 =$ **6372443 mm⁴ [Ans.]**

Moment of inertia about y-y axis passing through centre of gravity (G) of the section

$$I_{yy} = I_{yy1} + I_{yy2}$$

$$= \left[\frac{d_1 b_1^3}{12} + a_1 h_1^2\right] + \left[\frac{d_2 b_2^3}{12} + a_2 h_2^2\right]$$

$$= \left[\frac{140 \times 10^3}{12} + 1400 \times 0\right] + \left[\frac{10 \times 150^3}{12} + 1500(0)^2\right]$$

$= 11666.67 + 2812500 =$ **2824166.67 mm⁴ [Ans.]**

Problem 3: Find the moment of inertia of I-section, top flange 80 × 10 mm, web 10 × 140 mm and bottom flange 10 × 100 mm about x-x axis and y-y axis passing through centre of gravity of the section:

Solution : Fig. 7.14 shows I-section with given dimentions.

Position of centre of gravity (G): As the section is symmetrical about y-y axis.

$\therefore \quad \bar{x} = 100/2 = 50$ mm

To find the position of $\bar{y}$ proceed as under:

Areas: $a_1 = 80 \times 10 = 800\ mm^2$

$a_2 = 10 \times 140 = 1400\ mm^2$

$a_3 = 100 \times 10 = 1000\ mm^2$

$y_1 = 10 + 140 + 5 = 155$ mm

$y_2 = 10 + 70 = 80$ mm

$y_3 = 5$ mm

$$\therefore \quad \bar{y} = \frac{a_1 y_1 + a_2 y_2 + a_3 y_3}{a_1 + a_2 + a_3}$$

$$= \frac{800 \times 155 + 1400 \times 80 + 1000 \times 5}{800 + 1400 + 1000}$$

$= 75.3$ mm

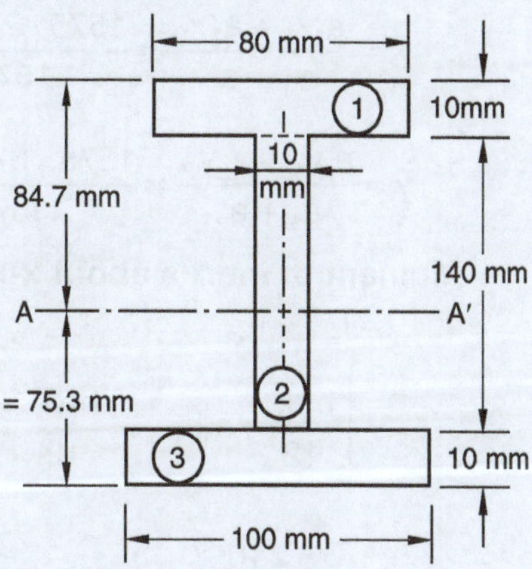

Fig. 7.14

Moment of inertia about x-x axis:

$$I_{xx} = I_{xx1} + I_{xx2} + I_{xx3}$$

$$= \left[\frac{b_1 d_1^3}{12} + a_1 h_1^2\right] + \left[\frac{b_2 d_2^3}{12} + a_2 h_2^2\right] + \left[\frac{b_3 d_3^3}{12} + a_3 h_3^2\right]$$

$$= \frac{80 \times 10^3}{12} + 800 \times 79.7^2$$

$$+ \frac{10 \times 140^3}{12} + 1400 \times 4.7^2 + \frac{100 \times 10^3}{12} + 1000 \times 70.3^2$$

$= [6667 + 5081672] + [2286667 + 30926] + [8333 + 4942090]$

$= 5088339 + 2317592.7 + 4950423$

= 12356355 mm⁴ [Ans.]

Moment of inertia about y-y axis:

$$I_{yy} = I_{yy1} + I_{yy2} - I_{yy3}$$

$$= \left[\frac{d_1b_1^3}{12} + a_1h_1^2\right] + \left[\frac{d_2b_2^3}{12} + a_2h_2^2\right] + \left[\frac{d_3b_3^3}{12} + a_3h_3^2\right]$$

$$= \frac{10\times 80^3}{12} + 800\times 0 + \frac{140\times 10^3}{12} + 1400\times 0 + \frac{10\times 100^3}{12} + 1000\times 0$$

$= 4266667 + 116667 + 833333 =$ **1271667 mm⁴ [Ans.]**

Problem 4: Find the moment of inertia of L-section 120 × 100 × 15 mm about x-x axis and y-y axis passing through the centre of gravity of the section. Also calculate the radius of gyration about x-x axis and y-y axis.

Solution: Fig. 7.15 shows the L-section of 120 × 100 × 15 cm

Position of centre of gravity (G): As the section is not symmetrical about any axis, therefore, to find the value of $\bar{x}$ and $\bar{y}$ proceed as under:

$a_1 = 15 \times 105 = 1575\ \text{mm}^2$ $\quad x_1 = 15/2 = 7.5\ \text{mm}$ $\quad y_1 = 15 + 105/2 = 67.5\ \text{mm}$

$a_2 = 100 \times 15 = 1500\ \text{mm}^2$ $\quad x_2 = 100/2 = 50\ \text{mm}$ $\quad y_2 = 15/2 = 7.5\ \text{mm}$

Position of $\bar{x}$ and $\bar{y}$ distances of centroids:

$$\therefore\ \bar{x} = \frac{a_1x_1 + a_2x_2}{a_1 + a_2} = \frac{1575\times 7.5 + 1500\times 50}{1575 + 1500} = 28.2\ \text{mm and}$$

$$\bar{y} = \frac{a_1y_1 + a_2y_2}{a_1 + a_2} = \frac{1575\times 67.5 + 1500\times 7.5}{1575 + 1500} = 38.2\ \text{mm}$$

Moment of inertia about x-x axis:

$$I_{xx} = I_{xx1} + I_{xx2}$$

$$= \left[\frac{b_1d_1^3}{12} + a_1h_1^2\right] + \left[\frac{b_2d_2^3}{12} + a_2h_2^2\right]$$

$$= \frac{15\times 105^3}{12} + 1575\times (29.3)^3$$

$$+ \frac{100\times 15^3}{12} + 1500\times (30.7)^2$$

$= 2799153 + 1441860 =$ **4241013 mm⁴**

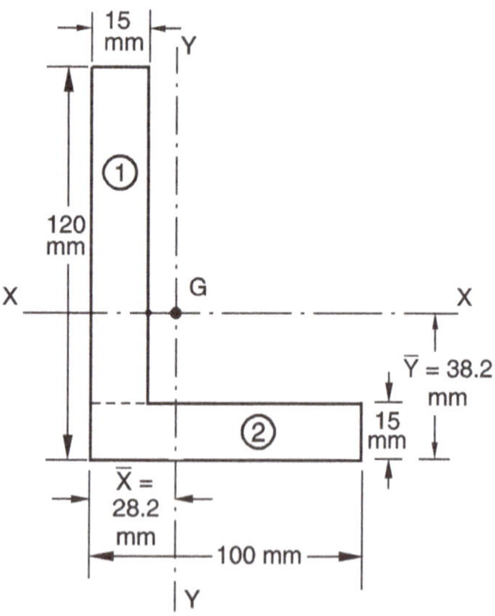

Fig. 7.15

Moment of inertia about y-y axis:

$$I_{yy} = I_{yy1} + I_{yy2}$$

$$= \left[\frac{d_1b_1^3}{12} + a_1h_1^2\right] + \left[\frac{d_2b_2^3}{12} + a_2h_2^2\right]$$

$$= \left[\frac{105\times15^3}{12}+1575\times(20.7)^2\right]+\frac{15\times100^3}{12}+1500\times(21.8)^2$$

$= 704403 + 1962860 =$ **2667263 mm^4 [Ans.]**

Radius of gyration about x-x axis [k_{xx}]:

$k_{xx} + \sqrt{I_{xx}/A}$ [$\therefore$ $A = a_1 - a_2 + 1575 + 1500 = 3075$ mm^2]

$= \sqrt{\dfrac{4241013}{3075}} =$ **37.14 mm [Ans.]**

Radius of gyration about y-y axis:

$k_{yy} = \sqrt{I_{yy}/A} = \sqrt{\dfrac{2667263}{3075}} =$ **29.45 mm [Ans.]**

Problem 5: Find the moment of inertia of the channel section 175 × 75 × 10 mm about the horizontal and vertical axes passing through the centre of gravity of the section.

Solution: Fig. 7.16. shows channel of 175 × 75 × 10 mm

Position of centre of gravity (G): As the section is symmetrical about x-x axes.

$\therefore\ \bar{y} = \dfrac{175}{2} = 87.5$ mm

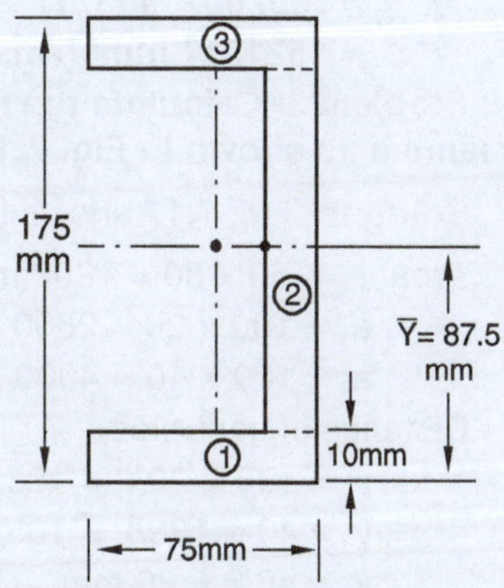

Fig. 7.16

To find $\bar{x}$ proceed as under:

Areas:

$a_1 = 75 \times 10 = 750$ mm^2

$a_2 = 155 \times 10 = 1550$ mm^2

$a_3 = 75 \times 10 = 750$ mm^2

Distance of centroids

$x_1 = 75/2 = 37.5$ mm

$x_2 = 75 - 10/2 = 70$ mm

$x_3 = 75/2 = 37.5$ mm

$$\therefore\ \bar{x} = \frac{a_1x_1 + a_2x_2 + a_3x_3}{a_1 + a_2 + a_3}$$

$$= \frac{750\times37.7+1500\times70+750\times37.5}{750+1550+750}$$

$$= \frac{164750}{3050} = 53.6 \text{ mm}$$

Moment of inertia of section about x-x axis [I_{xx}]:

$I_{xx1} = I_{xx1} + I_{xx2} + I_{xx3}$

$I_{xx1} = \dfrac{75\times10^3}{12}+750\times(87.5-10/2)^2$ $\left[\because I_{xx} = \dfrac{bd^3}{12}+ah^2\right]$

$= 6250 + 5104687 = 5110937$ mm^4

$I_{xx2} = \dfrac{10\times155^3}{12}+1550\times0 = 3103229$ mm^4

$$I_{xx3} = \frac{75 \times 10^3}{12} + 750(87.5 - 10/2)^2$$

$= 6250 + 5104687 = 5110937\ mm^4$

$\therefore\ I_{xx} = 5110937 + 3103229 + 5110937\ mm^4$

$= \mathbf{13325103\ mm^4}$ **[Ans.]**

Moment of inertia of section about y-y axis [I_{yy}]:

$I_{yy} = I_{yy1} + I_{yy2} + I_{yy3}$

$$I_{yy1} = \frac{10 \times 75^3}{12} + 750(54 - 75/2)^2 \qquad \left[\therefore I_{yy} = \frac{db^3}{12} + ah^2\right]$$

$= 351562 + 204188 = 555750\ mm^4$

$$I_{yy2} = \frac{155 \times 10^3}{12} + 1550(21 - 10/2)^2$$

$= 12917 + 396800 = 409717\ mm^4$

$$I_{yy3} = \frac{10 \times 75^3}{12} + 750(54 - 75/2)^2$$

$= 351562 + 204188 = 555750\ mm^4$

$\therefore\ I_{yy} = 555750 + 409717 + 555750$

$= \mathbf{1521217\ mm^4}$ **[Ans.]**

Problem 6: Calculate the moment of inertia about the centroidal axis x-x for the lamina as shown in Fig. 7.17.

Solution: Fig. 7.17 shows the given lamina. To find $\bar{y}$ proceed as under:

Area: $a_1 = 60 \times 30 = 1800\ mm^2$

$a_2 = 130 \times 20 = 2600\ mm^2$

$a_3 = 100 \times 40 = 4000\ mm^2$

Distance of centroids

$y_1 = 170 + 30/2 = 185\ mm$

$y_2 = 40 + 130/2 = 105\ mm$

$y_3 = 40/2 = 20\ mm$

$$\therefore\ \bar{y} = \frac{a_1 y_1 + a_2 y_2 + a_3 y_3}{a_1 + a_2 + a_3}$$

$$= \frac{800 \times 185 + 2600 \times 105 + 4000 \times 20}{1800 + 2600 + 4000}$$

$$= \frac{686000}{8400} = 81.7\ mm$$

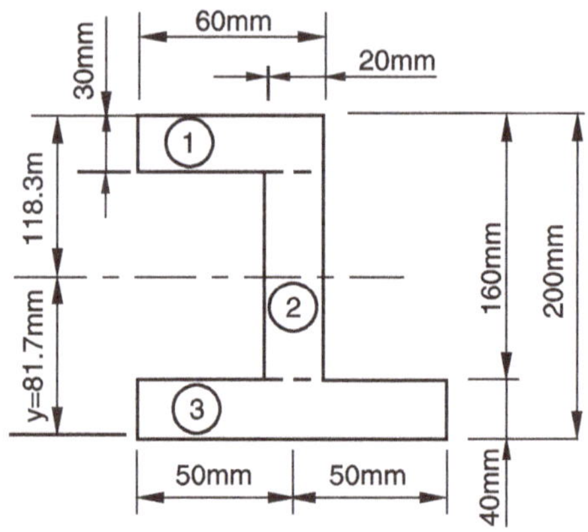

Fig. 7.17

$\therefore$ Centroidal axis x-x is 81.7 mm from base of lamina as shown in Fig. 7.17

Moment of inertia about x-x axis [I_{xx}]:

$$I_{xx} = I_{xx1} + I_{xx2} + I_{xx3} \qquad \left[\because I_{xx} = \frac{bd^3}{12} + ah^2\right]$$

$$I_{xx1} = \frac{60 \times 30^3}{12} + 1800(118.3 - 30/2)^2$$

$$= 135000 - 19207602 = 19342602 \text{ mm}^4$$

$$I_{xx2} = \frac{20 \times 130^3}{12} + 2600(105 - 81.7)^2$$

$$= 3661667 + 1411514 = 5073181 \text{ mm}^4$$

$$I_{xx3} = \frac{100 \times 40^3}{12} + 4000(81.7 - 40/2)^2$$

$$= 533333 + 15227560 = 15760893 \text{ mm}^4$$

$$\therefore \; I_{xx} = 19342602 + 5073181 + 15760893$$

$$\mathbf{= 40176676 \text{ mm}^4} \text{ [Ans.]}$$

Problem 7: Find the moment inertia of the section as shown in fig. 7.18 about x-x axis and y-y axis.

Solution: Fig. 7.18 shows the given section.

Position of centre of gravity (G): As section is symmetrical about y-y axis:

$$\therefore \quad \bar{x} = \frac{120}{2} = 60 \text{ mm}$$

To find the position of $\bar{y}$ proceed as under:

Area: $a_1 = 20 \times 120 = 2400 \text{ mm}^2$

$a_2 = 80 \times 20 = 1600 \text{ mm}^2$

$a_3 = 20 \times 120 = 2400 \text{ mm}^2$

$a_4 = 80 \times 20 = 1600 \text{ mm}^2$

Distance of centroids:

$y_1 = 120/2 = 60 \text{ mm}$

$y_2 = 100 + 20/2 = 110 \text{ mm}$

$y_3 = 120/2 = 60 \text{ mm}$

$y_4 = 40 + 20/2 = 50 \text{ mm}$

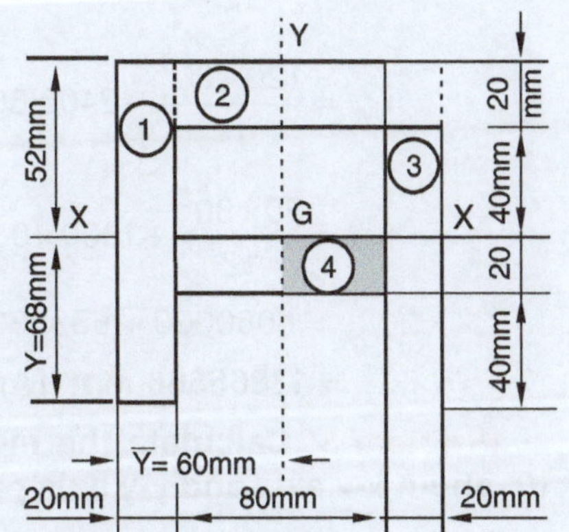

Fig. 7.18

$$\therefore \quad \bar{y} = \frac{a_1 y_1 + a_2 y_2 + a_3 y_3 + a_4 y_4}{a_1 + a_2 + a_3 + a_4}$$

$$= \frac{2400 \times 60 + 1600 \times 110 + 2400 \times 60 + 1600 \times 50}{2400 + 1600 + 2400 + 1600}$$

$$= \frac{544000}{8000} = 68 \text{ mm}$$

Moment of inertia about x-x axis [I_{xx}]:

$$I_{xx} = I_{xx1} + I_{xx2} + I_{xx3} + I_{xx4} \qquad \left[\because I_{xx} = \frac{bd^3}{12} + ah^2\right]$$

$$\text{Now, } I_{xx1} = \frac{20 \times 120^3}{12} + 2400(68 - 60)^2 = 3033600 \text{ mm}^4$$

$$I_{xx2} = \frac{80 \times 20^3}{12} + 1600(52-10) = 2875733 \text{ mm}^4$$

$$I_{xx3} = \frac{20 \times 120^3}{12} + 2400(68-60)^2 = 3033600 \text{ mm}^4$$

$$I_{xx4} = \frac{80 \times 20^3}{12} + 1600(68-50)^2 = 571733 \text{ mm}^4$$

$\therefore$ $I_{xx} = 3033600 + 2875733 + 3033600 + 571733$

$= \mathbf{9514666\ mm^4}$ **[Ans.]**

Moment of inertia about y-y axis [I_{yy}]:

$$I_{yy} = I_{yy1} + I_{yy2} + I_{yy3} + I_{yy4} \qquad \left[\because I_{yy} = \frac{db^3}{12} + ah^2\right]$$

Now, $I_{yy1} = \frac{120 \times 20^3}{12} + 2400(60-10)^2 = 6080000 \text{ mm}^4$

$$I_{yy2} = \frac{20 \times 80^3}{12} + 1600 \times 0 = 853333 \text{ mm}^4$$

$$I_{yy3} = \frac{120 \times 20^3}{12} + 2400(60-10)^2 = 6080000 \text{ mm}^4$$

$$I_{yy4} = \frac{20 \times 80^3}{12} + 1600 \times 0 = 853333 \text{ mm4}$$

$\therefore$ $I_{yy} = 6080000 + 853333 + 6080000 + 853333$

$= \mathbf{13866666\ mm^4}$ **[Ans.]**

Problem 8: Calculate the moment of inertia of z-section as shown in fig. 7.19. about x-x axis and y-y axis passing through centre of gravity of the section.

Soluton: Fig. 7.19 shows the given z-section.

Position of centre of gravity (G): As the section is not symmetrical about any axis, therefore, to find $\bar{x}$ and $\bar{y}$ proceed as under:

Area:	Distance of centroids:	
$a_1 = 100 \times 10$;	$x_1 = 70 + 100/2$;	$y_1 = 120 + 10/2$
$= 1000 \text{ mm}^2$	$= 120$ mm	$= 125$ mm
$a_2 = 120 \times 10$;	$x_2 = 70 + 10/2$;	$y_2 = 120/2$
$= 1200 \text{ mm}^2$	$= 75$ mm	$= 60$ mm
$a_3 = 70 \times 10$; $= 700 \text{ mm}^2$	$x_3 = 70/2$; $= 35$ mm	$y_3 = 10/2$ $= 5$ mm

$$\therefore \quad \bar{x} = \frac{a_1x_1 + a_2x_2 + a_3x_3}{a_1 + a_2 + a_3}$$

$$= \frac{1000 \times 120 + 1200 \times 75 + 700 \times 35}{1000 + 1200 + 700}$$

$$= \frac{234500}{2900} = 80.9 \text{ mm}$$

Now, $\bar{y} = \frac{a_1y_1 + a_2y_2 + a_3y_3}{a_1 + a_2 + a_3}$

$$= \frac{1000 \times 125 + 1200 \times 60 + 700 \times 5}{1000 + 1200 + 700}$$

$$= \frac{200500}{2900} = 69.1 \text{ mm}$$

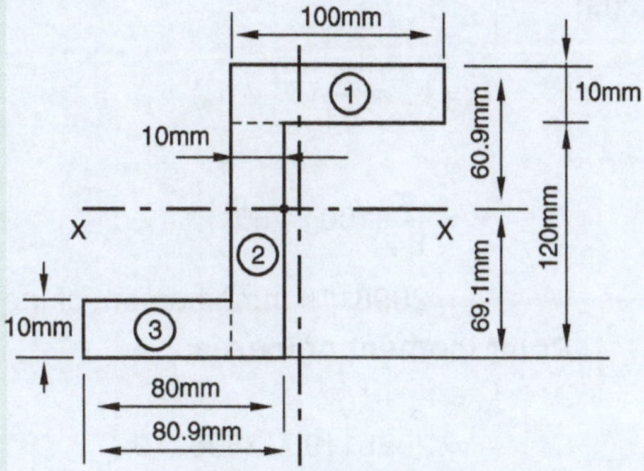

Fig. 7.19

Moment of inertia about x-x axis (I_{xx}):

$$I_{xx} = I_{xx1} + I_{xx2} + I_{xx3} \qquad \left[\because I_{xx} = \frac{bd^3}{12} + ah^2\right]$$

$$\therefore \quad I_{xx1} = \frac{100 \times 10^3}{12} + 1000(60.9 - 5)^2 = 3133143 \text{ mm}^4$$

$$I_{xx2} = \frac{10 \times 120^3}{12} + 1200(69.1 - 60)^2 = 1539372 \text{ mm}^4$$

$$I_{xx3} = \frac{70 \times 10^3}{12} + 700(69.1 - 5)^2 = 2882000 \text{ mm}^4$$

$$\therefore \quad I_{xx} = 3133143 + 1539372 + 2882000$$

$$= \mathbf{7554515 \text{ mm}^4} \text{ [Ans.]}$$

Moment of inertia about y-y axis (I_{yy}):

$$I_{yy} = I_{yy1} + I_{yy2} + I_{yy3} \qquad \left[\because I_{yy} = \frac{db^3}{12} + ah^2\right]$$

$$\therefore \quad I_{yy1} = \frac{10 \times 100^3}{12} + 1000(50 - 10.9)^2 = 2362143 \text{ mm}^4$$

$$I_{yy2} = \frac{120 \times 10^3}{12} + 1200(10.9 - 5)^2 = 51772 \text{ mm}^4$$

$$I_{yy3} = \frac{10 \times 70^3}{12} + 700(80.9 - 35)^2 = 1760600 \text{ mm}^4$$

$$\therefore \quad I_{yy} = 2362143 + 51772 + 1760600 = \mathbf{4174515 \text{ mm}^4} \text{ [Ans.]}$$

Problem 9: Find the polar moment of inertia of a hollow circular section whose external diametre is 100 mm and thickness is 10 mm.

Solution: Fig. 7.20 shows a hollow circular section

Moment of inertia about x-x axis and y-y axis:

$$I_{xx} = I_{yy} = \frac{\pi}{64}(D^4 - d^4)$$

$$= \frac{\pi}{64}(100^4 - 80^4)$$

$= 2898119\ mm^4$ moment of inertia:

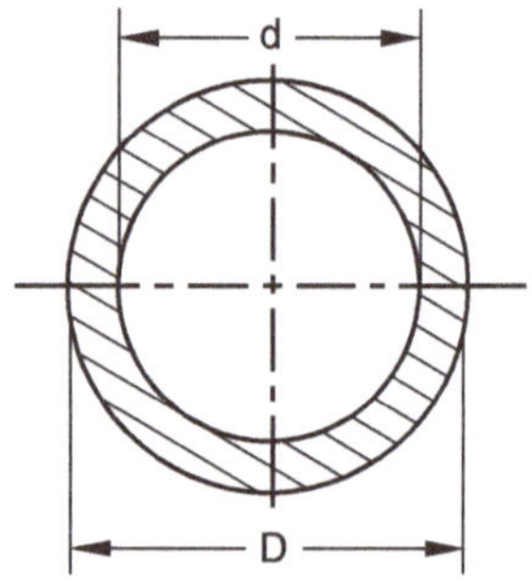

Fig. 7.20

Polar moment of inertia:

$$I_{zz} = I_{xx} + I_{yy}$$

$= 2898119 + 2898119$

$=$ **$5796238\ mm^4$ [Ans.]**

Problem 10: Calculate the moment of inertia about an axis passing through centroid and parallel to top flange plate of the composite section shown in fig. 7.21.

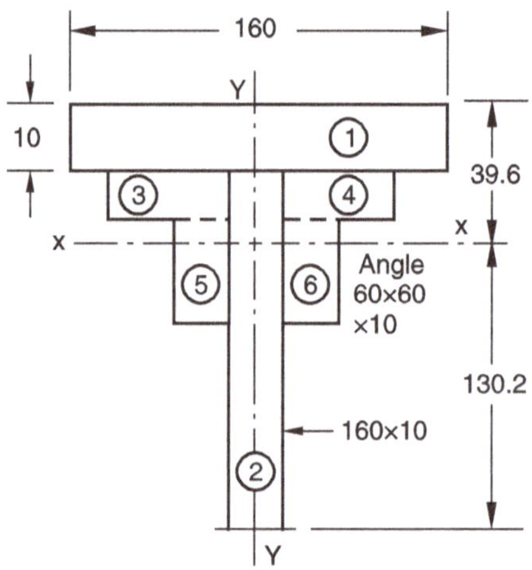

Fig. 7.21

Solution: Fig. 7.21 shows a composite section. As the section is not symmetrical about x-x axis, therefore, to locate its x-x position proceed as under:

Areas: $a_1 = 160 \times 10 = 1600\ mm^2$

$a_2 = 10 \times 160 = 1600\ mm^2$

$a_3 = a_4 = 60 \times 10 = 600\ mm^4$

$a_5 = a_6 = 10 \times 50 = 500\ mm^2$

$y_1 = 160 + 10/2 = 165\ mm$

$y_2 = 160/2 = 80\ mm$

$y_3 = y_4 = 160 - 10/2 = 155\ mm$

$$y_5 = y_6 = 160 - 10 - 50/2 = 125 \text{ mm}$$

$$\therefore \quad \bar{y} = \frac{a_1y_1 + a_2y_2 + a_3y_3 + a_4y_4 + a_5y_5 + a_6y_6}{a_1 + a_2 + a_3 + a_4 + a_5 + a_6}$$

$$= \frac{1600 \times 165 + 1600 \times 80 + 2(600 \times 155) + 2(500 \times 125)}{1600 + 1600 + 2 \times 600 + 2 \times 50} = 130.2 \text{ mm}$$

Moment of inertia about x-x axis (I_{xx}):

$$I_{xx} = I_{xx1} + I_{xx2} + I_{xx3} + I_{xx4} + I_{xx5} + I_{xx6} \qquad [\because I_{xx3} = I_{xx4}; I_{xx5} = I_{xx6}]$$

$$I_{xx1} = \frac{160 \times 10^3}{12} + 1600(39.8 - 5)^2 = 1950997$$

$$I_{xx2} = \frac{10 \times 160^3}{12} + 1600(130.2 - 80)^2 = 7445397.3$$

$$I_{xx3} = I_{xx4} = \frac{60 \times 10^3}{12} + 600(39.8 - 10 - 5)^2 = 374024$$

$$I_{xx5} = I_{xx6} = \frac{10 \times 50^3}{12} + 500(130 - 115)^2 = 216666.7$$

$$\therefore \quad \mathbf{I_{xx} = 10577775.7 \text{ mm}^4 \text{ [Ans.]}}$$

Problem 11: Find the moment of inertia of the section shown in fig. 7.22. about x-x axis.

Solution: Moment of inertia about x-x axis:

$$I_{xx} = I_{xx1} - I_{xx2}$$

$$I_{xx_1} = [I_{G_1} + a_1h_1^2]$$

$$= \frac{\pi}{64}D^4 + a_1h_1^2$$

$$= \frac{\pi}{64} \times 80^4 + 0 = 2010619 \text{ mm}^4$$

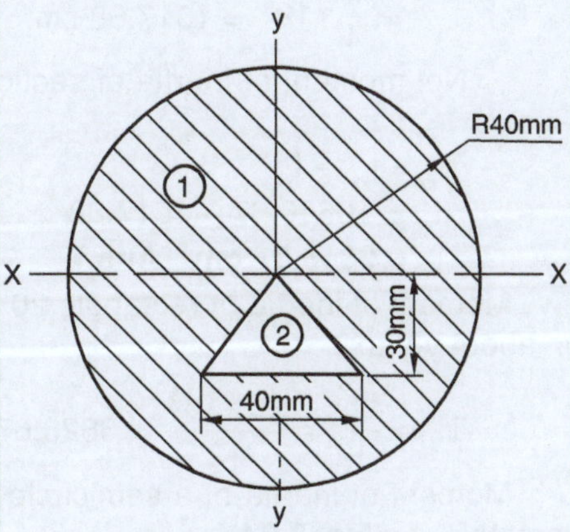

Fig. 7.22

$$Ixx_2 = [I_{G2} + a_2h_2^2] = \left[\frac{bh^3}{36} + \frac{1}{2} \times b \times h(k)^2\right]$$

$$- \quad \frac{40 \times 30^3}{36} + \frac{1}{2} \times 40 \times 30 \quad \frac{2}{3} \times 30^{\,2}$$

$$\mathbf{= 270000 \text{ mm}^4}$$

$$I_{xx1} = \frac{\pi}{64} \times 80^4 + 0 = 2010619 \text{ mm}^4$$

$$\therefore \quad I_{xx} = 2010619 - 270000 = \mathbf{1740619 \text{ mm}^4 \text{ [Ans.]}}$$

Problem 12: Calculate the Moment of inertia of the section shown in fig. 7.23 about AB Axis.

Solution: Moment of inertia about AB axis:

$$I_{AB} = I_{AB1} + I_{AB2} - I_{AB3}$$

$$I_{AB1} = \frac{bh^3}{36} = \frac{120 \times 120^3}{36}$$

$$= 17280000 \text{ mm}^4$$

$$I_{AB2} = \frac{1}{2} \times \frac{\pi}{64} D^4 = \frac{1}{2} \times \frac{\pi}{64} (120)^2$$

$$= 5089380 \text{ mm}^4$$

$$I_{AB3} = \frac{\pi}{64} (d)^4 = \frac{\pi}{64} (60)^4 = 636172 \text{ mm}^4$$

$$\therefore \quad I_{AB} = 17280000 + 5089380 - 636172$$

$$= \mathbf{21733208 \text{ mm}^4} \text{ [Ans.]}$$

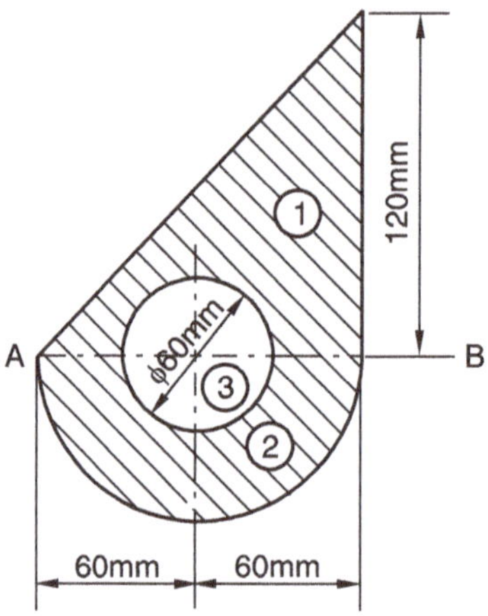

Fig. 7.23

Problem 13: Find the moment of inertia of the cross-section of a cast iron block about x-x axis and y-y axis as shown in Fig. 7.24.

Solution: Moment of inertia of rectangle 16 cm × 20 cm about x-x axis.

$$I_{xx1} = \frac{bh^3}{12} = \frac{16 \times 20^3}{12} = 10666.67 \text{ cm}^4$$

Moment of inertia of circle of 12 cm about x-x axis:

$$I_{xx2} = \frac{\pi}{64} \times d^4$$

$$= \frac{\pi}{4} \times 12^4 = 1017.88 \text{ cm}^4$$

∴ Net moment of inertia of section about x-x axis

$$I_{xx} = I_{xx1} - I_{xx2}$$

$$= 10666.67 - 1017.88$$

$$= \mathbf{9648.67 \text{ cm}^4} \text{ [Ans.]}$$

Moment of inertia of rectangle 20 cm × 16 cm about y-y axis

$$I_{yy1} = \frac{bh^3}{12} = \frac{20 \times 16^3}{12} = 6826.67 \text{ cm}^4$$

Moment of inertia of a semicircle about a diameter (see Fig. 3.24)

$$= \frac{1}{2} \times \frac{\pi}{64} d^4 = \frac{1}{2} \times \frac{\pi}{64} \times 12^4$$

$$= 508.94 \text{ cm}^4$$

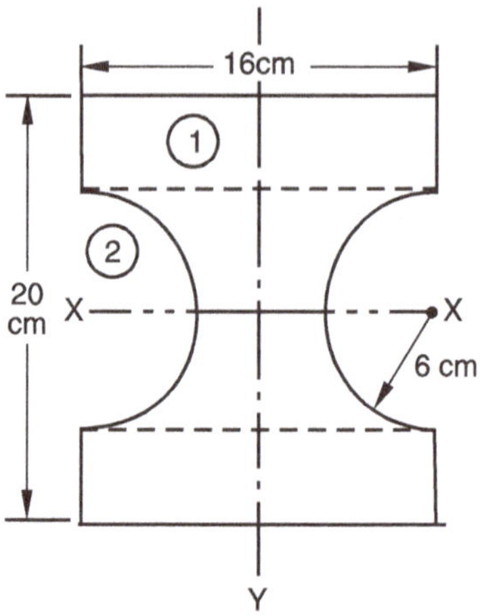

Fig. 7.24

Distance of C.G. of semicircle from the diameter = 0.424 r

[r = radius of semicircle]

$$= 0.424 \times 6 = 2.54 \text{ cm}$$

Area of semicircle = $\frac{1}{2} \pi r^2 = \frac{1}{2} \times \pi \times 6^2 = 56.55 \text{ cm}^2$

Moment of inertia of the semicircle about an axis passing through its C.G. and parallel to diameter

$= 508.94 - 56.55\,(2.54)^2$ [applying parallel axis theorem]

$= 508.94 - 365.98 = 142.96 \text{ cm}^4$

Distance of C.G. of semicircle from y-y axis,

$= 8 - 2.54 = 5.46$ cm

Moment of inertia of semicircle about y-y axis,

$I_{yy2} = 142.96 + 56.55 \times (5.46)^2$

$= 142.96 + 1685.85 = \mathbf{1828.81\ cm^4}$

$\therefore$ Moment of inertia about y-y axis

$I_{yy} = I_{yy1} - 2I_{yy2}$

$= 6826.67 - 2 \times 1828.81$

$= \mathbf{3169.05\ cm^4}$ **[Ans.]**

HIGHLIGHTS

1. The product of the area (or mass) about an axis and square of the distance of the C.G. of the area (or mass) from that axis is called moment of inertia. It is represented by I.

   $\therefore\ I = Ak^2$

2. Radius of gyration, $k = \sqrt{I/A}$

   Radius of gyration about x-x axis y-y axis.

   $k_{xx} = \sqrt{I_{xx}/A}$ and $k_{yy} = \sqrt{I_{yy}/A}$

3. According to theorem of perpendicular axis

   $I_{zz} = I_{xx} + I_{yy}$

4. According to the theorem of parallel axis

   $I_{AB} = I_G + Ah^2$

   where I_G = Moment of inertia of a given area about an axis passing through C.G. of the area.

   I_{AB} = Moment of inertia of the given area about an axis AB, which is parallel to the axis passing through C.G.

   A = Area of the section.

   h = Distance between the axis passing through C.G. of section and axis AB.

5. Moment of inertia of a circular section = $\pi D^4/64$

6. Moment of inertia of a rectangular section:

   (i) about horizontal axis passing through C.G. = $bd^3/12$

   (ii) about horizontal axis passing through base = $bd^3/3$

7. Moment of inertia of a triangular section:

   (i) about the base = $bh^3/12$

   (ii) about an axis passing through C.G. and parallel to the base = $bh^3/36$

MODEL TEST QUESTIONS

1. Define moment of inertia and explain why it is called second moment of area?
2. Define the terms: Radius of gyration and modulus of section.
3. State and prove the theorem of perpendicular axis.
4. State and prove the theorem of parallel axis.
5. Derive the expression for moment of inertia of rectangular section about horizontal axis passing through the base of the section.
6. Derive the expression for the moment of inertia of a triangular section about an axis passing through the C.G. of the section and parallel to the base.

OBJECTIVE TYPE QUESTIONS

1. Fill in the blanks:
   (i) Moment of inertia of a circular section is ....
   (ii) Moment of inertia of any section is .... (maximum, minimum) about an axis, passing through its C.G.
   (iii) Moment of inertia of a triangular section about its base is..................
   (iv) Polar moment of inertia of a circular section is................ times the moment of inertia about an axis passing through C.G.
   (v) Moment of inertia of a rectangular section about an horizontal axis passing through C.G. is....................

Ans. 1. (i) $\pi D^4/64$ (ii) Maximum (iii) $bh^3/12$ (iv) Two (v) $bd^3/12$

2. Pick up the correct words:
   (i) Moment of inertia of a rectangle about its x-x axis is given by:
   (a) $bd^3/12$ (b) $db^3/12$ (c) $bd^2/6$ (d) $bd^3/6$.
   (ii) Moment of inertia of a semi-circle about its x-x axis is given by:
   (a) $0.22\ r^4$ (b) $0.11\ r^4$ (c) $0.14\ r^4$ (d) $0.2\ r^4$
   (iii) Moment of inertia of a quadrant about its x-x axis is:
   (a) $0.055\ r^4$ (b) $0.04\ r^4$ (c) $0.06\ r^4$ (d) r^4.
   (iv) The moment of inertia about a principle axis is called
   (a) mass moment of inertia (b) second moment of area
   (c) principal moment of inertia (d) None of the above.

Ans. 2. (i) $bd^3/12$ (ii) $0.11\ r^4$ (iii) $0.055\ r^4$ (iv) principal moment of inertia.]

PROBLEMS FOR PRACTICE

1. Determine the moment of inertia of a T-section of dimensions 100 × 100 × 20 mm about the horizontal and vertical axes, passing through the centre of gravity of the section.

   [**Ans.** $I_{xx} = 314.2 \times 10^4\ mm^4$, $I_{yy} = 172 \times 10^4\ mm^4$]

2. Find the moment of inertia about the centroidal axes of the angle section of 80 × 200 × 20 mm

   [**Ans.** $2.91 \times 10^6\ mm^4$, $1.63 \times 10^6\ mm^4$]

3. A channel section of 45 mm × 100 mm × 5 mm. Find the moment of inertia of the section about the horizontal axis and vertical axis passing through centre of gravity.

[**Ans.** 132×10^4 mm^4, 166875 mm^4]

4. Calculate the moment of inertia of I-section of the following dimensions about X-X axis.

Top flange = 200 × 80 mm

Bottom flange = 100 × 40 mm

Central web = 280 × 60 mm

[**Ans.** 60328×10^4 mm^4]

5. The horizontal beam of channel section as shown in fig. 3.25 is simply supported. Calculated the moment of inertia about X-X axis.

[**Ans.** 1131×10^4 mm^4]

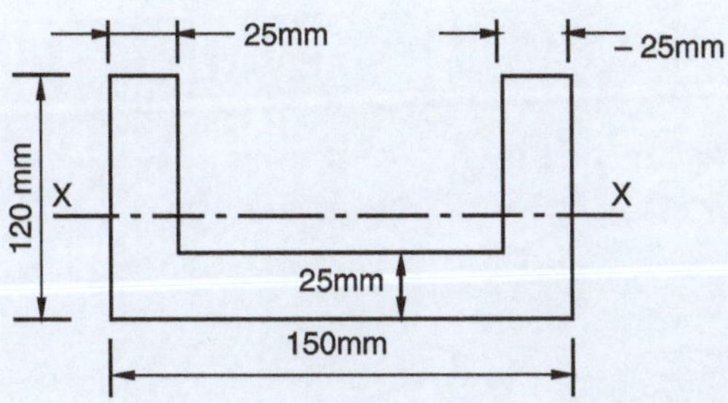

Fig. 7.25

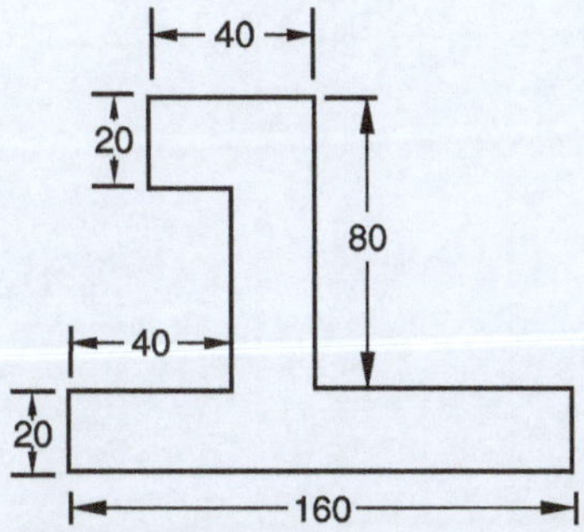

Fig. 7.26

6. Find the moment of inertia about the horizontal axis passing through the centroid of the section shown in Fig. 3.26.

[**Ans.** 2373322.22 mm^2]

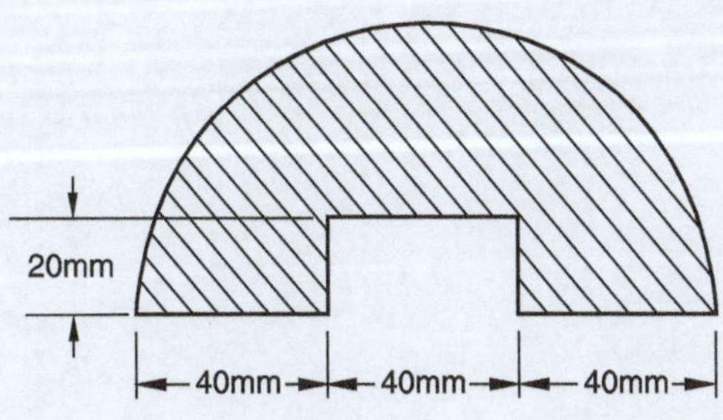

Fig. 7.27

7. Calculate the moment of inertia of the shaded area shown in Fig. 7.27 about horizontal and vertical centroidal axis.

[**Ans.** I_{xx} = 1354300 mm^4, I_{yy} = 4983000 mm^4]

This Page has been intentionally left blank

LABORATORY EXPERIMENTS

This Page has been intentionally left blank

Experiment No. 1

Objective: To verify the law of polygon of forces.

Apparatus: Universal force table apparatus, pulleys, a ring with five strings, weights (see Fig. 1-1).

Materials required: Paper sheet, drawing pins or fixing tape, set squares, pencil, scale cards, etc.

Theory: The Law of polygon of forces states that "if a number of coplanar concurrent forces acting on a body be represented in magnitude and direction by the sides of a polygon, taken in order, then their resultant Is represented in magnitude and direction by the closing side of the polygon, taken in the apposite; order.

Procedure :

1. Clamp the pulleys to the graduated disc of the force table and make it horizontal by adjusting the screws provided at its base.

2. Tie each of the five strings to the circumference of the small ring provided with the apparatus and place it around the axle of the apparatus. The other ends of the strings are to be attached to the weight hangers which are to carry small slotted weights, hanging over the pulleys as shown in fig. 1-1.

3. Put small weights on the weight hangers in such a manner that the ring is positioned symmetrically around the axle and it does not touch the axle of the apparatus.

4. Note the position of any one string on the disc and then find the angles between all the strings as $\theta_1, \theta_2, \theta_3, \theta_4$ and θ_5.

5. Draw the space diagram of the forces $F_1 = W_1$, $F_2 = W_2$, $F_3 = W_3$, $F_4 = W_4$ and $F_5 = W_5$ acting on the ring as shown in fig. 1-2.

6. Draw the force diagram abcde as shown in flg. 1-3 and verify that the polygon abcde is completed. !t the last force $F_5 = W_5$ represented by the side ea falls short or greater than the side which would complete the polygon, then find the percentage error between the force EA (R) and ea (R_1) required to complete the polygon as given in the observation chart.

7. Similarly, measure the angle o a b ;.i.e. α_1 (see fig. 1 -3) which' should be equal to θ_5 and if they are different, then find the percentage error between these angles as given In the observation chart.

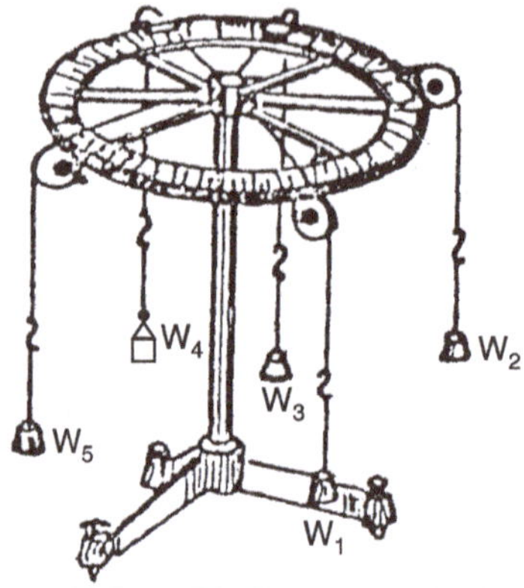

Fig. 1.1. *Universal force table apparatus*

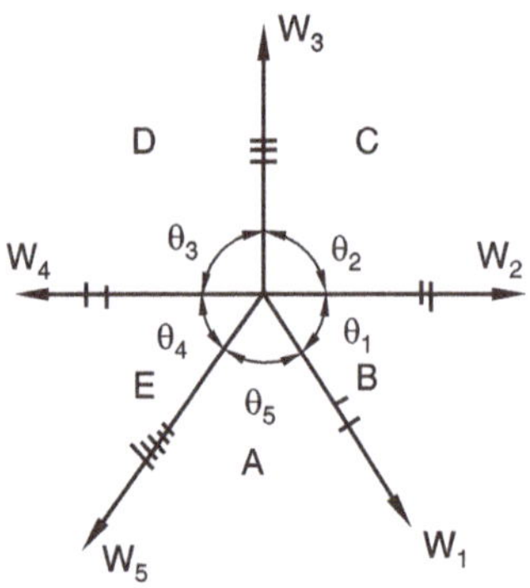

Fig. 1.2. *Space diagram*

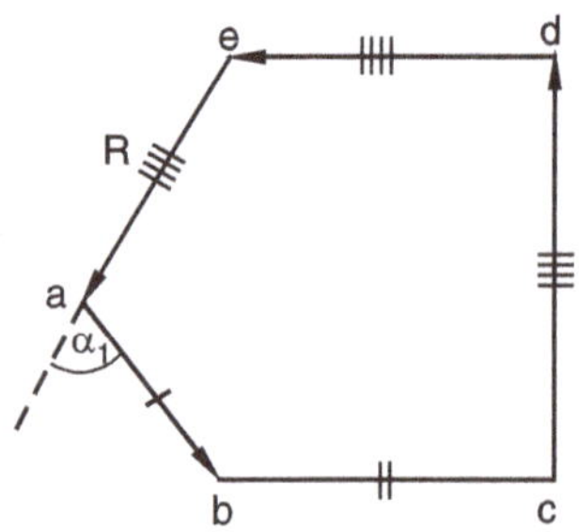

Fig. 1.3. *Vector diagram*

Observations:

S. No.	Weights suspended or forces					Angles					Calculated/ graphical value of 'R' & 'α'		Percentage Errors		Remarks
													In Magnitude	In Direction	
	$F_1 = W_1$	$F_2 = W_2$	$F_3 = W_3$	$F_4 = W_4$	$F_5 = W_5$	θ_1	θ_2	θ_3	θ_4	$\theta_1 = \alpha_1$	R_1	α_1	$\frac{R - R_1}{R} \times 100$	$\frac{\alpha - \alpha_1}{\alpha} \times 100$	
1.															
2.															
3.															

8. Change the weight and repeat the experiment at least thrice as before and get the mean value.as before and get the mean value.

Precautions:

1. The graduated disc i.e., the force table should perfectly be made horizontal by adjusting screws provided at its base. This can be checked with the help of a spirit level.

2. The pulleys should be frictionsless.

3. The knots of the strings should not be tight so that they may slide over the ring.

4. The ring should not touch the axle or the disc of the apparatus.

5. The position of the strings should be carefully noted only after the system has come to rest.

Specimen calculations:

$$\text{Percentage error in the magnitude} = \frac{R - R_1}{R} \times 100$$

$$\text{Percentage error in the direction} = \frac{\alpha - \alpha_1}{\alpha} \times 100$$

QUESTIONS FOR VIVA-VOCE

1. What is the "law of polygon of forces"?
2. How will you verify the "law of polygon of force"?
3. Why the force table is kept horizontal?
4. What can be sources of error in this experiment?
5. Name apparatus which can be used in place of universal force table apparatus for verifying the "polygon law of forces."

Experiment No. 2

Objective: To verify the forces in the different mambers of a jib crane.

Apparatus: Jib crane apparatus, weights, metre rod, thread, etc. (see fig. 2-1).

Material required : Paper sheet, drawing pins or fixing tape, set squares, pencil, scale cards, etc.

Theory: The experiment is based upon the triangle law of forces which states that "If two force acting upon a body can be represented in magnitude and direction by the two sides of a triangle, taken in order, then their resultant is given by the third side of the triangle, taken in the opposite order".

Procedure:

1. Note down the initial reading of spring balance in the tie and the compression balance in the jib.

2. Attach a weight 'W' at the hook as shown in fig. 2-1.

3. Note down the final readings on the balances of tie and jib.

4. Subtract the initial readings from the final readings to get the observed values of the forces, (say F_1 & F_2) in tie and jib respectively.

5. Measure the lenght of jib, tie and post.

6. Draw the space diagram to some convenient scale. The diagram represents the three members of the crane and also the load hanging at R. Name the members by bow's notation A, B and C in space diagram as shown in fig. 2-2.

7. Draw the force diagram to some convenient scale, For this purpose, draw ab = load 'W' through a, draw line ac parallel to member PR and through b draw line bc parallel to member QR. Thus, the line ac represents the force (F_3) in member PR (tie) and line bc represents the force (F_1) in member QR (jib) (see fig. 2-3.)

8. Repeat the experiment thrice (minimum) by varying the load. The forces are than verified by comparing two values i.e., observed values and calculated value.

Precautions:

1. The weights should be suspended gently without any jerk.

2. The axle of the pulley whould be lubricated before taking the observations.

3. The initial and final readings of the spring balance and compression balance be taken carefully.

4. The lenght of post, tie and jib members should be measured with a fine inextensible thread.

5. The space and force diagram should be drawn carefully and accurately.

Diagrams:

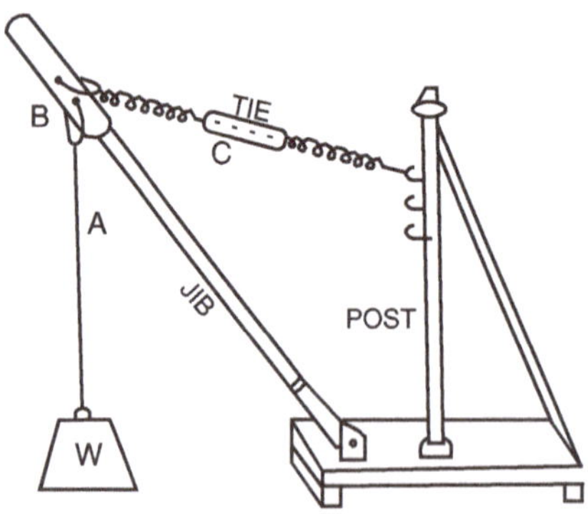

Jib crane apparatus

Fig. 2.1

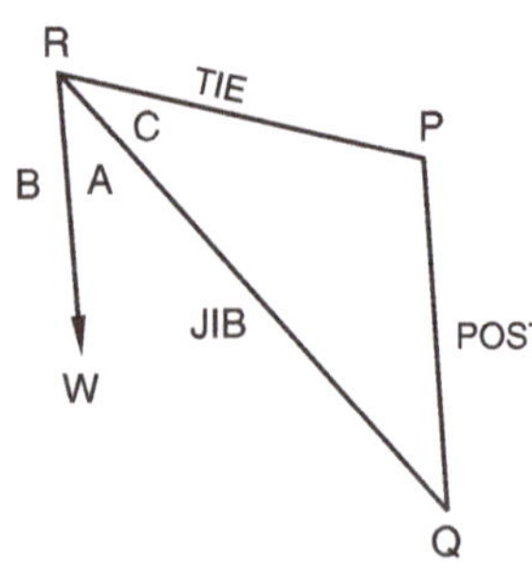

Space diagram

Fig. 2.2

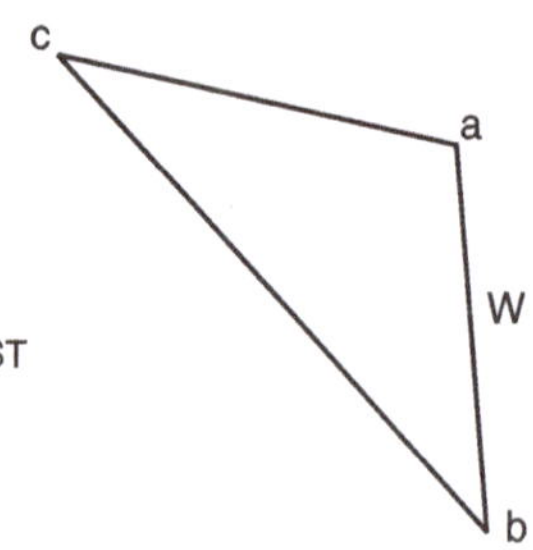

Force diagram

Fig. 2.3

S. No.	Weight, W	Initial Readings on Balances		Final Readings on Balances		Observed Forces		Calculated Forces Graphically		Percentage Errors		Remarks
		Tie	Jib	Tie	Jib	Tie F_1	Jib F_2	Tie F_3	Jib F_4	Tie $\frac{F_1 - F_3}{F_1} \times 100$	Jib $\frac{F_2 - F_4}{F_2} \times 100$	
1.												
2.												
3.												

Observations:

Height of post (PO) =

Length of tie (PR) =

Length of jib (RQ) =

Specimen calculations :

$$\text{Percentage error in the force of jib} = \frac{F_1 - F_3}{F_1} \times 100$$

$$\text{Percentage error in the force of tie} = \frac{F_2 - F_4}{F_2} \times 100$$

Result : Here, give the mean value of the percentage erros in the foces of jib and Tie.

QUESTIONS FOR VIVA-VOCE

1. Where do we use jib crane?
2. Name the different parts of the jib crane aparatus.
3. How do you know that the jib is in compression and the tie is in tension?
4. What can be the sources of error in this expriment?
5. State "triangle law of forces"

Experiment No. 3

Objective: **To verify the reactions at the supports of a simply supported beam.**

Apparatus: Parallel forces apparatus also known as Salter beam apparatus (consisting of a graduated beam supported on two compression balance springs as shown in Fig. 3-1), sliding hooks, weights, metre rod.

Material: Paper sheet, set squares, pencil, etc.

Theory: This experiment is based upon the principle of moments which states that if a body is subjected to a number of coplanar forces remains in static equilibrium, the algebraic sum of all the forcces and their moments about any point in their plane is zero. Mathematically:

$\Sigma F = 0$, which means that algebraic sum of all are forces (both the horizontal and vertical separately) is zero i.e., $\Sigma F_x = 0$, $\Sigma F_y = 0$ and $\Sigma M = 0$, which means that algebraic sum of moments of all the forces about a point is zero.

Procedure:

1. Take the initial readings of the spring balances at both the ends A and B in order to find the weight of beam.

2. Suspend two known weights 'W_1' and 'W_2' on the beam with the help of sliding hooks as shown in Fig. 3.1.

3. Take the final reading of the balances. The difference between the final and inital readings of the balances give the observed values of the reactions at the ends A and B.

4. Now measure the distances of each weight from one support and also note down the distance between the two supports.

5. Make a line diagram of the beam, showing the weights, reactions and the distance as shown in Fig. 3-2.

6. Take moments of all the forces, say the left hand end and find the reaction R_B.

7. Calculate the reaction R_A on the other end of beam.

8. Repeat the experiment thrice (minimum) by changing the loads in different positions.

Precautions:

1. The apparatus should be lying horizontal which can be changed with the help of a spirit levle.

2. The initial reading of spring balances should be read carefully.

3. The hooks should be place in grooves provided on the beam.

4. The weights should be suspended gently from hooks.

5. Before taking final readings, the beam should be slighty pressed downwards, so as to avoid any friction at the support.

6. The distances of weights from one end of the beam should be measured and noted down accurately.

Digram:

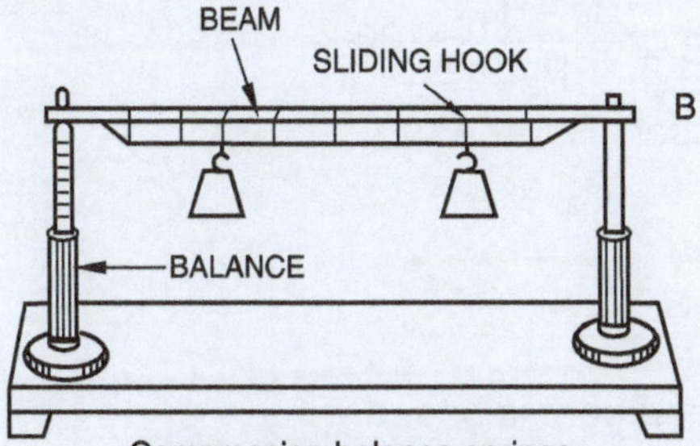

Compression balance springs
(Salter beam apparatus)

Fig. 3.1.

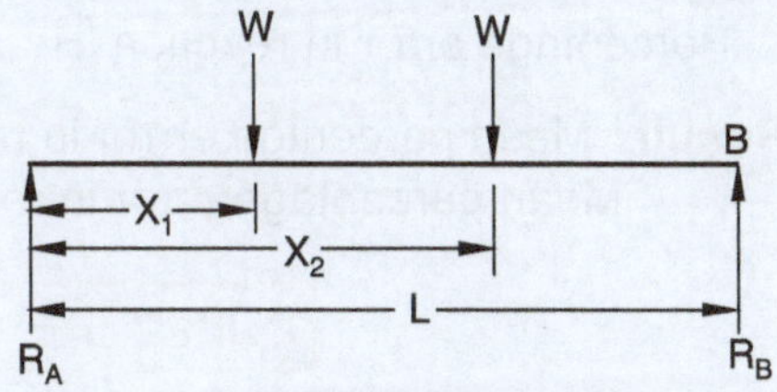

Line diagram

Fig. 3.2

Observations:

S. No.	Initial readings of the balance at		Final readings of the balance at		Observed value of reactions		Weights Suspended		Distance of Load from left end support		Moment of Load from left end support		Sum of Moments	Calculated Value of reactions		Percentage Error in reactions at	
	A	B	A	B	R_A	R_B	W_1	W_2	X_1	X_2	$W_1 \times X_1$	$W_2 \times X_2$	$W_1 \times X_1 + W_2 X_2$	R_{A1}	R_{B1}	A	B
1.																	
2.																	
3.																	

Mean Values:

Specimen calculations: By taking moments of all forces about the left end support A, we get

$$W_1 x_1 + W_2 x_2 - R_B \times L = 0$$

$$\therefore \quad R_B = \frac{W_1 x_1 + W_2 x_2}{L} \qquad \text{...(i)}$$

Also $R_A + R_B = W_1 + W_2$

$\therefore$ $R_A = (W_1 + W_2) - R_B$ ...(ii)

$\therefore$ Percentage error in reaction 'R_A' at A = $\dfrac{R_A - R_{A1}}{R_A} \times 100$ and

Percentage error in reaction 'R_B' at B = $\dfrac{R_B - R_{B1}}{R_B} \times 100$

Result : Mean percentge errror in reaction 'R_A' at A =

Mean percentage error in reaction 'R_B' at B =

QUESTIONS FOR VIVA-VOCE

1. State the principle on which this experiment is based.
2. What is the principle of moment?
3. How the weight of the beam has been taken into account in this experimnet.
4. What can be the sources of error in this experiment.
5. Can you test the accuracy of the compression balance with this experiment?
6. What do you mean by a simply supported beam?

Experiment No. 4

Objective: To find mechanical advantage, velocity ratio and effieicency of the inclined plane.

Apparatus: Inclined plane apparatus, wooden block, weight box, pan, thread and metre rod, etc (see Fig. 4-1).

Materials requuried: Paper sheet, set squares, pencil, etc.

Theory: Let a body of weight W moves on an inclined plane at an angle θ to the horizontal by an effort P. The effort P moves a vertical distance y, the weight of the body W also moves a distance y along the surface but only a distance $\times$ veritcally as shown in Fig. 4-1.

$$\text{M. A.} = \frac{W}{P}, \qquad \text{V. R.} = \frac{y}{x} = \frac{y}{y\sin\theta} = \frac{1}{\sin\theta}$$

$$\text{Efficiency, } \eta = \frac{\text{M.A.}}{\text{V.R.}} = \frac{W\sin\theta}{P}$$

Procedure:

1. Set the plane to the desired inclination and not down the angle of inclination (θ)

2. The wooden block and the pan of known weight be connected by means of a thread.

3. Note down the weight of the block (say W) and that of the pan (say W_1) in the observation chart.

4. Place the wooden block on the inclined plane and pass the thread over the fictionless guide provide with the apparatus.

5. Now put the extra weights in the pan gradually, till the wooden block just tends to side up over the inclined plane.

6. Note down the extra weight (say W_2) placed on the pan and adding to it the weight of the pan (W_1). Find out the total weight ($W_1 + W_2$) which will be the effort P applied to move the wooden block up the inclined plane.

7. Finally, calculate the value M. A., V. R. and η

8. Repeat the experimnet thrice (minimum) by varying the angle of inclination (θ) of the plane.

Precautions:

1. The guide pulley should move freely. It should be lubricated to make it frictioniess.

2. The inclined plane should be clean and smooth.

3. The thread should be free from knots.

4. The direction of thread applying the effort should be parallel to the inclined plane.

5. The extra weight should be gently put on the pan.

6. The extra weights should be increased so gradually to such a limit that the wooden block just to move slowly and not abruptly.

Diagram:

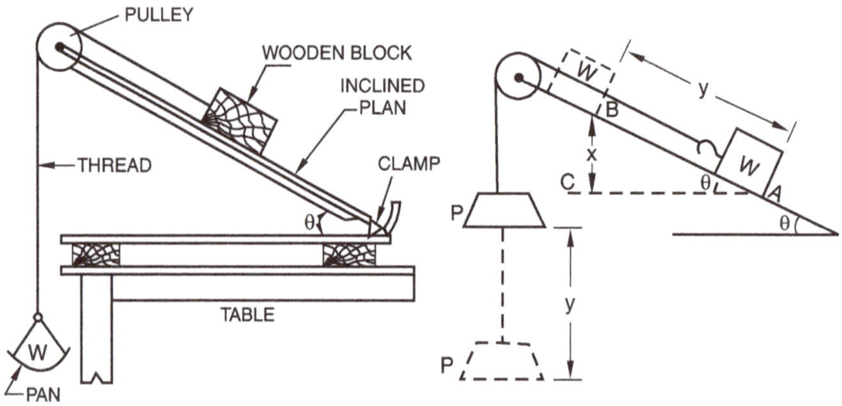

Fig. 4.1. *Inclined plane*

S. No.	Weight of wooden block 'W' in N	Weight of pan 'W'$_1$ in N	Weight added in the pan 'W'$_2$ in N	Total effort applied P = ('W'$_1$+ W'$_2$) in N	Angle of Inclination (θ)	M.A.= $\frac{W}{P}$	V.R.= $\frac{1}{\sin\theta}$	η.= $\frac{W \sin\theta}{P}$	Remarks
1.									
2.									
3.									

Mean values:

Specimen calculations:—

$$\text{M. A.} = \frac{W}{P} = \ldots\ldots, \qquad \text{V. R.} = \frac{1}{\sin\theta} = \ldots\ldots\ldots .$$

$$\text{Efficiency, } \eta = \frac{\text{M.A.}}{\text{V.R.}} = \ldots\ldots.$$

QUESTIONS FOR VIVA-VOCE

1. What do you mean by friction?
2. Define an inclined plane and co-efficient of friction.
3. Does the co-efficient of friction depends on the areas of surfaces in contact?
4. What is angle of friction? Does it depend upon the weight of the slider (wooden block) placed on the inclined plane?
5. What is the equation of motion of a body when it moves (i) upwards, (ii) downwards on a rough inclined plane?
6. Define M.A. of an inclined plane.
7. What can be the sources of error in this experiment?

EXPERIMENT NO. 5

Objective: **To determine mechanical advantage, velocity ratio and % age efficiency in case of a simple screw jack.**

Apparatus: Screw jack apparatus, weights, vernier calliper, metre rod, etc.

Theroy: The screw jack is a simple machine used to lift heavy load by a small effort. It is commonly used for lifting vehicles like cars, buses, trucks while repairing or replacing their punchered wheels. It works on the principle of screw and nut. The nut is rotated with the help of a tommy bar at the end of which effort is applied. By doing so, the screw is raised and thus the load placed on its head gets lifted.

The screw jack consists of a square threaded spindle (known as screw) which carries a double flanged circular table at its top to carry loads to be lifted. Two pulley are fixed on the heavy base through them the thread passes. The effort is applied from the weights provided on the apparatus as shown in fig. 5-1.

The required relations for finding mechanical advantage, velocity ratio and percentage efficiency of the jack can be derived as follows:

Let p = pitch of the screw.
D = Diameter of the double flanged table,
W = The load to be lifted.
P = The effort applied to lift the load

$$\text{M. A.} = \frac{\text{Load lifted}}{\text{Effort applied}} = \frac{W}{P}$$

and $$\text{V. R.} = \frac{\text{Distance moved by the effort}}{\text{Distance moved by the load}} = \frac{\pi D}{p}$$

also $$\%\text{ age } \eta = \frac{\text{M.A.}}{\text{V.R.}} \times 100$$

Procedure:

1. Masure the pitch of the screw (p) with the help of a vernier calliper.

2. Measure the circumference of the flanged table with an inextensible thread and a metre rod.

3. Warp the thread around the flanged table passing over the guide pulleys and lie their ends with a pan or hook as shown in Fig. 5-1.

4. Note down the weight of each hook.

5. Place a known weight (say W) on the top of flanged table and start hanging weights with the hooks gradually till the load starts lifting.

6. Find M. A., V. R. and percentage efficiency and record the same in the observation chart.

7. Repeat the experiment thrice (minimum) by increasing the loads and noting down the corresponding efforts applied.

Precautions:

1. The circuference of the disc and the pitch of the screw should be measured accurately.

2. The guide pulleys should be free from friction.

3. The screw should be well lubricated to reduce the friction.

4. Use both the pulleys to find the total effort applied.

5. The weight should be placed in the hooks very gently.

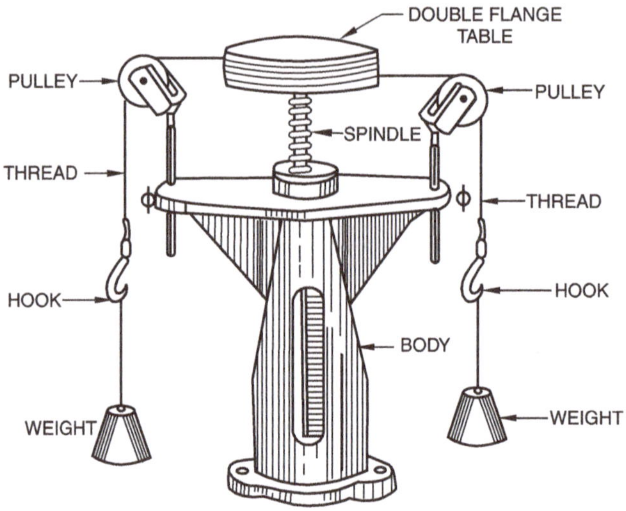

Fig. 5.1. *Simple screw jack apparatus*

Observations:

S. No.	Circumference of the table i.e. πD in mm	Pitch 'p' of the screw in mm	Velocity ratio V.R. = $\frac{\pi D}{p}$	Weight W to be lifted in N	Total effort 'P' applied to lift the wt. in N ($P_1 + P_2$ + wt. of hook)	M.A.= $\frac{W}{P}$	% $\eta = \frac{M.A.}{V.R.} \times 100$	Remarks
1.								
2.								
3.								

Mean values:

Specimen calculations:

$$\text{Velocity ratio, V. R.} = \frac{\pi D}{p} = \ldots\ldots$$

$$\text{Mechanical advantage, M. A.} = \frac{W}{P} = \ldots\ldots$$

$$\text{Also Efficiency, } \eta = \frac{M.A.}{V.R.} = \ldots\ldots$$

QUESTIONS FOR VIVA-VOCE

1. Where a simple screw jack is used?
2. How will you increase the efficiency of the screw jack?
3. Does the efficiency of this apparatus increase, if the load lifted is increased? If so, how?
4. Why the effort is applied by using both the pulleys?
5. What can be the sources of error in the experiment?
6. Are square threads on spindle area more efficient than V threads?
7. What do you mean by pitch of the thread?

Experiment No. 6

Objective: To find the mechnical advantage, velocity ratio and efficiency in case of worm and worm wheel.

Apparatus: Worm and worm wheel apparatus, weight, metre rod, thread, etc (see Fig. 6-1).

Theory: The apparatus consists of a toothed wheel (known as worm wheel) carrying load drum. A rope wound round the load drum carrying weight W to be lifted. The worm wheel is meshed with the worm. To the end of the worm spindle, a pulley is attached. Another rope passes round the pulley and at the end of which a scale pan is tied, on which the weights are placed to find the value of effort P as shown in Fig. 6-1.

Let Radius of the load drum $= r$

Radius of the pulleys $= R$

No. of teeth of the worm wheel $= T$

$$\text{Mechanical advantage, M. A.} = \frac{W}{P}$$

For one revoluation of the pulley,

Distance moved by the pulley (effort) $= 2\pi R$

$$\therefore \text{ Velocity ratio, V. R.} = \frac{\text{Distance moved by the pulleys (effort)}}{\text{Distance moved by the drum (load)}}$$

$$= \frac{2\pi R}{\frac{2\pi r}{T}} = \frac{R}{r} \times T \quad \text{and} \quad \text{M.A.} = \frac{W}{P}$$

$$\text{Also Precentage efficiency, \%} = \frac{\text{M.A.}}{\text{V.R.}} \times 100$$

$$= \frac{W \times r}{P \times R \times T} \times 100$$

Procedure:

1. Measure the circumference of the drum and the pulley with a thread and metre rod.

2. Wrap the rope, round the drum and attach a load (W) with it.

3. Wrap another rope round the pulley and attach a pan with it, as showing in Fig. 6.1.

4. Place sufficient weights on to the pan till the load W just starts moving upwards.

5. Note down the weights palced in the pan and by adding weight of the pan, find the value of effort P.

6. Calculate M. A., V. R. and percentage efficiency as explained above.

7. Repeat the experiment thrice (minimum) with the different loads and tabulate the results.

Diagram:

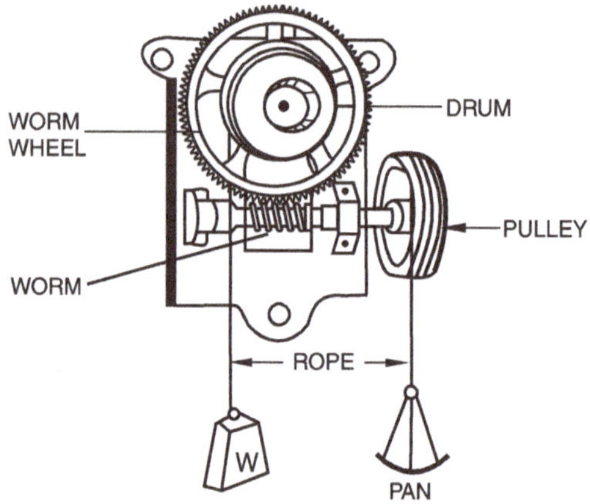

Fig. 6.1. *Worm and worm wheel aparatus*

Observation:

S. No.	Load 'W' in N	Effort 'P' in N	M.A.= $\frac{W}{\rho}$	V.R.= $\frac{R}{r}\times T$	$\% \eta = \frac{M.A.}{V.R.} \times 100$	Remarks
1.						
2.						
3.						

Mean Values:

Mean values:

Specimen calculation:

$$V.R. = \frac{R}{r}\times T = ...$$

$$M.A. = \frac{W}{P} = ...$$

and
$$\%\eta = \frac{W\times r}{P\times R\times T}\times 100 = ...$$

Precautions:

1. The bearings of the worm spindle, worm teeth and that of the wheel should be well lubricated to minimise friction.
2. The circumference of the drum and the pulley should be measured accurately.
3. Weight should be placed gently on the effort pan.
4. There should be no overlapping of the strings on the pulley and the drum.
5. The pan and load W should not touch the wall.
6. The load W and effort P should move slowly and not abruptly.

QUESTIONS FOR VIVA-VOCE

1. Where the worm and wrom wheel apparatus is used?
2. What is the difference between worm and worm wheel?
3. What is the principle working of worm and worm wheel?
4. How will you increase the efficiency of the apparatus?
5. Does the efficiency increase as load lifted is increased? If so, how?
6. What can be the sources of error this experiment?

EXPERIMENT No. 7

Objective: **To find to mechanical advantage, velocity ratio and efficiency in case of single purchase winch crab.**

Apparatus: Single purchase winch carb apparatus, weights, pan, rope, thread, metre rod, etc (see Fig. 7-1).

Theory: It consists of two toothed wheels, smaller and larger one, the smaller wheel is known as the pinion and the larger one as the gear. The gear is attached to the load drum. A rope is wound round the drum which carries load W to be lifted. Another rope passes round the pulley to which is attached a scale pan in which weights can be put to find the effort P.

Let T_1 = No. of teeth on the pinion.
T_2 = No. of teeth on the gear or spur wheel
R = Radius of the wheel
r = Radius of the drum

Mechanical adventage, M.A. = $\frac{W}{P}$

For one revolution of the effort pulley,

Distance moved by effort = $2\pi R$

and distance moved by load = $2\pi r \times \frac{T_1}{T_2}$

$$\therefore \text{ V. R.} = \text{Distance moved by load} = \frac{2\pi R \times T_2}{T_1 \times 2\pi r} = \frac{R}{r} \times \frac{T_2}{T_1}$$

$$\text{Also } \%\eta = \frac{\text{M.A.}}{\text{V.R.}} \times 100 = \frac{W \times r \times T_1}{P \times R \times T_2} \times 100$$

Procedure:

1. Measure the circumference of the drum and the pulleys with the help of thread rod metre rod.

2. Count the number of teeth of pinion and the gear.

3. Wrap a rope round the effort pulley and attach a scale pan to its free end.

4. Wrap the other rope round the drum in such a manner that when the effort is applied, the load gets lifted.

5. Hang the load W to the rope passing round the drum.

6. Now place sufficient weights in the pan gently till the load just starts moving upwards.

7. Note down the weights added in the pan and by adding weight of the pan, find the value of effort P.

8. Calculate the M.A., V. R. and η as explained above.

9. Repeat the experiment thrice (minimum) and tabulate the result.

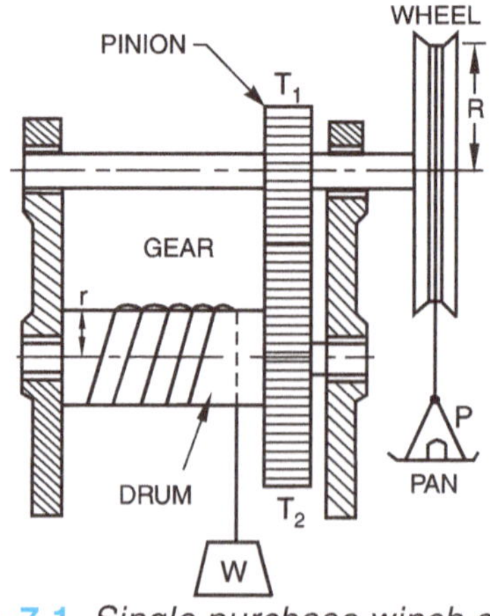

Fig. 7.1. *Single purchase winch crab*

Precautions:

1. Bearings of axles, the pinion and the gear (spur wheel) should be properly lubricated so as to reduce friction.

2. The weights should be put gently in the scale pan.

3. Number of teeth on the pinion end and the gear should be carefully counted.

4. The strings should not overlap.

Obeservations:

No. of teeth on pinion, T_1 = ...
No. of teeth on the gear, T_2 = ...
Diameter of effort pulley, D = ...
Diameter of drum, d = ...
Weight to the scale pan, W = ...

S. No.	Load 'W' in N	Effort 'P' in N	$M.A. = \frac{W}{P}$	$V.R. = \frac{R}{r} \times T$	$\% \eta = \frac{M.A.}{V.R.} \times 100$	Remarks
1.						
2.						
3.						

Mean value:

Mean Values:

$$M.\ A. = \frac{W}{P} = \ldots$$

$$V.\ R. = \frac{R}{r} \times \frac{T_2}{T_1} = \ldots$$

and $$\%n = \frac{W \times r \times T_1}{P \times R \times T_2} \times 100$$

QUESTIONS FOR VIVA-VOCE

1. What is a single purchase winch crab?
2. Where the winch crab is used?
3. How will you find the efficiency of winch crab?
4. How will you increase the efficiency of winch crab?
5. Does the efficiency increase as the load lifted is increased?
6. What can be the sources of errors in this experiment?

Experiment No. 8

Objective: **To find the centre of gravity of regular lamina.**

Material required : Paper sheet, drawing pin or tape, set squares, scale cards, etc.

Theory: Every body is attracted towards the centre of the earth due to gravity. The force with which the body is attracted towards earth is called weight of the body. The whole weight of the body is supposed to act at a point in the body and this point is known as centre of gravity. Similarly, the point where the whole area of a lamina is supposed to act is also known as centroid or centre of gravity.

Procedure:

The centre of gravity of any lamina may be found out by adopting the following steps:

1. Draw the two refernce axes ox and oy so that the whole figure lies on one side of both the refernece axes.

2. Check whether the section is symmertrical about any centroidal axes. If it is then C. G. will lie on that axis.

3. Divide the composite section into convenient parts' such as rectangle, triangle etc.

4. Calculate the areas a_1, a_2 and a_3 of the various sections.

5. Find the distances of their centroids from the respective axes ox and oy as y_1, y_2 and y_3 and x_1, x_2 and x_3.

6. Then find the position of C.G. and calculate $\overline{X}$ and $\overline{Y}$ form the follwoing relations:

Diagram:

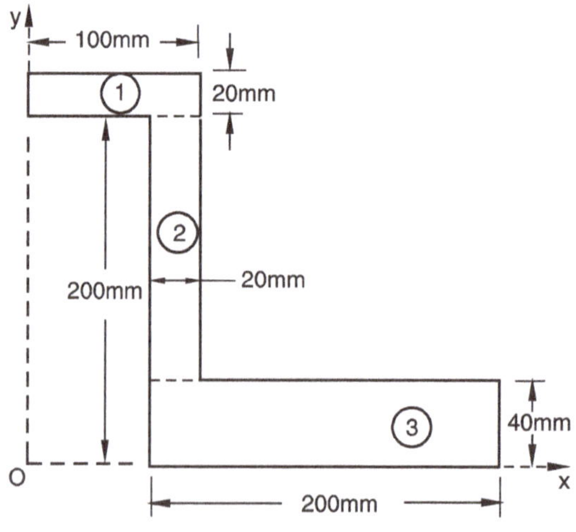

Fig. 8.1.

Observation:

S. No.	Areas in mm^2	Distance from OY axis in mm	Distance from OX axis in mm	Position of C.G. from OY axis = $\overline{X}$	Position of C.G. from OX axis = $\overline{Y}$
1.	$a_1 =$	$x_1 =$	$y_1 =$		
2.	$a_2 =$	$x_2 =$	$y_2 =$	$\overline{X} = \dfrac{a_1x_1 + a_2x_2 + a_3x_3}{a_1 + a_2 + a_3}$	$\overline{Y} = \dfrac{a_1y_1 + a_2y_2 + a_3y_3}{a_1 + a_2 + a_3}$
3.	$a_3 =$	$x_3 =$	$y_3 =$		

$$\overline{X} = \frac{a_1x_1 + a_1x_2 + a_3x_3}{a_1 + a_2 + a_3}$$

$$\overline{Y} = \frac{a_1y_1 + a_1y_2 + a_3y_3}{a_1 + a_2 + a_3}$$

If any part of the section is cut out or removed, then the remaining portion of the section is called remainder. For determineing the C.G. of a remainder, the portion removed is imagined to be intact at its position, but its mass or area is considered negative.

Precautions:

1. Calculate the areas a_1, a_2, a_3 etc, accuratly.
2. $\overline{X}$ and $\overline{Y}$ should be find out carefully with standard relations.

Note: *The centre of gravity of regular lamina can also be find out as stated above for irregular lamina.*

This Page has been intentionally left blank

MODEL TEST PAPERS

MODEL TEST PAPER – 1

Duration: 3 hrs. *M. Marks: 75*

SECTION A

Q.1. **(A) Fill in the blanks**

(a) In a couple, the lines of action of the forces are ____

(b) Friction is divided into two types viz. static friction and ____ friction.

(c) ____ is the ratio of load lifted to the effort applied.

(d) The maximum frictional force which comes into play when a body just begins to slide over the surface of another body, is known as ____.

(e) The velocity ratio of a first system of pulleys with four pulleys as movable pulleys, is ____

(B) State True or False

(a) Those machines, in which friction is zero, are called ideal machines.

(b) The C.G. of a hollow right circular cone of radius 'r' and height 'h' lies at (h/4) from the base.

(c) The force of friction developed between the surfaces of contact of two bodies in motion is known as dynamic friction.

(d) If arm of couple is double, its moment will not alter.

(e) An body in which distance between any two particles always remain constant is known as rigid body.

SECTION B

Note: Attempt any five questions

Q.2. (i) Differentiate between equilibriant and resultant force.

(ii) Discuss polygon law of forces.

(iii) Discuss any three properties of a couple.

(iv) Find C.G. of an angle section 100 mm × 60 mm × 20 mm.

(v) Write laws of solid friction.

(vi) There are four pulleys in a third system of pulleys. An effort of 320 N is required to lift an unknown weight. If efficiency of machine = 75°, find the weight lifted.

(vii) Differentiate between reversible and irreversible machines.

SECTION C

Note: Attempt any three questions

Q.3. A uniform beam AB of length 2 m is placed on 2 supports (simple) A an B. The beam carries point loads of 1 KN, 2 KN and 3 KN at distance of 0.5 m, 1 m and 1.5 m respectively. Find reactions developed at A and B.

Q.4. A rod of length 5 m is bent 80 as to form five sides of a regular hexagon. Find distance of its C.G. from either end of the rod.

Q.5. A body resting on a rough horizontal plane required a pull of 180 N inclined at 30° to the plane just to move it. It was that a push of 220 N inclined at 30° to the plane just moved the body. Determine the weight of the body and coefficient of friction.

Q.6. What load will be lifted by an effort of 12 N if the velocity ration is 18 and efficiency of machine at this load is 60°. If the machine has a constant frictional resistance, determine the law of the machine and find the effort required to run the machine at a load of 90 N.

Model Test Paper – 2

Duration: 3 hrs. *M. Marks: 75*

Section A

Q.1. (A) State True or False:

(i) Velocity is a vector quantity.

(ii) The method of determination of the resultantt of some forces acting simultaneously on a particle is called composition of forces.

(iii) A couple means two equal and like forces.

(iv) With the help of lever we can get more output than input.

(v) Angle of friction is always greater than 90°.

(vi) Center of gravity is a lern related to bodies having area.

(vii) newton's first law is called Law of Inertia.

(viii) A reversible machine has efficiency more than 50%

(B) Fill in the blanks.

(i) An ideal machine is one whose efficiency is equal is equal to ...

(ii) Large force acting for a short time is called ...

(iii) The C.G. of a solid cone lies at a distnace of ... its height.

(iv) Static friction is always ... than dynacmic friction.

(v) The resultant of two like parallel force P and Q is ...

(vi) The sense of a force is indicated by ...

(vii) The fundamental quantities are ...

Section B

Note: Attempt any five questions

Q.2. Explain the elements of force.

Q.3. Write SI units in which the following quantities are measured:

(i) Force (ii) Moment

(iii) Torque (iv) Acceleration

Q.4. Where do the centroid of the following lamina lies from their base:

(i) Triangle of height 'h'

(ii) Semicircle of radius 'r'

(iii) Square of side 'a'

Q.5. In a second system of pulleys there are three pulleys in the upper block and two pulleys in the lower block. If its efficiency is 75% find its Mechanical advantage.

Q.6. What is a couple? How are these classified? Explain.

Q.7. A body weighing 50 Newton is lying on a horzontal plane for which $\mu = 0.5$ determine:

(i) Normal Reaction

(ii) Limiting force of friction

(iii) Horizontal force required for moving it.

SECTION C

Attempt any two questions from this section

Q.8. Find the centroid of the area obtained after removing a semicircle of radius 10 cm from a quadrant of a circle of radius 20 cm.

Q.9. A body of mass 10 kg, lying on a horizontal plane, is attached to one end of a string. The string passes over a smooth pulley and carries at its other end a mass of 5 kg. If the co-efficient of friction between the mass of 10 kg and plane is 0.25, determine:

(a) Acceleration of the masses

(b) Tension in the string.

Q.10. In single winch crab number of teeth on the pinion is 12 and that on the main gear 48. The diameter of the load drum is 20 cm and the lenght of the handle is 40 cm. Determine:

(a) Velocity Ratio.

(b) Efficiency of the machine when a load of 1000 N is lifed by an effort of 200 N. Effort lost in friction while lifting a load of 100 N.